Human Teeth - From Function to Esthetics

Edited by Lavinia Cosmina Ardelean and Laura-Cristina Rusu

Published in London, United Kingdom

Human Teeth - From Function to Esthetics
http://dx.doi.org/10.5772/intechopen.104146
Edited by Lavinia Cosmina Ardelean and Laura-Cristina Rusu

Contributors
Tharanikumar Sivakumar, Chandrasekaran Krithika, Nandikha Tharanikumar, Lavinia Cosmina Ardelean, Laura-Cristina Rusu, Codruta Victoria Tigmeanu, Daniela Elisabeta Jumanca, Divya Krishnamoorthi, Alona Melnyk, Valerii Filonenko, Yasmine N. Alawaji, Marut Phuphaniat, Maria Fatima Martins do Rosário Dias, Valter Pedroso Alves, Gunel Mammadova Kizi, Ana Sintra Delgado, Meghna Bhandary, Rachaita Chhabra, K. Ananya Rao, Mohammed Shahid, Robert Ćelić, Hrvoje Pezo, Stanislava Senzel, Gracia Ćelić, Monisha P. Khatri, Shreya Kishore, Srujana Hemmanur, Milos Stepovic, Maja Vulovic, Ivona Bankovic, Miroslav Misic, Radisa Vojinovic, Edith Lara-Carrillo, María Teresa Hernández-Solís, Víctor Hugo Toral -Rizo, Ronell Eduardo Bologna-Molina, Wael Hegazy-Hassan, José Edgar Garduño-Mejía, Ana Miriam Santillán-Reyes, Norma Guadalupe Ibáñez-Mancera, Nachum Samet, Diego Muñoz, Nicole Manzur, Sergio Cáceres, Christian Rochefort, Seyide Tugce Gokdeniz, Mehmet Eray Kolsuz, Arda Buyuksungur, Lawrence O. Igbinosa, Richard Evbuomwan, Martins A. Okoromu, Uwa S. Osarenkhoe, Tanvi Satpute, Emanuela Lidia Crăciunescu, Meda-Lavinia Negrutiu, Mihai Romînu, Andreea Codruţa Novac, Cristina Modiga, Borislav-Duşan Caplar, Cosmin Sinescu, Daniela-Maria Pop, Andréa Correia Miranda Dolores Correia Miranda Valdivia, Maria de los Angeles Vázquez Sánchez, Diana Elizabeth Aguirre Corté, Evangelina Gutierrez-Cortés, Fares Kablan, Ali K. Refai

First published in London, United Kingdom, 2023 by IntechOpen
IntechOpen is the global imprint of INTECHOPEN LIMITED, registered in England and Wales, registration number: 11086078, 5 Princes Gate Court, London, SW7 2QJ, United Kingdom

British Library Cataloguing-in-Publication Data
A catalogue record for this book is available from the British Library

Additional hard and PDF copies can be obtained from orders@intechopen.com

Human Teeth - From Function to Esthetics
Edited by Lavinia Cosmina Ardelean and Laura-Cristina Rusu
p. cm.

This title is part of the Dentistry Book Series, Volume 15
Topic: Oral Health
Series Editor: Sergio Alexandre Gehrke
Topic Editor: Sandra Marinho and Esra Guzeldemir-Akcakanat

Print ISBN 978-1-83768-658-2
Online ISBN 978-1-83768-659-9
eBook (PDF) ISBN 978-1-83768-660-5
ISSN 2631-6218

IntechOpen Book Series

Dentistry

Volume 15

Aims and Scope of the Series

This book series will offer a comprehensive overview of recent research trends as well as clinical applications within different specialties of dentistry. Topics will include overviews of the health of the oral cavity, from prevention and care to different treatments for the rehabilitation of problems that may affect the organs and/or tissues present. The different areas of dentistry will be explored, with the aim of disseminating knowledge and providing readers with new tools for the comprehensive treatment of their patients with greater safety and with current techniques. Ongoing issues, recent advances, and future diagnostic approaches and therapeutic strategies will also be discussed. This series of books will focus on various aspects of the properties and results obtained by the various treatments available, whether preventive or curative.

Meet the Series Editor

Dr. Sergio Alexandre Gehrke is a doctorate holder in two fields. The first is a Ph.D. in Cellular and Molecular Biology from the Pontificia Catholic University, Porto Alegre, Brazil, in 2010 and the other is an International Ph.D. in Bioengineering from the Universidad Miguel Hernandez, Elche/Alicante, Spain, obtained in 2020. In 2018, he completed a postdoctoral fellowship in Materials Engineering in the NUCLEMAT of the Pontificia Catholic University, Porto Alegre, Brazil. He is currently the Director of the Postgraduate Program in Implantology of the Bioface/UCAM/PgO (Montevideo, Uruguay), Director of the Cathedra of Biotechnology of the Catholic University of Murcia (Murcia, Spain), an Extraordinary Full Professor of the Catholic University of Murcia (Murcia, Spain) as well as the Director of the private center of research Biotecnos – Technology and Science (Montevideo, Uruguay). Applied biomaterials, cellular and molecular biology, and dental implants are among his research interests. He has published several original papers in renowned journals. In addition, he is also a Collaborating Professor in several Postgraduate programs at different universities all over the world.

Meet the Volume Editors

Professor Lavinia Cosmina Ardelean, DMD, Ph.D., is a professor and head of the Department of Technology of Materials and Devices in Dental Medicine, "Victor Babes" University of Medicine and Pharmacy, Timisoara, Romania. She has authored/co-authored 16 books, 13 book chapters, and more than 120 peer-reviewed papers. She is an editorial board member of numerous journals and a member of reviewer/topic boards of the journals *Materials*, *Prosthesis*, *Metals*, and *Polymers*. She has guest-edited eight special issues in different journals and four books. She is also involved in an ongoing book editing project. Dr. Ardelean is an active reviewer with more than 200 reviews to her credit who was recognized as a top reviewer in "cross-field" and "materials science" in 2019. She currently holds two patents. Her research interests include most areas of dentistry, with a focus on dental materials/biomaterials, dental alloys, resins, ceramics/bioceramics, CAD/CAM milling, 3D printing/bioprinting in dentistry, laser welding, scanning, and oral health care.

Professor Laura Cristina Rusu, DMD, Ph.D., is the mother of two lovely boys and a full-time professor and head of the Oral Pathology Department, Faculty of Dental Medicine, "Victor Babes" University of Medicine and Pharmacy, Timisoara, Romania. Her Ph.D. thesis was centered on allergens in dental materials. In 2017 she obtained a Dr. Habil and was confirmed as Ph.D. coordinator in the field of dental medicine. She took part in 10 research projects, including FP7 COST Action MP 1005, and authored more than 140 peer-reviewed papers. She has published ten books and book chapters as author and co-author. She has guest-edited four special issues in different journals and edited or co-edited four books. She is also involved in an ongoing book editing project. She currently holds three patents. Dr. Rusu is a member of the editorial board of the *Journal of Science and Art* and *Medicine in Evolution*. She is a topic editor for the journal *Materials*. Her main scientific interests are oral pathology and oral diagnosis in dental medicine, focusing on oral cancer.

Contents

Preface **XIII**

Section 1
Dental Anatomy and Morphology **1**

Chapter 1 **3**
Dental Anatomy and Morphology of Permanent Teeth
by Emanuela Lidia Crăciunescu, Meda-Lavinia Negruțiu, Mihai Romînu, Andreea Codruța Novac, Cristina Modiga, Borislav-Dușan Caplar, Cosmin Sinescu and Daniela-Maria Pop

Chapter 2 **27**
All We Need to Know about Normal and Abnormal Human Teeth
by Milos Stepovic, Maja Vulovic, Ivona Bankovic, Miroslav Misic and Radisa Vojinovic

Chapter 3 **47**
Frequency and Distribution of Pulpal Calcifications in Teeth Involved in Jaw Tumors
by María Teresa Hernandez-Solis, Edith Lara-Carrillo, Víctor Hugo Toral-Rizo, Ronell Eduardo Bologna-Molina, José Edgar Garduño-Mejía, Norma Guadalupe Ibáñez-Mancera, Wael Hegazy-Hassan and Ana Miriam Santillán-Reyes

Chapter 4 **57**
Aesthetic Smile Designing
by Monisha P. Khatri, Shreya Kishore and Srujana Hemmanur

Section 2
Prevention and Oral Health **75**

Chapter 5 **77**
Oral Health: Fundamentals, Importance, and Perspectives
by Andréa Dolores Correia Miranda Valdivia, Maria de los Angeles Vázquez Sánchez, Diana Elizabeth Aguirre Cortés and Evangelina Gutiérrez Cortés

Chapter 6 93
Oral Health: A Doorway to General Well-being
by Lawrence O. Igbinosa, Richard Evbuomwan, Martins A. Okoromu and Uwa S. Osarenkhoe

Chapter 7 103
The Role of Sugar-Free Chewing Gum in Dental Caries Prevention
by Marut Phuphaniat

Chapter 8 117
Periodontitis, Its Associations, and Prevention
by Yasmine N. Alawaji

Section 3
Prosthetics and Implantology **149**

Chapter 9 151
Development of Gallium Silicon Titanium Alloys for Dental Implants
by Tharanikumar Sivakumar, Chandrasekaran Krithika and Nandikha Tharanikumar

Chapter 10 181
Diagnosis and Treatment Planning for Multiple Implants in Esthetic Zone
by Divya Krishnamoorthi

Chapter 11 195
Simple Surgical Methods for Soft Tissue Management around Fresh Extraction Sockets during Socket Preservation
by Fares Kablan

Chapter 12 209
Dental Implants: Immediate Placement and Loading in the Esthetic Zone
by Nachum Samet

Chapter 13 223
Implantogenomic: Conceptualizing Osseointegration toward Personalized Dental Implant Therapy
by Ali K. Refai

Chapter 14 239
The Relationship between Dental Occlusion and "Prosthetic Occlusion" of Prosthetic Restorations Supported by Natural Teeth and Osseointegrated Dental Implants
by Robert Ćelić, Hrvoje Pezo, Stanislava Senzel and Gracia Ćelić

Chapter 15 269
Elimination of Retainers in Anterior Zone in Removable Partial Dentures
by Diego Muñoz, Christian Rochefort, Nicole Manzur and Sergio Cáceres

Chapter 16 281
Novel Prosthetic Solutions for High-Quality Aesthetics
by Lavinia Cosmina Ardelean, Laura-Cristina Rusu, Daniela Elisabeta Jumanca and Codruta Victoria Tigmeanu

Section 4
Orthodontics 299

Chapter 17 301
From the Orthodontic Smile to the Perfect Smile: A New Categorization
by Maria do Rosário Dias, Valter Pedroso Alves, Gunel Mammadova Kizi and Ana Sintra Delgado

Chapter 18 315
Clinical and Phonetic Features of Dentognathic Deformations, Their Orthodontic Treatment
by Alona Melnyk and Valerii Filonenko

Section 5
Advances and Future Perspectives 335

Chapter 19 337
Practical and Theoretical Considerations for Dental Restorative Materials
by Tanvi Satpute

Chapter 20 349
Human Teeth Is Useful Even after Its SHED! So, Why Discard It?
by Meghna Bhandary, Rachaita Chhabra, K. Ananya Rao and Mohammed Shahid

Chapter 21 379
Artificial Intelligence in Dentistry
by Seyide Tugce Gokdeniz, Arda Buyuksungur and Mehmet Eray Kolsuz

Preface

The teeth play an important role in daily living, being involved in chewing and digesting food and speaking. They also have an integral role in facial esthetics. Well-shaped and aligned white teeth are considered essential to an appealing smile and have a high impact on self-esteem, school or job performance, and other areas of life.

Oral health issues, which range from cavities to oral cancer, cause different degrees of pain and disability. Oral diseases have a high prevalence; nearly 100% of the adult population has at least one cavity.

Edentulism, the partial or total absence of teeth, has debilitating consequences and is considered the ultimate stage of oral suffering. Therefore, maintaining proper oral hygiene is crucial to prevent dental issues. Preventive methods include proper oral hygiene, regular dental checkups and cleaning, a healthy diet, limited sugar intake, and no smoking. Routine dental checkups help detect the early stages of dental illness. Most oral diseases are preventable and treatment in the early stages is easier, cheaper, and faster than at the later stages of disease.

To achieve a perfect balance between function and esthetics, dental treatments have evolved to be capable of solving pretty much any dental problem with a high rate of success. Dental treatment, depending on the oral health issue, involves using a wide range of materials, including smart materials. Current available dental techniques include lasers, microscopy, and 3D bioprinting as well as digital solutions that are gaining popularity. Nowadays, even edentulism is a condition that can be treated effectively by means of various modern prosthetic devices or dental implants.

This book includes twenty-one chapters and discusses different aspects of human teeth function and esthetics, including dental anatomy and morphology, smile design, oral health and prevention of oral disease, prosthetic solutions for both functional and esthetic issues, advances in dental implants and orthodontics, and future perspectives in dentistry.

Lavinia Cosmina Ardelean
Department of Technology of Materials and Devices in Dental Medicine,
"Victor Babes" University of Medicine and Pharmacy,
Timisoara, Romania

Laura-Cristina Rusu
Department of Oral Pathology,
"Victor Babes" University of Medicine and Pharmacy,
Timisoara, Romania

Section 1

Dental Anatomy and Morphology

Chapter 1

Dental Anatomy and Morphology of Permanent Teeth

Emanuela Lidia Crăciunescu, Meda-Lavinia Negruțiu, Mihai Romînu, Andreea Codruța Novac, Cristina Modiga, Borislav-Duşan Caplar, Cosmin Sinescu and Daniela-Maria Pop

Abstract

The present chapter is proposing a detailed and illustrated description of dental morphology of permanent dentition. The main topics are related to nomenclature, age of emergence, a description of teeth's tissues (pulp, dentin, enamel, and cement), and morphology of all permanent teeth. The main focus of this chapter is the description of individualized morphology and specific variations of each permanent tooth. The goal of all treatment phases in dental medicine is to restore the function, integrity, and morphology of the oral cavity, and all these achievements are reached through deep knowledge of dental morphology. Cavities are restored with direct dental materials, which need to be carved according to the natural shape, outlines, occlusal and proximal contacts of teeth's morphology, reproducing also the shade and translucencies of natural teeth. The same goal dominates the prosthodontic field. It is well known in dental medicine that shape, size, and position assure the optimal function and preserve the self-maintenance of dental arches and dento-maxillary system. For esthetic, function, and self-preservation, all dental treatment fields have to first consider the dental morphology.

Keywords: teeth, crown, root, morphology, contact areas, occlusion, esthetics, oral cavity

1. Introduction

In the past decades, dental medicine has evolved according to the technological progress, development of high quality and esthetic dental materials, and new and conservative attitudes in clinical practice. Dental anatomy and morphology are the basic components of the skills and knowledge needed in all dental treatment phases. The knowledge related to function, shape, color, phonetics, position on the dental arch, and the relations with the opposing arch is as important as dental anatomy and morphology. The study of dental anatomy, morphology, physiology, and occlusion is mandatory in all the curriculum of dental schools and provides the basic and the most important component in developing a successful treatment plan for all phases of dental treatments, which include odontotherapy, endodontics, prosthodontics, implantology, orthodontics, periodontology, and pediatric dentistry.

Dental practitioners, during the inspection of the oral cavity, will see the clinical crowns, the attachment of gingival tissue, the shape, size, position and angulations of the teeth, the relation of proximal contact areas, the occlusal contacts, and the evidence of parafunctions and esthetics. Having this picture, the basic knowledge, and considering all the factors and all related anatomical structures, a correct treatment plan will be made.

2. Introduction to dental anatomy and morphology

2.1 Nomenclature, numbering system, and dental tissues

For studying dental anatomy, a common language is required. Humans have two dentitions in their lifetime that support the anatomical structures and orofacial functions such as mastication, speech, and give shape and beauty to the face. Teeth are highly calcified structures with individualized tissues, supported in the upper and lower jaw by *bony sockets* also called *alveolus*. Teeth are comprised of four types: incisors, canines (cuspids), premolars (bi-cuspids), and molars. The first set of teeth is the *primary or deciduous dentition*, which begins to form prenatally at about 14 weeks and is completed postnatal at about 3 years of age. In total, 20 teeth, 10 maxillary and 10 mandibular teeth, and 5 teeth on each side of the jaw are consisting the deciduous dentition. The teeth in deciduous dentition are grouped on maxillary and mandibular jaw as follows: four incisors, two canines, and four molars (**Table 1**). World Dental Federation proposed a numbering system for deciduous and permanent dentitions, and the system was adopted by the World Health Organization and accepted by other organizations such as the International Association for Dental Research [1].

Morphological characteristics of deciduous teeth are different from permanent ones. Deciduous teeth have smaller crowns and roots, more prominent cervical ridges, and narrower "neck" or cervical diameter. The roots are narrower and flare, and the buccolingual diameters are smaller than the permanent's diameters. Crowns of deciduous anterior teeth have a mesiodistal diameter higher in comparison with the cervical-incisal diameter. The roots of deciduous anterior are long and narrower, and the roots of deciduous molars are longer, slender, and flare to allow more room for the development of permanent crowns. The cervical ridges of the deciduous anterior teeth are more prominent, the crowns and roots of deciduous molar are more slander mesiodistally, the buccal and lingual surfaces of deciduous molars are flatter, and the shade of deciduous is whiter [2].

The teeth supported by the upper jaw are called *maxillary* teeth and the ones supported by the lower jaw are called *mandibular* teeth. The deciduous dentition lasts until about 6 years of age when the first succedaneous or permanent teeth are emerging into the oral cavity. In this stage, the transitional dentition or mixed dentition is formed, which is present in the oral cavity until 12 or 13 years of age and ends when all the deciduous teeth are lost and all the permanent ones have emerged in the oral cavity. Mixed dentition, present between 6 and 12 years of age, can be a difficult period of time because of missing teeth, different shades and hues of the recently emerged teeth, malposition, and crowning, which may need orthodontic treatment.

In total, 32 teeth, 16 maxillary and 16 mandibular teeth, and 8 teeth on each side of the jaw are consisting the permanent dentition. Teeth in permanent dentition are grouped on each jaw as follows: four incisors, two canines, four premolars, and six

DOI: http://dx.doi.org/10.5772/intechopen.110223

			Upper right					Upper left		
Tooth	**Second molar**	**First molar**	**Canine**	**Lateral incisor**	**Central incisor**	**Central incisor**	**Lateral incisor**	**Canine**	**First molar**	**Second molar**
Nomenclature	5.5	5.4	5.3	5.2	5.1	6.1	6.2	6.3	6.4	6.5
Age of emerge	25–33 months	13–19 months	16–22 months	9–13 months	8–12 months	8–12 months	9–13 months	16–22 months	13–19 months	25–33 months
Age of emerge	23–31 months	14–18 months	17–23 months	10–16 months	6–10 month	6–10 month	10–16 months	17–23 months	14–18 months	23–31 months
Nomenclature	8.5	8.4	8.3	8.2	8.1	7.1	7.2	7.3	7.4	7.5
Tooth	Second molar	First molar	Canine	Lateral incisor	Central incisor	Central incisor	Lateral incisor	Canine	First molar	Second molar
			Lower right					Lower left		

Table 1.
Nomenclature and age of emergence of deciduous dentition.

molars. The evolution of permanent dentition ends at about 14–15 years of age when the roots have completed their development. The age of emergence of permanent teeth is described in **Table 2**. The exception is related to third molars, which are emerging at 18–25 years of age [1, 3].

Teeth have two primary components, one *crown*—exposed in the oral cavity—and one or more *roots* supported by the alveolar socket. The tooth has four primary tissues —*enamel, dentine, cement, and pulp*. The main component of a tooth is a bone-like tissue called *dentine*.

Enamel covers the visible part of the crown in the oral cavity and is a highly calcified tissue (95.5% inorganic, 0.5% organic, and 4% water). The upper incisors reach the highest density of enamel, which is increasing progressively during development. The thickness of the enamel layer varies from about 2.5 mm in the cusps area, 2.0 mm in the incisal edge, and 0.5 mm at the cervical enamel. The color is semitranslucent and depends on the enamel thickness. In the thick opaque area, the color appears bluish-white and yellow-white in the area where the enamel layer is thinner. Enamel is the hardest structure of the body, having a value of 5–8 on the Moch scale (diamond-10 Moch). The high hardness influences the tensile strength and compressibility, which indicate that enamel is extremely brittle [4, 5].

The enamel is structured in *enamel rods (prisms), rod sheaths, and cementing interred substance*. Generally, enamel rods are aligned perpendicular to dentine-enamel junction and usually in cervical area the rods become twisted. The rods are aligned in planes best suited to withstand the occlusal forces as long as the forces are perpendicular to the tooth surface. Rod sheath is an area identified in histologic sections of a tooth, found where enamel rods, the functional unit of enamel, meet interrod enamel. Both types of enamel meet at sharp angles and form the appearance of a space called the rod sheath. The rod sheath consists of more protein and the rod sheath is characterized as being hypomineralized in comparison to the rest of the highly mineralized enamel. The rod sheath is Inorganic matrix tying the enamel rods together. The interred substance cement the rods together. Enamel forms junctions with dentine and cement. The *dentine-enamel junction* (DEJ) has a pitted aspect and in the rounded pits fit the enamel rod, which generates a strong bond between the two tissues [3, 4].

Underlying the enamel tissue is the dentine root, a bone-like tissue that consists of a spongy bone-like tissue, which consists of the root, and is covered by *cementum* which is present only on the tooth surface. The crown delimited by the enamel and the root delimited by the cement join in a junction called cementoenamel junction (CEJ), which is also called the cervical line. The cementum is specially adapted to anchor and support the tooth in the bony socket and covers the dentine root. Both enamel and cement are covering the dentine, and three variations in their link may exist: the cervical enamel is covered by cementum (65%), contact line between enamel and cementum (25%), or enamel and cement do not touch, and in this situation, dentin is exposed (10%) [3, 6].

Dentin is the largest tissue of a tooth, gives the basic shape of the crown and root, and forms the walls of the pulp cavity in the crown and root area. The color is light yellow in deciduous teeth and yellow in permanent teeth, less hard than enamel, and harder than the cementum. In contrast to the enamel, dentine is highly elastic, can support the non-resilient enamel, and is highly permeable. In its chemical composition, dentine has 70% inorganic matter, 17% organic matter (collagen, proteins, and citric acid), and 13% water. Dentine is structured into dentinal tubules that are ending beneath the enamel and contain the Tome's fibers also called processes of odontoblasts cells present in the dental pulp. For this reason, dentine is a living and sensitive hard tissue of the tooth. Dentine tubes have an "S" curved shape and start perpendicular to

M3	M2	M1	P2	P1	C	LI	CI	CI	LI	C	P1	P2	M1	M2	M3
1.8	1.7	1.6	15	1.4	1.3	1.2	1.1	2.1	2.2	2.3	2.4	2.5	2.6	2.7	2.8
17–21 years	12–13 years	6–7 years	10–12 years	10–11 years	11–12 years	8–9 years	7–8 years	7–8 years	8–9 years	11–12 years	10–11 years	10–12 years	6–7 years	12–13 years	17–21 years
17–21 Years	11–13 years	6–7 years	11–12 years	10–12 years	9–10 years	7–8 years	6–7 years	6–7 years	7–8 years	9–10 years	10–12 years	11–12 years	6–7 years	11–13 years	17–21 years
4.8	4.7	4.6	4.5	4.4	4.3	4.2	4.1	3.1	3.2	3.3	3.4	3.5	3.6	3.7	3.8
M3	M2	M1	P2	P1	C	LI	CI	CI	LI	C	P1	P2	M1	M2	M3

CI—central incisor, LI—lateral incisor, C—canine, P1—first premolar, P2—second premolar, M1—first molar, M2—second molar, and M3—third molar.

Table 2.
Nomenclature and age of emerge for permanent dentition.

the pulp chamber surface but are straight under the cusps and in the root area. The dentine encloses the *dental pulp*, the soft tissue formed by blood vessels, nerves, and specialized tissue, all responsible for the tooth's vitality [3, 6].

Dental pulp is contained within the pulp chamber of the crown, close to the cervical line and in the cervical third part of the crown, and is followed by the pulp canal along the root. The pulp chamber and root canal form together the pulp cavity. Pulp has the main role in forming the dentine, has a sensory function through which the tooth is felt, and also has a nutrient function by supplying nutrition to the dentine through the blood vessels and odontoblastic processes with an important role in maintaining the tooth's vitality. The last function is the defensive one, through which the pulp is producing secondary dentine to maintain the vitality of the tooth in case of damaging factors such as cavities, occlusal overloading, and aging. In young teeth, the pulp chamber is bigger and follows the shape of the crown through the extension of pulp chamber into the cusps. These extensions are called pulp horns. By aging, the pulp chamber is reducing its size because of secondary dentine deposition. The root canal and apical foramen are wide in case of young teeth and become narrow by aging. Root canals can be straight or curved, single or not, with accessory lateral canals usually in the apical foramen or at the floor of the pulp chamber. *Apical foramen* or *root apex* is a small foramen through which the blood vessels and nerves enter the pulp chamber and may be located on the lateral side of the apex. A tooth can have more than one foramina. Through apical foramen, there is a communication with periodontal space. Endodontic and periodontal pathology are interconnected and influenced one by another through apical foramen or accessory foramens and root canals [7].

The three hard tissues of the tooth, enamel, dentine, and cementum, and the soft tissue represented by dental pulp have to be considered in relation with the oral-facial structures, oral cavity, and all surrounded soft tissues and anatomical structures.

2.2 Surfaces, ridges, and landmarks in dental morphology

The crowns of anterior teeth have four axial surfaces and one incisal ridge. The premolars and molars have five surfaces, four axial, and one occlusal. The name of the surface is given after the anatomical position and relation to the adjacent anatomical structures. For upper and lower incisors and canines, the surfaces toward the lips are called *labial* surface. The surfaces that face the median line of the face or toward adjoining teeth are called *mesial* and opposite or the ones distant from the median line are called *distal* surfaces. Mesial and distal surfaces are also called *proximal* surfaces [1, 3].

The surfaces facing the tongue are called *lingual.* In case of premolars and molars, the name of the surfaces is maintained with one exception, the surfaces facing the cheeks are called *buccal* surfaces. Buccal and labial surfaces, when spoken collectively, are called *facial* surfaces. The surfaces of premolars and molars, which establish contact or occlusion with the ones positioned on the opposite jaw, are called *occlusal* surfaces and *incisal ridge* with respect to incisors and canines (**Figure 1**) [1, 3].

Beyond surfaces, teeth have other landmarks divided into positive and negative landmarks. The positive landmarks are cusps, tubercle, cingulum, marginal ridge, triangular ridges, transversal ridges, oblique ridges, and lobes. The negative landmarks are fossa, sulculus, pits, and developmental grooves [1, 3].

Cusps are elevations specific to premolar's and molar's crowns and divide the occlusal surfaces. Every cusps has a pyramidal shape with a quadrangular base with the exception of the mesial-lingual cusp of the upper first molar. Tubercles are small

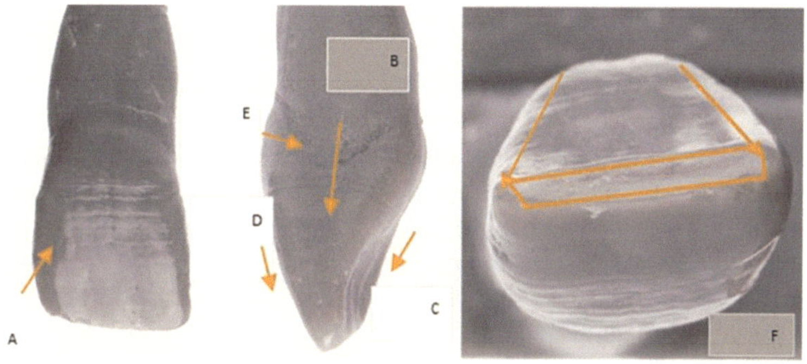

Figure 1.
Upper central incisor. A—Buccal or facial surface, B—mesial surface-proximal view, C—lingual surface from proximal view, D—facial surface from mesial view, E-cervical line—mesial view, and F—Incisal view—incisal ridge.

elevations, similar in shape to the cusps but are actually an extra formation of enamel. *Cingulum* is the lingual lobe of anterior teeth and gives a bulk aspect in the cervical third of the lingual surface. *Ridges* are any linear elevation on the surface of the tooth and receive the name after the position on the surface. *Marginal ridges* are the rounded borders of the enamel that form and limit the margins of the occlusal surface. Each premolar and molar has a mesial and distal ridge on the occlusal surface. The *transversal ridges* start from the tips of the cusps of laterals toward the center of the occlusal surfaces and are named after the cusps to which they belong, like triangular ridge of the mesial-buccal cusp of the first mandibular molar. The union of two triangular ridges crossing on transversal direction forms the *transversal ridges*. The *oblique ridge* is specific to maxillary molars and is formed by the union of the triangular ridges of the distal-buccal and mesial-buccal cusps. *Lobes* are the primary structures in the development of the crown and cusps and mamelons are representations of lobes. *Mamelons* are any rounded protuberances present on the incisal ridges of recently emerged teeth, and by aging, their contour will become less evident [1, 3].

Fossae are irregular depressions or concavities present on the occlusal surfaces of the premolars and molars and on the lingual surfaces of the incisors. The *central fossae*, present on the occlusal surfaces, are formed by the convergences of ridges with the grooves. The *triangular fossae* are present on the occlusal surface of lateral teeth, mesial or distal to the marginal ridge, and also on the lingual surface of the maxillary incisors where the lingual fossae and marginal ridge meet the cingulum. *Pits* are point depressions found at the junction of developmental grooves [1, 3].

For an accurate description and positioning of the landmarks, the crown and roots are divided after the third rule, which means that each surface of the crown is divided into three equal parts in mesiodistal direction and in the cervical-incisal/occlusal direction. Following this rule, the buccal surface of the central upper incisor has three-thirds on cervical-incisal distal direction, which are third incisal, third middle, and third cervical. The buccal surface of the central upper incisor has three-thirds in mesiodistal direction, which are third mesial, third middle, and third distal. By following this rule, a landmark can be placed more precisely, for example, the mesio-incisal angle of the maxillary central incisor is placed in the third incisal (in cervical-incisal direction) and in the third mesial part in mesiodistal direction. The same rule is valid for the root [8].

3. Dental anatomy of the anterior teeth

3.1 Common features of permanent incisors and canines teeth

The incisors are the cutting blades of human dentition and together with the canines form the group of anterior teeth with great impact on esthetic, eating, speech and facial expression, and harmony. Both deciduous and permanent dentitions have four incisors on each jaw, two central incisors, and two lateral incisors, all single root teeth. The central incisors are the closest to the midline, whereas the lateral is more distal to the midline of the jaw. The mandibular central incisors emerge around 7 years of age, followed by the maxillary central incisors, maxillary lateral and the last ones, mandibular lateral incisors around 8 years of age. Mandibular incisors have the smallest and shortest root and the weakest resistance due to their reduced size. The relation between upper and lower incisor is forming the overjet and overbite with great impact on esthetic and normal relation between the upper and lower jaw. The evaluation of overbite and overjet is done early and gives important information about the growth pattern of the jaws, the length of the dental arches, and optimal space for permanent teeth. The *overbite* is the overlap of the maxillary central incisors over the mandibular central incisors measured relative to the incisal ridges. Normal overbite is approximately 1/3–1/2 (30 and 50%) of the height of the mandibular incisors. *Overjet* is defined as the extent of horizontal overlap, anterior-posterior direction, of the maxillary central incisors over the mandibular central incisors. The normal value of the overjet should be maximum 2 mm. The relations between the upper and lower incisors indicated a physiologic occlusion and development of the jaws or a malocclusion [9, 10].

Canines are the third teeth in line, from the median line of the dental arch and are refer as cornerstones of the dental arch. Canine's roots are the longest among all the teeth and the crowns are usually as long as the crowns of central incisors. Canines are single root teeth and their "V" crown shape and position are important for the guidance of the lower jaw into the intercuspal position. These teeth have a great value for an efficient function, stability, and natural facial expression. Lower permanent canines emerge around 9–10 years of age and the upper permanent canines emerge at 11–12 years of age, being through the last permanent teeth that emerge between incisors and premolars. Because of their timing emerge, placement between incisors and premolars and molars, which already emerged, canine's space can be restricted and for this reason are predisposed for malposition, crowding and malocclusion. Being located between incisors and premolars, canines support these teeth due to their shape, size, and position on the dental arch. A quality that has to be overlooked is the canine eminence that along with the bone ridge over the labial roots has an esthetic value by contouring the normal facial expression at the corner of the lips. Canines value is manifested through stability, efficiency, and natural facial expression [4].

3.2 Maxillary and mandibular incisors

3.2.1 Maxillary incisors

Maxillary central incisors (1.1, 2.1) are single cone shape root teeth, next to the median line. In smile and speech, the central incisors are the most exposed teeth with a great impact on esthetics. Maxillary central incisors (1.1, 2.1) have four axial surfaces: labial, lingual, mesial and distal, and one incisal ridge.

The labial face is the widest, mesial-distally, and convex in the cervical incisal direction. It is less convex than the lateral incisal and canine and has a rectangular shape. The labial surface looks symmetrical and regularly formed for both centrals and is having a straight incisal edge. The height of the crown is 10–11 mm and mesial-distal measurement is 8–9 mm at the contact area. The mesial-distal diameter in the cervical area is 6.5–8 mm. The maximum convexities from the mesial and distal surfaces are consisting the contact area with its neighbors. The mesial outline is less convex than the distal one, and these two surfaces by continuing with the incisal ridge are forming two angles: mesial-incisal, more sharp than the distal-incisal. The incisal outline is usually regular and straight in a mesial-distal direction, but after a long function, it tends to become ascendant distally. The labial face has three lobes named mamelons, divided by two cervical-incisal groves. Due to function, the mamelons tend to become less evident. The cervical outline of the crown has a semicircular direction and is concave toward the incisal ridge (**Figure 2**). From proximal view, the facial surface is convex and the lingual surface is convex in the third cervical and concave in the third incisal area [11].

The lingual outline of the cervical line is placed below a smooth convexity called cingulum and has a similar shape with the labial cervical line. Cingulum is placed in the third cervical part of the lingual surface. Mesial and distal from the cingulum are the mesial and distal marginal ridges, which continue the incisal ridge. Below the cingulum is present a shallow concavity called lingual fossa, which is bordered mesially by the mesial marginal ridge, incisally by the lingual area of the incisal ridge, distally by the distal marginal ridge, and cervically by the cingulum. From the cingulum starts developmental grooves into the lingual fossa.

The mesial and distal surfaces have a triangular shape with the base in the cervical and the apex oriented toward the incisal ridge. The maximum convexities of both mesial and distal surfaces are joining with neighbor teeth into the contact area that is positioned into the third incisal and third labial of the proximal surface.

Viewed from incisal, the incisal ridge is positioned in the middle of the maximum labial-lingual diameter of the central incisor and in the same time, centered over the root (**Figure 3**).

The maxillary central incisor may show a wide range of particularities with regard to labial outline, labial profile curvature, labial lobe grooves, mamelons, and cingulum and are classified into three basic shapes: tapering when mesial and distal borders converge toward cervical line, square when mesial and distal borders are almost

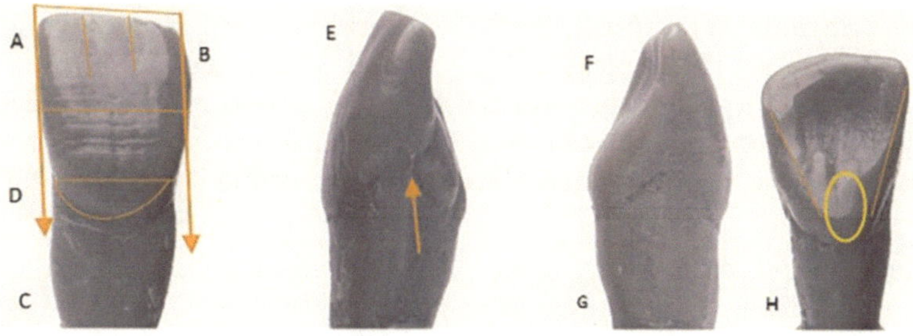

Figure 2.
A. Facial surface—the three mamelons divided by developmental grooves; B. Outline of distal ridge; C. Cervical line on labial surface; D. Outline of mesial ridge; E. Mesial surface with the view of labial surface; F. Distal surface —aspect of cervical line and outline of labial surface; G. Distal surface—aspect of cervical line and outline of lingual surface; and H. Lingual surface—view of cingulum, marginal ridges, and developmental grooves.

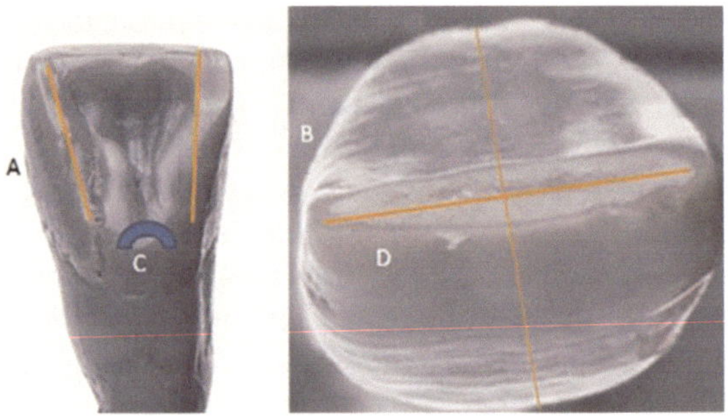

Figure 3.
A—Distal marginal ridge, B—Mesial marginal ridge, C—Cingulum, D—Incisal ridge from incisal view.

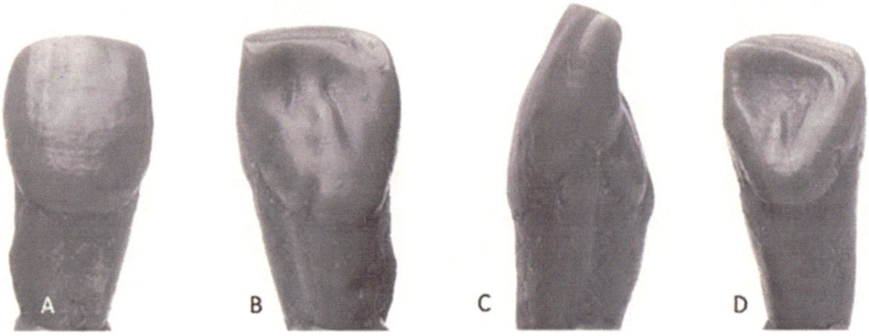

Figure 4.
Upper lateral incisor. A—Labial surface, B—lingual surface, C—mesial surface, and D—Lingual-distal surface.

parallel, and an ovoid shape when mesial and distal borders converge incisal and cervical [1, 3, 11].

Maxillary lateral incisor (1.2, 2.2) varies more than any tooth and is smaller in all dimensions except the root length that has a significant distal curvature of the apex (**Figure 4**). The root has one root canal, with the pulp chamber centered within the root. On cross-section, the root shows a large variation in shape and may be triangular, oval, or round. The crown measures from 2 to 3 mm shorter than the central incisor. The labial aspect is similar with the central incisor but has more accentuated curvature, rounded incisal ridge, and rounded incisal angles. The mesial outline of the labial face resembles with the central incisor but with a rounded mesial-incisal angle. The distal outline is always more rounded and shorter than the mesial one. On labial surface, the cingulum is prominent with deep developmental grooves within a more concave lingual fossa. The mesial surface is similar with the central incisors one, the distal surface appears thicker than the mesial one, and the cervical line is usually more cervically than it is on the mesial surface. The incisal aspect can be similar with the central incisor, but it also may resemble with a small canine [1].

3.2.2 Mandibular incisors

Mandibular central incisors (3.1, 4.1) are single-root anterior teeth, the smallest ones with regular and symmetric surfaces and outlines. The crown is about half of the mesial-distal diameter of the maxillary central incisor and the labial-lingual diameter is only 1 mm less. Because of its size, the mandibular central incisor is the only tooth that has occlusal contact with only one tooth, the upper central incisor. Except mandibular central incisor, if present, the maxillary third molars have occlusal contact only with the third mandibular molars. The permanent central incisors emerge at 6–7 years of age, after the permanent molars.

The labial surface is straight and its long axis is continued by the root, is wider in mesial-distally then the lingual surface, and is wider in the cervical third because of the presence of a smooth cingulum. From incisal view, more labial surface can be seen. The mesial and distal outline are parallel and slightly tapered toward cervical line. The labial surface is regular, convex, and flattened with a convexity in the middle third. The mesial-incisal angle formed by the incisal margin and mesial outline is straight with right angles and is characteristic of lateral lower incisor.

The lingual aspect is smooth with a concavity in the incisal third placed between the marginal ridges. In some cases, the marginal ridges are more prominent near the incisal edge and the concavity becomes more contoured. The lingual surface is flat in the incisal third and convex in cervical third. The cingulum is not marked by any developmental grooves.

The mesial surface is smooth in the incisal third, flatter in the middle third, and flat in the cervical third receiving a slightly convex line. Immediately below the middle third a concavity is present.

The distal surface is similar with the mesial one and has a developmental depression on the distal surface of the root with a deeper and defined groove. The cervical line is placed incisal with about 1 mm on the mesial surface.

Usually, the lower lateral incisor does not exhibit too many variations, however, labial surface may have a degree of labial inclination and over contoured mamelons separated by well-defined grooves (**Figure 5**). The pulp cavity and root canal are narrow but can also be very large in size. Usually, the tooth has one root canal narrow in mesial-distal section but wider on labial-lingual cross-section. This tooth may have the second root canal [1, 3, 11].

3.3 Maxillary and mandibular canines

The crown of upper canines is usually as long as the crowns of central incisors, but the single strong developed root is the longest than those of any teeth. Canine crowns have a single pointed cusp narrower more lingual than labial. The labial surface has a smaller mesial-distal diameter in comparison with central incisors with about 1 mm.

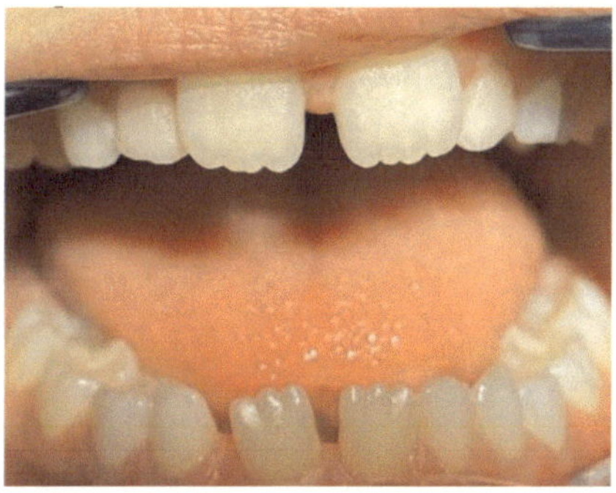

Figure 5.
Labial view of upper and lower central incisors in mixed dentition—the mamelons and developmental grooves are well defined.

The cervical line is concave toward incisal on facial and lingual surface and convex toward incisal on mesial and distal surface as it is for any tooth. The mesial outline is convex from the cervical line and forms the mesial contact area, which is approximately at the middle third to the incisal third junction. The distal outline is usually concave between the cervical line and distal contact area that is found in the center of the middle third of the distal surface.

The cusp tip is on the same line with the center of the root. The cusp has two slopes, the mesial one being shorter than the distal one. The labial surface of the cusp is smooth with ought any developmental lines, except the shallow depressions that divide the three lobes. The middle labial lobe is the most developed one and forms a ridge on the labial surface. The connection of the outline of the mesial slope of incisal ridge with the outline of the mesial surface forms a rounded angle positioned more incisal than the distal angle. The same link is present for the distal slope outline with the outline of the distal surface but in this case, the angle is more rounded and placed more cervically.

The lingual surface of the crown is narrower than the labial one and has a large cingulum which can look like a small cusp. In this morphology type, the ridges and developmental grooves of the lingual surface are well-defined. The marginal ridges are converging, and link with the cingulum and occasionally a well-defined lingual ridge is confluent with the cusp tip and extends near the cingulum. Between this lingual ridge and the marginal ones are evident shallow concavities called mesial and distal lingual fossae. In other cases, the lingual surface is smooth and the fossae and ridges are difficult to distinguish (**Figure 6**).

The lingual surface of the root is narrower than the labial one and the mesial and distal developmental depressions are evident on the root surface (**Figure 6**) [1, 3, 11].

The corwns of mandibular lower canines (3.3, 4.3) are narrower mesial-distal, less than 1 mm when compared with the maxillary ones but are longer with about 0.5–1 mm. The root is usually of same size or shorter and about 1 mm wider than lateral lower incisors. The labial surface has a pentagon shape, and the lingual surface is smoother with a less defined cingulum and marginal ridges being similar with the lingual surface of the lower lateral incisor. The cusp is less structured, and the cusp ridge is thinner labio-lingually in comparison with the upper canine. The cusp tip is usually on a line with the center of the root from mesial to distal, but it can also lie lingualy toward this line, same as mandibular incisors. Lower canines can have bifurcated roots, and this situation is not rare.

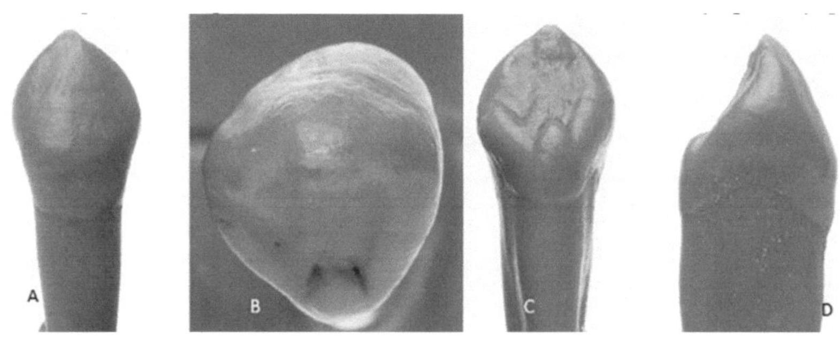

Figure 6.
A—"V" shape labial surface, labial ridge, mesio-incisal, and distal-incisal angles, B—incisal view of the cusp, labial, and lingual surface, C—lingual surface, cingulum, marginal ridges, developmental grooves, and fossae, and D—mesial surface with the high position of the cervical line.

The differences between the labial surfaces of maxillary and mandibular canines refer to a longer appearance of the lower one because of the narrower mesial-distal diameter. The height of the contact areas is placed above the cervix. The mesial outline is almost straight if the mandibular canine and contact area are positioned near the mesio-incisal angle, and the distal contact area is also placed toward incisal. The distal surface is very similar with the distal surface of the upper canine [1, 11].

4. Dental anatomy of lateral teeth

4.1 Common features of permanent premolars and molars

Premolars, referred as posterior or lateral teeth, are succedaneous teeth present only in permanent dentition and replace the deciduous molars. There are four premolars on each jaw and two on each quadrant. Premolars emerge between 10 and 12 years of age, before the permanent canines and second molars. First maxillary and mandibular premolars emerge earlier than second maxillary and mandibular premolars. These teeth are referred as bicuspids and are taking their place on the dental arch between canines and molars having a transitional morphology between the canines and molars.

Both maxillary premolars (1.4, 1.5, 2.4, 2.5) are developed from four lobes same as anterior teeth, but, comparing with the anterior, the lingual cusp is well-defined. The buccal cusps of first maxillary premolars are long and sharp similar with the canine's cusp. The second premolar has a shorter crown and root, the root being equal in height with the molar's roots though the crowns are slightly longer. Usually, the first maxillary premolar has two roots—one buccal and one lingual, and the second maxillary premolar has one root.

Permanent mandibular premolars (3.4, 3.5, 4.4., 4.5) have the same position on the dental arch as the maxillary ones, replace the mandibular deciduous molars, and are developed from four lobes, same as maxillary ones.

The first mandibular premolar has two cusps, the buccal one is larger and longer and the lingual one is much smaller, similar with the cingulum, and is nonfunctional, the morphology of this tooth being very similar with the mandibular canine's. Second premolars have three cusps, one buccal, and two small lingual cusps being more similar with a molar. The first mandibular premolar is smaller than the second one, but the first maxillary one is bigger than the second one.

Molar's most important function is mastication, though the crowns and roots are considerable in size, except the crown's height which is shorter than premolars. The two roots of mandibular molars and the three roots of maxillary molars are longer and curved.

Maxillary molars (1.6, 1.7, 1.8, 2.6, 2.7, 2.8) have three roots, two buccal and one lingual, one massive crown with well-developed two buccal cusps, and two lingual cusps. The first permanent maxillary molar emerges at 6 years old and soon after emerges the mandibular first molars. Both emerge distally by the second deciduous molars and do not replace any deciduous tooth. The first molar's occlusal relation is important for anticipating malocclusions.

Mandibular molars (3.6, 3.7, 3.8, 4.6, 4.7, 4.8) are the largest teeth, showing variations related to cusps number, from 3 to 5, size, occlusal surface, length of the crowns, and root. The outlines of the crowns are similar, quadrilateral, for all three molars, and each has two roots, a mesial, and a distal one. Third molars show a fusion of the two roots. The crowns are shorter cervical-occlusal, but the mesiodistal and buccal-oral diameters are much larger than those of the mandibular anterior teeth.

The roots are bifurcated, not so long but more bulky and thick, and assure a great anchorage and stability of the tooth in alveolar socket [1, 3, 11].

4.2 Maxillary and mandibular premolars

The first maxillary premolar (1.4, 2.4) emerges first after the canine at age of 10. Premolar's crowns have four axial surfaces (mesial, distal, buccal and lingual) and one occlusal surface. The two cusps, buccal and lingual, form the occlusal surface. Both are well shaped and defined. The buccal cusp is usually longer than the lingual one. The long axis of the crown is continuing the long axis of the root.

The buccal and the mesial aspect of this tooth is trapezoidal, and the crown shows a little curvature at the cervical line from a buccal point of view. Lingually, its gross outline is the reverse of the buccal gross outline.

When it comes to the distal aspect, the crown of the first maxillary premolar differs from the mesial aspect. The crown surface is convex at all points except for a small flat area, the curvature of the cervical line which is less on the distal than on the mesial surface, usually there is no groove crossing the distal marginal ridge of the crown, and the root is flattened on the distal surface above the cervical line.

The occlusal aspect of this tooth resembles a six-sided or hexagonal figure, which are called mesiobuccal, mesial, mesiolingual, distal, distolingual, and distobuccal. The two buccal sides are approximately equal, but the mesial side is shorter and the mesiolingual side is also shorter than the distolingual one. The occlusal surface is limited by the cusp and marginal ridges, which are in the same line with each other. The two ridges of the buccal cusp, the mesiobuccal, and distobuccal are in line, but their alignment is on distobuccal direction and for this reason the distobuccal cusp ridge is buccaly placed toward mesiobuccal cusp ridge. The two cusps are dived by the central groove placed in the center of the occlusal surface (**Figure 7**). Usually, the surface of the tooth has no supplemental grooves, which makes the surface relatively smooth but has two pits, a mesial and a distal.

In most cases, it has two roots and two pulp canals or even two buccal roots, similar with molars. In the less cases in which it only has one root, there are still two canals found. However, the first maxillary premolar root(s) may be irregularly curved or distally inclined in the apical third [1, 3, 11].

The maxillary second premolar (1.5, 2.5) is supplementing the first one in function. It is less angular, which gives a more rounded effect to the crown. Just like the first

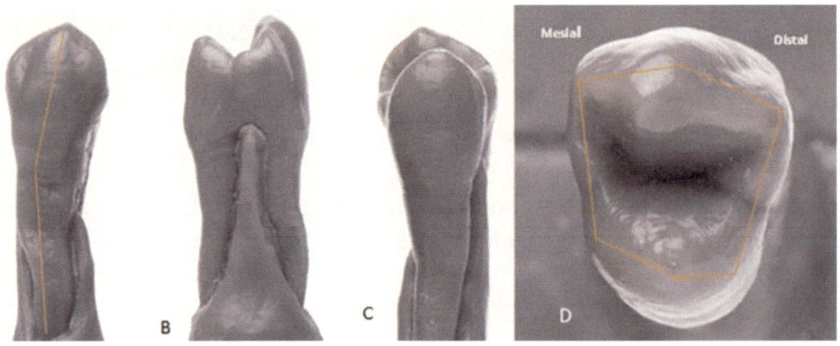

Figure 7.
Maxilary first premolar; A—Buccal surface—"V" shape buccal cusp, buccal ridge, and inclination toward the root, B—Medial surface—buccal cusp is higher than the lingual one, C—Lingual surface—lingual cusp is smaller than the buccal one, D—Occlusal surface—hexagonal shape, slopes of the cusps, marginal ridges, central groove, and mesial and distal pits.

one, it has two cusps, of the same height, but are less sharp. The buccal cusp is not as long as one of the first maxillary premolars.

From a buccal point of view, in many cases, the crown and the root of this tooth are thicker at their cervical area. Also, the buccal ridge of the crown may not be so prominent in comparison with the first maxillary premolar.

Lingual, the crown of this tooth has a trapezoidal form, and because the buccal and lingual cusps are almost the same length, the shape of it is generally symmetrical.

From mesial view, the buccal and lingual cusps are nearly equal as height, but the buccal one is slightly more prominent than the lingual one. The mesial marginal ridge is not crossed by a groove. Also, this surface of the crown is not marked by concavity, but it is evenly convex from the cervical line to the marginal ridge. Distally, the aspect of the second maxillary premolar is similar to the first one.

The occlusal aspect of this tooth is differentiated in many ways from the first maxillary premolar. The shape is ovoid instead of hexagonal, and the mesial and distal borders show little to any lingual convergence. This trait along with the equality of the cusps determines a rectangular shape of the tooth. The central groove is placed more lingual and as a result, the buccal cusp appears to be larger.

Most maxillary second premolars only have one canal and one root. In some exceptions, we can find two roots with two canals or one root with two canals [1, 3, 11].

4.2.1 The mandibular first premolar

The mandibular first premolar is the first tooth from the mandibular lateral group. It is situated between the canine and the second mandibular premolar, and it has characteristics from both.

The particularities that resemble those of the canine: the buccal cusp is long and sharp, and it is the only occluding cusp; the buccolingual distance is approximately equal with the canine one; the occlusal surface slopes sharply lingually, in cervical direction; the mesiobuccal cusp ridge is shorter than the distobuccal one; and the outline of the occlusal aspect resembles the outline of the incisal aspect of the canine.

The particularities that resemble those of the second mandibular premolar: except for the longer cusp, the outline of the crown and root resembles the second premolar; the contact areas, mesially and distally are similar; and the tooth has more than one cusp.

Buccal, the crown, is roughly trapezoidal. The cervical margin is represented by the shortest of the uneven sides. The middle buccal lobe is well-developed, creating a large, pointed buccal cusp. The distal cusp ridge is longer than the mesial one.

From a lingual point of view, the crown tapers toward the lingual, since the lingual measurement mesiodistally is less than that buccally. The lingual cusp is always small. A major part of the crown is made up of the middle buccal lobe. This makes it resemble the canine.

From the mesial aspect, the crown is roughly rhomboidal. This tooth shows an outline that is fundamental and characteristic to all posterior teeth when viewed from the mesial or distal aspect. The convexity of the outline of the lingual lobe is lingual to the outline of the root.

The distal aspect of the mandibular first premolar is different from the mesial one in some respects. The shape is spheroidal, and it has an unbroken curved surface.

Occlusal, we can observe considerable variation in the gross outline of the tooth. Both mandibular premolars exhibit more variations in form occlusally than the maxillary premolars. The common characteristics of all mandibular first premolars, regardless of type, from occlusal point of view are: the buccal lobe in the middle makes up the major bulk of the tooth crown; the buccal ridge is prominent; and the mesiobuccal and distobuccal line angles are prominent even though they are rounded.

In most cases, it has one root and one canal, and in very rare cases, two roots and two canals. The mesial and distal surfaces of the roots are wider than the buccal an lingual, so the root canal will follow the same pattern [1, 3, 11].

4.2.2 The mandibular second premolar

The mandibular second premolar is resembling the first premolar from the buccal aspect only. Although the buccal cusp is not so pronounced, the mesiodistal proportion of the crown and its general outline are similar. There are two common forms of this tooth: the first form, which probably occurs more often, is the three-cusp type, and the second is two-cusp type. The two types differ mainly from the occlusal point of view.

From the buccal aspect, the mandibular second premolar (**Figure 8**) presents a shorter buccal cusp than the first one, with mesiobuccal and distobuccal cusp ridges describing angulation of less degree. The contact area seems higher because of the short buccal cusp.

Lingual, this tooth's crown shows considerable variations from the crown portion of the first premolar: lingual lobes are developed to a greater degree; less of this occlusal surface is visible. In the three-cusp type, the lingual development creates the greatest variation between the two teeth. In the two-cusp type, the lingual cusp development attains equal height as the three cusp. This surface of the crown is smooth and spheroidal, showing a bulbous from above the constricted cervical portion.

Mesial, the second premolar differs from the first one as follows: the crown and root are wider buccolingually; the buccal cusp is not centered over the root trunk; the lingual lobe development is greater; the marginal ridge is at right angles at the long axis of the tooth; less of the occlusal surface is visible; no mesiolingual developmental groove on the crown; the root is longer and generally slightly convex on the mesial surface; and the aspect of the root is usually blunter on the second premolar.

The distal aspect of this tooth is similar to the mesial aspect, except that more of the occlusal surface may be visible. This is possible because the distal marginal ridge is at a lower level than the mesial marginal ridge when we look at the tooth vertically.

From the occlusal point of view, the outline form of each type shows variations. The square or three-cusp type is square lingual to the buccal cusp ridges when

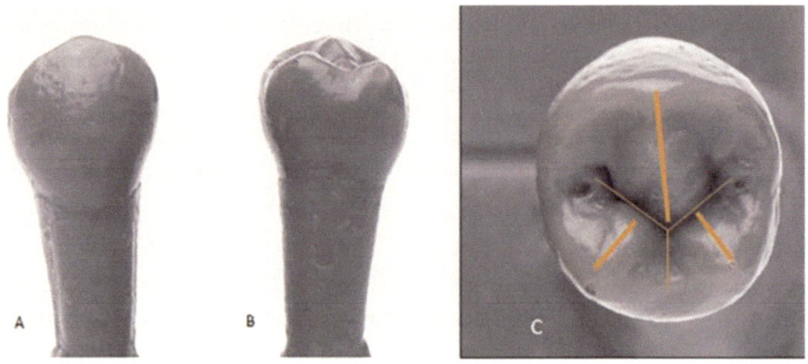

Figure 8.
Mandibular second premolar; A buccal surface and root, B lingual surface, and C occlusal surface.

developed. The three cusps are unequal: the buccal is the largest, the mesiolingual is next, and the distolingual is the smallest. The round or two-cusp type viewed from this point is much different than the three-cusp one: the outline of the crown is rounded lingual to buccal cusp ridges; there is some lingual convergence of mesial and distal sides, no more than in the square type; the mesiolingual and distolingual line angles are rounded; and there is one well-developed lingual cusp directly opposite the buccal one in a lingual direction.

For the majority of the population, the mandibular second premolar has a single root and a single root canal, but rarely it can also have two roots and two root canals [1, 3, 11].

4.3 Maxillary and mandibular molars

4.3.1 The maxillary first molar

The maxillary first molars (1.6, 2.6) are the largest and strongest teeth on the maxillary arch (**Figure 9**). It has four well-formed large cusps, and a small low functioning one called the Carabelli tubercle is placed on the mesio-lingual cusp. The main cusps are the mesiobuccal, distobuccal, mesiolingual, and distolingual, while the fifth, Carabelli tubercle, can take the form of a well-developed cusp, or it can downgrade to a series of depressions, grooves, and pits on the mesial portion of the lingual surface.

From a buccal aspect, the crown is roughly trapezoidal. It has cervical, and occlusal outlines representing the uneven sides, the cervical one being the shorter one. The buccal developmental groove dividing the buccal cusps is more or less equidistant between the mesiobuccal and distolingual line angles.

Lingually, the gross outline of the maxillary first premolar is the reverse of the buccal aspect. From this point, we can see the Carabelli tubercle situated on the mesiolingual cusp.

From a mesial point of view, we can observe the increased buccolingual dimensions and the cervical curvature of the crown outlines at the cervical third buccally and lingually. The cervical line is irregular, and it curves occlusaly. Distally, the gross outline of the first maxillary molar is the reverse of the mesial aspect.

The occlusal aspect of this tooth shows that its shape is roughly rhomboidal. An outline following the four major cusp ridges and the marginal ridges is especially so.

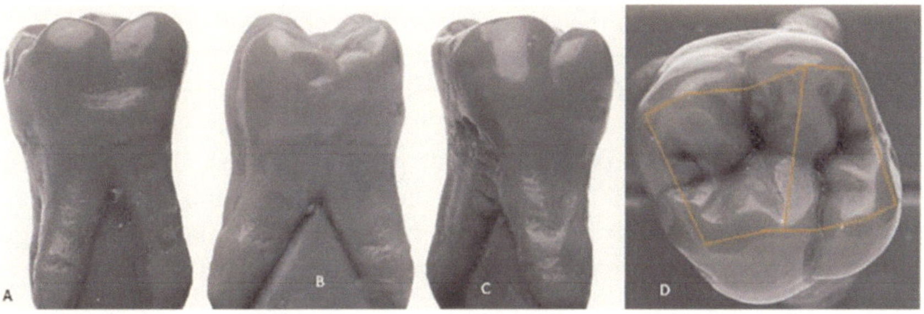

Figure 9.
A Buccal surface and buccal roots, B distal surface, distal-buccal and lingual root, C lingual surface and root, and D occlusal surface and oblique ridge.

Because of the mesiodistal and buccal-lingual diameter, the crown is wider mesially than distally and wider lingually than buccally. The largest cusp is the mesiolingual, followed by the mesiobuccal, distolingual, distobuccal, and Carabelli tubercle.

The occlusal surface has four fossae, two major fossae, the mesial one is triangular and the distal one is linear and two minors ones, a mesial and a distal one, both triangular in shape. The marginal ones are outlined by the marginal ridges, and the central ones are divided by the oblique ridge.

The oblique ridge is crossing the occlusal surface and makes the union of the triangular ridges of the distobuccal and mesiolingual cusps. The oblique ridge is reduced in size in the center of the occlusal surface and in some morphological variations is crossed by a developmental groove and the two major fossae are connected. The mesial and distal marginal ridges are irregular and confluent with the mesial and distal cusp ridges of the major cusps. The central fossa is concave and connected with developmental grooves and also short grooves, and central developmental pit.

There are three well-developed roots, two buccal and one lingual. They are well separated, which gives maximum anchorage against the forces that tend to unseat them. The lingual root is the largest, the mesiobuccal one is a little shorter but broader, and the distobuccal is the smallest. The percentage in which the first maxillary molars development deviates from the normal is small [1, 3, 11].

4.3.2 The maxillary second molar

The maxillary second molar supplements the first one in function. Its roots are as long or even longer than the first molar's. The distobuccal cusp is not as large and well-developed, and the distolingual cusp is smaller. The crown of this tooth is approximately 0.5 mm shorter cervico-occlusally than the one of the first molar, but buccolingually the dimensions are about the same.

Buccaly, the crown is slightly shorter cervico-occlusally and narrower mesiodistally than the maxillary first molar. The apex of the mesiobuccal root is on the same imaginary line as the buccal groove of the crown.

Lingual, there are a few important differences between the second and first molars. The distolingual cusp of the crown is smaller and there is no fifth cusp evident. Also, the distobuccal cusp may be seen through the sulcus between the mesiolingual and distolingual cusp.

From a mesial aspect, the buccolingual dimension is about the same as the one of the first molar, but the crown height is smaller. The roots do not spread as far buccolingually, staying within the confines of the buccolingual crown outline.

From a distal aspect, because the distobuccal cusp is smaller than the one of the first molar, more of the mesiobuccal cusp is visible from this angle. The mesiolingual cusp cannot be seen.

The occlusal aspect shows a rhomboidal shape in most cases, although in comparison with the first maxillary molar, the acute angles of the rhomboid are smaller, and the obtuse angles are greater. It is common to find supplemental or accidental grooves and pits on the occlusal surface of the second molar than are usually found on the first one and.

Most maxillary second molars have three roots and three canals. Although the presence of two canals in the mesiobuccal root of the maxillary second molar is not common, it may occur [1, 3, 11].

4.3.3 The maxillary third molar

All the third molars, mandibular and maxillary, show more development variations than any other teeth in the mouth. In describing the normal third maxillary molar, direct comparisons will be made with the second one, which he supplements in function. Their design is also similar. The maxillary third molar, known as "wisdom tooth," often appears as a developmental anomaly. It can vary considerably in size, contour, and relative position to the other teeth.

From a buccal aspect, the crown is shorter cervico-occlusally and narrower mesiodistally than the one of the second molar.

From the lingual point of view, there is only one large cusp and, therefore, no lingual groove. However, in some cases, the third molar with the same essential particularities has a poorly developed distolingual cusp with a developmental groove lingually.

Mesial, aside from the differences in size, the main features are the fused roots that have a bifurcation in the apical third. The root is considerably short in relation to the crown length. The crown and root portions are usually poorly developed, with irregular outlines.

From the distal point of view, most of the buccal surface of the crown is at sight. More of the occlusal surface may be seen than at the second molar, because of the more acute angulation of the occlusal surface in relation to the long axis of the root. The distance between the cervical line and the marginal ridge is short.

Occlusal, the third molar presents a heart-shaped outline. On this tooth, there are three functional cusps: two buccal and one lingual. It presents many accidental grooves unless the tooth is very much worn.

The number of roots can vary from one to five and the number of incased root canals from one to even six. However, in most cases, one- to three-rooted third molars are most frequent [1, 3, 11].

4.3.4 The mandibular first molar

The mandibular first molar (**Figure 10**) is the biggest tooth on the mandibular arch. They have five well-developed cusps: mesiobuccal, distobuccal, mesiolingual, distolingual, and distal. Although the crown is relatively short cervico-occlusally, the mesiodistal and buccolingual measurements provide a broad occlusal form.

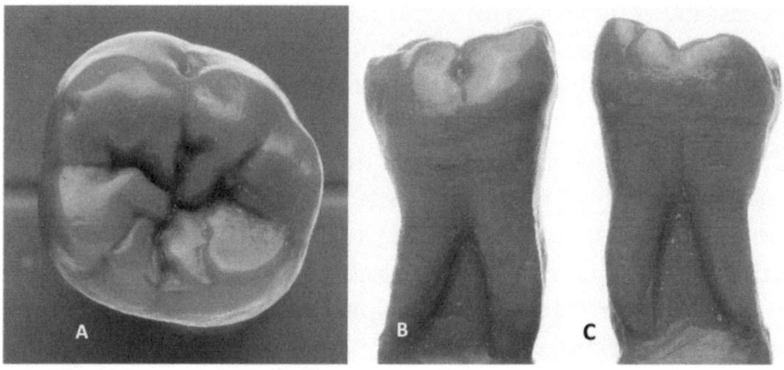

Figure 10.
Mandibular first molar; A—Occlusal surface with the three buccal cusps and two lingual, B—Buccal surface and three buccal cusps and the mesial and distal roots, and C—Lingual surface with two lingual cuspa and mesial and distal roots.

From a buccal point of view, the mandibular first molar's shape is roughly trapezoidal, and the cervical and occlusal outlines represent the uneven side of the trapezoid. If posed vertically, all fives cusps are visible. This side also shows two groves: the mesiobuccal developmental groove and the distobuccal developmental groove.

From the lingual aspect, there are three cusps visible: two lingual and the lingual portion of the distal one. The lingual ones are pointed, so the cusp ridges are high enough to hide the two buccal cusps' view. The lingual developmental grove serves as a line of demarcation between the lingual cusps. Some first molars show no groove on this side but show a depression lingual to the cusp ridges.

From the mesial point of view, if held with its mesial surfaces at right angles to the line of vision, two cusps and only one root are to be seen: the mesiobuccal and mesiolingual cusps, and the mesial root. The buccolingual height of the crown is greater at the mesial portion than it is at the distal portion. Also, the mesial root is longer than the distal one.

Distal, the gross outline of the tooth is similar to the mesial view. From this point of view, the distal cusp is in the foreground on the crown portion. The distal cusp is placed a little buccal to center buccolingually, and the distal contact area appears on its distal contour.

From the occlusal aspect, the mandibular first molar is somewhat hexagonal. The crown measurement is 1 mm greater mesiodistally than buccolingually. The crown converges lingually from the contact areas. The occlusal surface presents a major fossa and two minor fossae. The development grooves on the occlusal surface are the development groove, the mesiobuccal development groove, the distobuccal development groove, and the lingual development groove.

This tooth usually has two roots: mesial and distal. The mesial root usually has a more complicated root canal system because of the presence of two canals. The distal root usually has one canal, but often there can be two also [1, 3, 11].

4.3.5 The mandibular second molar

The mandibular second molar has four well-developed cusps: two buccal and two lingual, and it supplements the first molar in function. The anatomy differs in some details. Normally, the second molar is smaller than the first molar by a fraction of a millimeter, in all dimensions.

Buccal, comparing to the first molar, the crown is shorter cervico-occlusally and narrower mesiodistally. The crown and root have a tendency toward greater overall length, but they are not always longer. The only groove on this side is the buccal developmental groove, which acts like a line of demarcation between the mesiobuccal and distobuccal cusps.

From the lingual aspect, there are several differences between the second and first mandibular molars. The crown and root of the second one converge lingually but to a slight degree; the mesiodental calibration at the cervix lingually is always greater than the one of the first molar; the mesial and distal curvatures that describe the contact areas are more noticeable from the lingual aspect; they prove to be at a slightly lower level than those of the first molar.

From the mesial aspect, except for the dimension differences between the second and first mandibular molars, the differences are small: the cervical ridge buccally is less pronounced; the cervical line shows less curvature; and the mesial root is somewhat pointed apically.

From the distal aspect, the second molar is similar in shape to the first one, the only difference being the absence of a distal cusp and a distobuccal groove.

Occlusal, the mandibular second molar differs considerably from the first one. The small distal cusp of the first molar is not present at the second, and the distobuccal lobe development is sometimes more pronounced than the messiobucal one. Many of the second mandibular molars are rectangular from the occlusal point of view. Also, many of them show considerable prominence cervically on the mesiobuccal lobe only.

Most of the second mandibular molars have two roots with three root canals, two in the mesial root and one in the distal root. The proportions of the crown and roots are very similar to the first molar. The roots of the second molar may be straighter with less divergence from the furcation than in the first one and sometimes they are shorter [1, 3, 11].

4.3.6 The mandibular third molar

The mandibular third molar varies considerably in different individuals and presents many form and position anomalies. It supplements the second molar in function, but in general, it is showing irregular development of the crown portion, with undersized and malformed roots. However, its design conforms to the general plan of all mandibular molars. In many cases, third molars have five or more cusps, with the crown portions larger than the mandibular second molar. In these cases, the alignment and occlusion may not be normal because there is insufficient room available.

From a buccal aspect, mandibular third molars vary considerably in shape and outline. However, a well-developed third molar closely resembles the second molar. The crown is wider at contact areas mesiodistally than at the cervix, the buccal cusps are short and rounded, and the crest of contour, mesially and distally, is located a little more than at half of the distance from the cervical line to the tip of the cusps.

Lingual, the observations of the third molars coincide with the buccal aspect. When the tooth is well-developed, corresponds closely to the morphology of the second molar, except for size and root development. Same as for the mesial and distal points of view.

The occlusal aspect is in a big part similar to the second mandibular molar when the development facilitates good alignment and occlusion. They tend toward a more rounded outline and a smaller buccolingual measurement distally.

Most mandibular third molars have two roots and three canals. They are usually shorter, with a poor development. The roots may be separated with a definite point of bifurcation, or they may be fused in all parts of their way [1, 3, 11].

5. Conclusion

In conclusion, the present chapter summarizes the most essential terminology and dental anatomy of permanent dentition, though the domain is much more extended. Terminology is the basis for communication in the domain, and this aspect cannot be minimized. Dental practitioners need an accurate communication about the dental morphology with dental technicians and laboratories for performing highly esthetic and functional prosthodontic restauration. The highest esthetic in prosthodontics and odontotherapy is given by the finest details of dental morphology.

This chapter is presenting the nomenclature of permanent and deciduous teeth, age of emerge, and replacement of the deciduous teeth which are essential for clinical evaluation and interception of any possible malposition or malocclusion. Each tooth has individualized morphology given by curves, lines, ridges, and angles all having a functional aim and basis. For this reason, it is important to be accurately reproduced in all dental practical domains. Dental morphology can vary for each individual because of the invariable norm in nature. The shape, size, and angles of the teeth are related to the sex and constitutional type and therefore it is necessary to have a starting point in the study of dental anatomy.

Dental anatomy is a teach in the first year of study of any dental school, being the basic for the study of dental medicine. Having strong knowledge about dental anatomy, future subjects such as endodontics, restorative dentistry, prosthodontics even extraction, surgery, implantology, and periodontics find their basics in dental morphology. Composite resins, high quality, and highly esthetic dental materials used in odontotherapy cannot reach highly esthetic fillings if are not modeled and laired according to the morphology of the teeth and the optical parameters of dentin and enamel. The biggest challenge is found for the restauration of the anterior teeth.

Restorative dentistry is approaching and considering all the dental concepts and details of dental morphology in elaborating the treatment plan. A successful clinician or a successful dental technician should be able to mentally create the picture of the teeth from any aspect. This mental picture should be correlated with the patience's appearance, to esthetics, and with natural appearance but in the same time to support the function of the dental-maxillary system.

Conflict of interest

The authors declare no conflict of interest.

Author details

Emanuela Lidia Crăciunescu, Meda-Lavinia Negruțiu*, Mihai Romînu, Andreea Codruța Novac, Cristina Modiga, Borislav-Dușan Caplar, Cosmin Sinescu and Daniela-Maria Pop
Research Center in Dental Medicine Using Conventional and Alternative Technologies, Department of Prostheses Technology and Dental Materials, Faculty of Dental Medicine, Victor Babeș University of Medicine and Pharmacy, Timișoara, Romania

*Address all correspondence to: meda_negrutiu@yahoo.com

References

[1] Nelson SJ, Ash MM. Wheelers Dental Anatomy, Physiology and Occlusion. 9th ed. Sauders Elsevier St. Louis; 2010. ISBN: 978-1-4160-6209-7

[2] Tsai HH. Morphological characteristics of the decisous teeth. The Journal of Clinical Pediatric Dentistry. 2001;**25**(2):95-101. DOI: 10.17796/jcpd.25.2.l871896721408pt6

[3] Mosby's. Mosby's Dental Dictionary, 3rd ed. St Louis, Missouri, US: Elsevier; 2014. ISBN: 978-0-323-10012-03251

[4] Berkovits BK, Holland GR, Moxham BJ. Oral Anatomy, Histology and Embryology, 3rd ed. Mosby, Edinburgh: Mosby; 2002. ISBN 978-0-7234-3181-7

[5] Goldberg M, Smith AJ. Cells and extracellular matrices of dentin and pulp: A biological basis for repair and tissue engineering. Critical Reviews in Oral Biology and Medicine. 2004;**15**(1):13-27. DOI: 10.1177/154411130401500103

[6] Sakaguchi R, Ferracane J, Powers J. Craig's Restorative Dental Materials. 14th Edition. St Louis, Missouri, US: Elsevier; 2018. ISBN: 9780323478212

[7] Tjäderhane L, Paju S. Dentin-Pulp and Periodontal Anatomy and Physiology. In: Orstvik D. Essential Endodontology: Prevention and Treatment of Apical Periodontitis. 3rd ed. Hoboken, NJ, US: John Wiley & Sons Ltd.; 2020. pp. 11-58. DOI: 10.1002/9781119272014

[8] Division of teeth into thirds. 2022. Available from: https://www.globehealth.net/dental-anatomy/division-of-teeth.html

[9] Okeson JP, ed. Management of Temporomandibular Disorders and Occlusion. 6th ed. St Louis, Missouri, US: Mosby; 2008. ISBN: 9780323676748

[10] Proffit WR. Fields HW. Sarver DM. Contemporary Orthodontics 5th ed. Oxford: Elsevier Mosby. 2012. ISBN: 9780323096270

[11] Nelson SJ. Wheeler's Dental Anatomy, Physiology and Occlusion, 11th Edition. St Louis, Missouri, US: Sauders Elsevier; 2010. ISBN: 9780323638784, 2020

Chapter 2

All We Need to Know about Normal and Abnormal Human Teeth

Milos Stepovic, Maja Vulovic, Ivona Bankovic, Miroslav Misic and Radisa Vojinovic

Abstract

Type of dentition, number of teeth in primary and permanent dentition, and appearance of the teeth in both dentitions have been extensively described in the literature. There are 20 teeth in primary dentition and 32 in permanent dentition. Teeth typically exhibit normal appearance, although small variations may appear. Abnormal teeth can be detected via careful clinical or radiographical observation. Variations in appearance, structure, and eruption of teeth can be influenced by factors such as genetics, trauma, certain drugs, and periapical lesions. These factors can alter teeth shape, number, size, and position. Teeth with anomalies can have a negative impact on a person's appearance as well as functions such as eating and can even lead to psychological problems. They can be detected with different X-ray techniques, but cone beam computed tomography (CBCT) is the most precise.

Keywords: normal human teeth, abnormal human teeth, dental anomalies, tooth anatomy, cone beam computed tomography

1. Introduction

There is a specific period in which humans develop primary dentition and then permanent dentition. Primary dentition consists of four incisors, two canines, and four molars in each jaw, whereas permanent dentition includes all of this plus an additional group of teeth called premolars and one more set of molars. These groups of teeth have a well-established description and appearance and any deviation in these areas is classified as an abnormality. Dentists must have thorough knowledge of normal tooth anatomy, as they are usually the first provider one visits for clinical or radiographical examination of one's teeth. Abnormal teeth often have an unusual appearance, structure, or position in the jaw, which can cause problems during growth as well as make dental therapy challenging. Genetic disorders and external risk factors contribute equally to the abnormal appearance of teeth and even to their absence.

2. Types of dentition and periods of teeth eruption in humans

At birth, humans often do not have any teeth in their upper and lower jaws; this is considered normal and is a precursor to the development of what is hoped to be normal-looking teeth. However, sometimes teeth develop abnormally due to trauma or genetic anomalies. These teeth may have abnormal appearance or inadequate strength or function. These are called rudimental teeth. Periods of teeth eruption can be divided into primary (milk) dentition and secondary (permanent) dentition [1]. Periods during which humans have both primary and permanent teeth in their jaws is called mixed dentition. This usually occurs when the first permanent molar finds its place in the dental arch, usually at the age of six years [2].

The first teeth that erupt into the oral cavity are called primary teeth. There are 20 primary teeth, 10 per jaw. In each arch of the mouth, there are two central incisors, two lateral incisors, two canines, and four molars. These teeth will be exfoliated by permanent teeth eruption, so-called replacement teeth [3, 4]. Primary first and second incisors as well as the canines are replacements for their counterpart permanent teeth, but the first and second primary molars are replacement teeth for first and second permanent premolars, whereas permanent molars do not have replacement teeth and they grow at the end of the primary teeth arch. The first teeth that will appear in the dental arch are mandibular incisors followed by maxillary incisors. By the end of the first year, all anterior teeth are usually visible. At the second year, all primary teeth, except for the second molars, are erupted. In the third year, all primary teeth are present and functional in both jaws [5, 6] (**Table 1**). Eruption of primary teeth is often called teething and it is followed by increased salivation and local gingival irritation at the site of teeth eruption, which can vary in intensity and usually ends when the tooth appears in the oral cavity [7].

Primary teeth have an important role in developing the jaw as well as jaw and facial muscles, which are important for speaking, chewing, biting, and so on; thus, primary teeth must be kept healthy until the time of their replacement. Prematurely lost primary teeth can cause the absence of space for permanent teeth, which can lead

Maxillary tooth	**Eruption Date (average)**	**Exfoliation Date (average)**
Central Incisor	8–12 Months	6–7 Years
Lateral Incisor	9–13 Months	7–8 Years
Canine	16–22 Months	10–12 Years
1st Molar	13–19 Months	9–11 Years
2nd Molar	25–33 Months	10–12 Years
Mandibular tooth	**Eruption Date (average)**	**Exfoliation Date (average)**
Central Incisor	6–10 Months	6–7 Years
Lateral Incisor	10–16 Months	7–8 Years
Canine	17–23 Months	9–12 Years
1st Molar	14–18 Months	9–11 Years
2nd Molar	22–31 Months	10–12 Years

Table 1.
Average time of eruption and exfoliation in primary dentition, both jaws.

to malposition of the permanent teeth. Mispositioned teeth have greater chances of developing caries and may not be esthetically pleasing [8].

Permanent dentition begins at about six years of age with the growth of central incisors and first molars of the upper and lower jaws and ends at about 12 years of age with the growth of second molars. Third molars, which can vary in their appearance or even be absent, appear between 18 and 30 years of age; sometimes they never erupt due to missing embryonic tooth germ (**Table 2**). While growing, permanent teeth exfoliate primary teeth and grow in their place or behind them. The phase during which a person has both primary and permanent teeth is called mixed dentition. Once the second primary molar and primary canine are replaced with permanent teeth, dentition becomes fully permanent. At the end of development, there is a total of 32 permanent teeth in both jaws [9]. The permanent and primary teeth can be affected by different anomalies, which can alter their appearance. Permanent teeth have greater risk of developing anomalies.

Tables 1 and **2** show the time of teeth eruption from both dentitions. Delayed eruption (6–8 months from the approximate average times) indicates a problem and may be a sign of tooth anomalies (number, shape, or position). More rarely, delayed eruption is associated with Down syndrome, hypothyroidism, hypopituitarism, achondroplasia, osteopetrosis, or chondroectodermal dysplasia as well as the possible development of cysts [10]. A significant variation affecting a single tooth or only a few teeth should be carefully investigated.

Tooth	Eruption Date(Avg.)
Maxillary	
Central Incisor	7–8 Years
Lateral Incisor	8–9 Years
Canine	11–12 Years
1st Premolar	10–11 Years
2nd Premolar	10–12 Years
1st Molar	6–7 Years
2nd Molar	11–13 Years
3rd Molar	17–21 Years
Mandibular	
Central Incisor	7–8 Years
Lateral Incisor	8–9 Years
Canine	9–10 Years
1st Premolar	10–12 Years
2nd Premolar	11–12 Years
1st Molar	6–7 Years
2nd Molar	11–13 Years
3rd Molar	17–21 Years

Table 2.
Average time of eruption in permanent dentition, both jaws.

3. Anatomical structure of tooth

The anatomy of a tooth divides into the crown and the root. The crown of the tooth is visible in the oral cavity, and the roots of the tooth are located in the alveolus of the maxilla and mandible and attached to the periodontal ligament. The anatomic crown is covered by enamel, while the anatomic root is covered by cementum. A clinical crown is not necessarily the same length as the anatomical crown and clinical root, which may differ because of periodontal diseases or chirurgical procedures. The gingiva covers the border of the alveolar process that is adjacent to the teeth. The anatomical boundary between the enamel-covered crown and cementum-covered root is the cementoenamel junction. The roots of the tooth vary depending on the type of tooth. The number of roots often is the same for the same group of teeth. The maxillary molars typically have three roots: a palatal root, mesiobuccal root, and distobuccal root. First premolars of the maxilla have two roots and the second premolars and all anterior teeth have typically one root. In the mandible, all anterior teeth and both premolars typically have one root, while molars have two roots: mesial and distal. The only exceptions are the third molars, which can have different numbers of roots. The longest root is considered to have maxillary canines. The periodontal ligament is composed of connective tissue fiber that connects the tooth to the alveolar bone. At the upper part of the root is the apical foramen where the neurovascular structures enter the tooth [11].

The blood supply of the teeth originates from the maxillary artery, which is the largest terminal branch of the external carotid artery. The specific arteries carrying blood to the teeth travel through the root canal and have the following names: anterior superior alveolar arteries, posterior superior alveolar arteries, and inferior alveolar arteries. The veins of the teeth follow the arteries, having similar names. The nerves supplying the teeth also accompany the arteries through the root canal and originate from the maxillary and mandibular branches of the trigeminal cranial nerve. Near the teeth, these major nerves branch into superior alveolar nerves and inferior alveolar nerves [12].

4. Surfaces of teeth and appearance of teeth in both dentitions

Permanent teeth are divided into two groups: anterior teeth and posterior teeth. Anterior teeth include the teeth toward the front of the mouth, including the central incisors, the lateral incisors, and the canines, while posterior teeth are the teeth toward the back of the mouth and include the premolars and molars [13].

Anterior teeth show four surfaces and one cutting ridge: vestibular surface, lingual surfaces in bottom teeth, palatinal surfaces in upper teeth, mesial surfaces in teeth closer to the middle line of the jaw, distal surfaces in teeth far from the middle line of the jaw (mesial and distal surfaces are also called proximal surfaces) and the incisal ridge.

Posterior teeth show five surfaces: vestibular surface, lingual surfaces in bottom teeth and palatinal surfaces in upper teeth, mesial surface in teeth closer to the middle line of the jaw, distal surfaces in teeth far from the middle line of the jaw, and the occlusal surface also called the masticator surface.

Special anatomical structures of teeth include the mound-ridge complex (cuspis, tuberculum, cingulum, crista, cumulus) and fissure complex (fissures, fossae, pits).

The area of the tooth that contacts the adjacent tooth is called the contact area. The number of cusps varies among anterior and posterior teeth. The incisors do not have cusps, canines have one cusp, premolars usually have two cusps (sometimes they can show molarization when they have an additional cusp), and molars have four cusps (with occasional appearance of a fifth cusp called the cusp of Carabelli). Cingulum is characteristic of anterior teeth of both jaws (incisors and canines) and it represents the portion of enamel forming a convex protuberance in the cervical third of the palatal and lingual surfaces of anatomic crowns. Cumulus is also characteristic of anterior teeth, mostly seen on early erupted permanent incisors, and they are located on the incisal surfaces in the form of three dents. The fissure complex is characteristic of posterior teeth. A tooth usually has one main central fissure that goes from the mesial to distal side of the occlusal surface of the tooth. It continues with additional fissures on each side going toward the proximal buccal and oral surfaces (mesiobucal and mesiooral, and distobuccal and distooral) and together they form the borders of the triangular fossae on the mesial and distal proximal side of occlusal surfaces. Central fossa is located on the middle of the central fissure. There is also a lateral fissure in the molar region that divides buccal and oral cusps on the mesial and distal cusps [14].

Tafti and Clark described the primary teeth in detail.

a. Teeth in Primary Dentition

1. Incisors - The maxillary incisors are essentially smaller morphological versions of permanent teeth incisors. The incisors are used for cutting food and have sharp edges. They consist of the central incisors and the lateral incisors. The central incisor is larger than the lateral incisor, and the maxillary central incisor is the largest of all the incisors. They also prominently figure into the esthetic of the oral region. Their labial surface is convex and their palatal surface is concave. Incisors of the mandible are smaller than the incisors of the maxillary teeth. The incisor crown is trapezoidal in the labial view and contains three tubercled incisal edges called cumulus.

2. Canines – The maxillary canines are also morphologically similar to permanent teeth canines. The canines have a trapezoidal crown with one labial cusp. The labial surface is convex and the palatal surface is concave. The canines serve to support the incisors. The canines have a characteristic pointed incisal edge and a large cone-shaped crown. Canines of the mandible have a trapezoidal crown with a single labial cusp.

3. Molars – The molars are the largest of the primary teeth and provide significant function in mastication. The crowns of the primary molars are more bulbous in morphology. The buccal, lingual, mesial, and distal surfaces of the maxillary and mandibular molars are all convex. The occlusal surface is rectangular in both jaws.

    - First Molars – The maxillary first molars morphologically resemble both a molar and a premolar. The occlusal surface of a first molar consists of the mesiobuccal, distolingual, mesiolingual, and distobuccal cusps. The largest cusp is the mesiolingual cusp, which is also the sharpest cusp.

The distolingual cusp is small and rounded in morphology. The primary first molar has a less pronounced mesiobuccal cusp when compared to the permanent first molar. The mandibular first molars have four cusps: two buccal and two lingual. These include the distobuccal, distolingual, mesiolingual, and mesiobuccal cusps. The occlusal surface is narrow secondary to mesiobuccal and mesiolingual cusp convergence. There are three pits.

- Second Molars – Primary maxillary second molars resemble the permanent maxillary first molar. Rhomboidal in shape, the maxillary second molar has four cusps, two on both the buccal and lingual aspects. A fifth cusp can be present and is called the cusp of Carabelli. Two well-defined cusps can be seen on the buccal view, and the lingual surface has three cusps: the mesiolingual, distolingual, and the cusp of Carabelli, much less than in permanent dentition. The buccal surface of the maxillary second molar is divided into the mesiobuccal, distobuccal, and distal cusps. The lingual cusps include the distolingual and the mesiolingual cusps [15].

b. Teeth in Permanent dentition

1. Incisors – The primary function of incisors is to cut food. They have a sharp incisal edge. When first erupted, the incisors have three cumulus, which often disappear when biting food. The cingulum is located on the oral surfaces and can have a pit formed in it where it meets the lingual surface of the tooth. Central incisors have sharper and more acute incisor angles than lateral incisors. The maxillary central incisors are also unique in that they are larger than mandibular central incisors. Maxillary lateral incisors often vary the most in their shape. Sometimes maxillary lateral incisors can have a peg-like shape. The maxillary lateral incisors may also be congenitally missing. The mandibular central incisor is the smallest tooth of all the incisors.

2. Canines – The main function of canines is to tear food. They have a single, pointed cusp. They also serve to form the corners of the mouth. The canines have very prominent cingulum, but the maxillary cingulum is more prominent than the one on the mandibular canines and it rarely has pits.

3. Premolars (bicuspids) – These teeth are located behind and adjacent to the canines and are designed to crush and grind food. These teeth can have two–three cusps (the mandibular second premolar typically has three). The buccal cusp is typically longer and wider than the lingual cusp. In the mandibular first premolar, the lingual cusp is very small and usually not functional. Often the first premolar is the tooth that is extracted in orthodontic procedures to make room for crowded teeth to move into place.

4. Molars – The most posterior teeth in the mouth are the molars. They have broader and flatter surfaces with four–five cusps and have the largest crowns of any other teeth. They are designed to grind food. They are named starting with those closest to the midline as first molars, second molars, and third molars. Third molars can vary in size and shape or even be absent. They are commonly referred to as wisdom teeth and often have fused roots. The mandibular molar is the first permanent tooth to erupt in the mouth [16].

Indicated differences between primary and permanent teeth include:

1. Primary teeth are smaller in all dimensions than analogous permanent dentition teeth.

2. Crowns of primary teeth look shorter (lower) compared to permanent teeth.

3. Primary teeth always show a larger md width compared to the natural height of the crown ("chucky" appearance of the crown of all primary teeth).

4. The class of incisors and canine teeth is characterized by the presence of a cervical ridge. Members of the primary dentition molar class show an analogous structure, but only to the cervical third of the buccal surface (buco-cervical ridge).

5. The labial surface is completely smooth, without the presence of cumulus, lobes, and developmental depressions.

6. The root tree in the primary teeth is absent.

7. The root branches of the primary molar are longer and more graceful with a greater degree of angulation and divergence.

8. Primary teeth are milky or blue-white, unlike the crowns of permanent dentition teeth, which are mainly yellow-gray or gray-white.

9. Enamel and dentin are thinner and less transparent.

10. The pulp chamber of primary teeth is voluminous.

5. Normal tooth structure

Nanci elucidated that tooth enamel, the protective outer layer of the dental crown, is the hardest and most mineralized tissue in the human body. Enamel is an avascular, hard material that protects the outer tooth and gives the tooth its whitish color. Enamel color is determined by enamel thickness. Enamel consists of more than 95% apatite, a calcium phosphate mineral that can be found in all mineralized tissues. Apatite crystals grow predominantly along their c axis and align parallel to one another, effectively forming an enamel rod; each rod is enveloped with organic matter, which makes up only 1–2% of enamel. The role of organic matter in providing a scaffold for enamel minerals to grow has long been recognized [17]. Enamel is formed by cells called ameloblasts, which are active only until the tooth is in developmental stage. Once the tooth appears in the oral cavity, those cells are no longer active and this is why the enamel is the only structure of the tooth that cannot form by itself [18].

Dentin is less mineralized than enamel but more than bone or cementum. Its characteristic feature is a regular pattern of microscopic dentinal tubules, 3 μm in diameter, which extend from the pulpal surface to the enamel–dentine junction. The tubules show lateral and terminal branching near the enamel–dentine junction and may project a short distance into the enamel [11]. The diameter of the dentine tubule is narrowed by deposition of peritubular dentine. This is different from normal dentine (intertubular dentine) because it is more mineralized and lacks a collagenous matrix. Dentine is formed slowly throughout life and thus there is always an unmineralized zone of predentine at the surface of the mineralized dentine, adjacent to the odontoblast layer at the periphery of the pulp [19].

Cementum covers the root surface and is an important tissue to maintain dental and periodontal attachment. It is light yellow in appearance, thinner near the tooth neck, and wider at the root apex. Cementum is excreted by cells called cementoblasts, which develop from undifferentiated mesenchymal cells in the connective tissue of the dental sac. Cementum is slightly softer than dentin and consists of about 45–50% inorganic mineral. Cementum is a mineralized connective tissue similar to bone except that it is avascular; the mineral is also apatite and the organic matrix is largely collagen. The two main types of cementum are cellular and acellular. The cementum attached to the root dentin and covering the cervical part of the root is acellular, or primary, cementum. The lower apical part of the root is covered by cellular, or secondary, cementum [20].

The dental pulp resides in a rigid chamber comprising dentine, enamel, and cementum, which provide strong mechanical support and protection from the microbial-rich oral environment. The mature pulp bears a strong resemblance to the embryonic connective tissue, with a layer of highly specialized cells, the odontoblasts, along its periphery. The physical confinement of the dental pulp, its high incidence of sensory nerve innervation, and the rich microcirculatory components make the dental pulp a unique tissue [21]. A recent study on the bacterial invasion into dentinal tubules of human teeth with or without viable pulp has shown that teeth with pulps are much more resistant to bacterial invasion into the dentinal tubules than teeth with root canal fillings. Considering this, the pulp plays an important role in this defense process. In teeth with pulps, the dentinal tubules are occupied by dentinal fluid and odontoblastic processes, which may behave collectively as a positively charged hydrogel. The pulp's specialized cells, the odontoblasts, and perhaps undifferentiated mesenchymal cells retain the ability to form dentine throughout life. This enables the healthy pulp to partially compensate for the loss of enamel or dentine caused by dental caries. Odontoblasts may also form sclerotic dentine, reactionary dentine, and reparative dentine in response to different stimulus, such as thermal stimulus (warm and cold), caries, or operative procedures. Hence, the pulp is a sensory organ involved in the immune defense reaction by processing antigens. Regardless of the nature of the sensory stimulus, such as thermal change, mechanical deformation, or trauma, the pulp registers different impulses as a common sensation like pain. Such pain-registering ability is important as part of the defense mechanisms of the pulp [22].

6. Tooth anomalies

There are different types of tooth abnormalities and they can appear in both dentitions but are more common in permanent teeth rather than primary teeth. Anomalies can affect the number of teeth (hyperdontia, hypodontia, anodontia), the size of teeth (macrodontia, microdontia), the position of teeth (retention (impact), clenched teeth, rotation, ectopia (dystopia), transposition), and the shape of teeth (taurodontism, dilaceration, flexion, conjoined teeth, abnormalities in the shape of the tooth root, aplasia, Hutchinson's teeth, evagination) [23]. Sometimes trauma can affect the appearance of teeth or cause their retention in the bone. Certain drugs and minerals can affect tooth color and some genetic disorders can affect the structure and appearance of teeth [24].

6.1 Anomalies in the number of teeth

A supernumerary tooth is defined as "any tooth or odontogenic structure that is formed from embryonic tooth germ in excess of usual number for any given region of the dental arch." Küchler et al. (2011) concluded that hyperdontia is a condition characterized by additional teeth (mesiodens) within the normal dentition. Supernumerary teeth are frequently located in the anterior maxilla, between the central incisors. In permanent dentition, the prevalence of supernumerary teeth ranges between 0.5% and 5.3% [25]. Rajab and Hamdan (2001) found that it affects males twice as much as females. This phenomenon is often associated with syndromes such as Ehlers–Danlos, cleidocranial dysplasia, and Gardner syndrome. However, it may also be seen in non-syndromic patients [26]. It has also been reported that males are commonly affected in midline and premolar regions and females are commonly affected in the incisor and canine regions. Supernumerary teeth are more likely to be rudimental looking but can sometimes resemble regular teeth [27]. King et al. (2007) found that formation of supernumerary teeth in any part of the dental arch can lead to complications, including delayed eruption, malposition, impaction, diastemas, crowding, and poor esthetics [28]. Supernumerary tooth can be of different forms (conical, tuberculate, supplemental, and odontoma) and occur in different locations (mesiodens, paramolar, and distomolar). The mesiodens is located between the two central incisors and is mostly conical in shape. Distomolars are located distally to the third molar, while paramolars are located palatally or labially next to a molar [29].

Al-Ani et al. (2017) found that tooth agenesis may involve the primary or permanent dentition and is a developmental anomaly where at least one tooth is absent. Total anodontia is complete tooth agenesis. Missing teeth can be further classified as hypodontia, involving one to five absent teeth, and oligodontia, involving six or more missing teeth [30]. Ritwik and Patterson (2018) found that when excluding third molars (wisdom teeth), tooth agenesis has a prevalence between 3% and 10% in the permanent dentition. Third molars are the most common congenitally missing permanent teeth, with a reported prevalence of 23%. These are followed by mandibular second premolars, maxillary lateral incisors, and, less frequently, the maxillary second premolars [31]. Dinckan et al. (2018) found that syndromes such as ectodermal dysplasia or cleft lip and palate are frequently associated with tooth agenesis and factors such as trauma, infection, or drugs. Complications of tooth agenesis include malocclusion, poor esthetics, reduced masticatory function, and speech difficulties [32].

Hypodontia is an inherited condition, but it can also be caused by trauma, infection, radiation, and endocrine disturbances. It is characterized by developmentally missing teeth. It is very rare in the primary dentition and more common in females than males, with a 3:2 ratio. In addition to missing teeth, people with hypodontia may have rather small or very conical teeth. The tooth most commonly missing is the permanent maxillary lateral incisor (74% incidence) followed by the maxillary and mandibular second premolars [30]. Hypodontia most commonly affects last or most distal teeth in the teeth group, decreasing in incidence from posterior to anterior teeth, in the following order: third molars, second premolars, and lateral incisors. Absent third molars are considered a normal variation nowadays and may not be considered to be part of hypodontia [33].

6.2 Anomalies in the size of teeth

Macrodontia and microdontia are rare shape anomalies of dentition, where teeth are larger or smaller than average.

Microdontia is the term given to describe a tooth very much reduced in size. Microdontia generally affects individual teeth, usually the maxillary second incisor and the third molar. Occasionally, however, many teeth in the same dentition may be affected, in which case the teeth may be spaced apart [34]. Microdontia may accompany clefts of the lip or palate, Ehlers–Danlos syndrome, hypopituitarism, and ectodermal disorders. A frequent variation in shape is the peg-shaped maxillary lateral incisor. In generalized microdontia, all teeth in the dentition appear smaller than normal and enlarge the spaces between teeth, which is aesthetically unpleasing. Teeth may be measurably smaller than normal, as in pituitary dwarfism, or they may be relatively small in comparison with a large mandible and maxilla (acromegaly). In isolated microdontia, a single tooth is smaller than normal. This is mostly seen in permanent dentition in lateral incisors that can be cone- or peg-shaped and very much affects the appearance, and third molars, more common in maxilla [35].

Macrodontia refers to teeth that are physically larger than normal and could clinically be confused with other conditions such as fusion (two separate follicles fusing to form one tooth) and gemination (two teeth that form from one follicle but are not separated) [36]. Three different types of macrodontia can be found: true generalized, where many teeth in the mouth are affected (very rare); relative generalized, where all the teeth are affected and the teeth can either be of normal size in a very small jaw creating the illusion of macrodontia or all the teeth may be slightly enlarged; and isolated macrodontia, where only a single tooth is affected (very rare). The prevalence of macrodontia in permanent dentition is 0.03% to 1.9%, with a higher incidence in males [37]. Macrodontia in the anterior region poses an esthetic problem for patients, leading to crowding, plaque accumulation, and so on.

6.3 Anomalies in the shape of teeth

Gemination (twinning) is an abnormality where a single tooth forms two crowns. This is a consequence of partial division of the tooth, resulting in a bifid crown but retaining the normal number of expected teeth. Gemination is a rare abnormality in which a single tooth tries to divide. Partial division may produce a shared pulp canal and root. Complete division produces a normal tooth along with a supernumerary tooth [38]. Gemination usually affects the primary teeth, but the permanent dentition can be involved as well. The incisor region is the most commonly affected area with no apparent gender predilection. Although the prevalence rate is variable in individual reports, the overall prevalence appears to be approximately 0.5% in the primary dentition and 0.1% in the permanent dentition [39]. Gemination is more prevalent in the anterior maxillary region, affecting incisors and canines, although it can also affect molars and bicuspids. Bilateral cases are seen less frequently, with a prevalence of 0.02% in both dentitions. They are very rare in posterior teeth. it is more frequently found in Mongolians (5%) than in Caucasians (0.5%) [40].

Different from germination, fusion leads to one less tooth within the dental arch; it involves two separate developing tooth follicles that join together [41]. Clinically, a fused tooth looks like a larger tooth and sometimes is mixed with gemination because both types of teeth have an incisure on the middle of the crown, but radiographic difference is much clearer. Fused teeth have separate pulp chambers for each tooth, while gemination has one larger pulp chamber and there are two roots or two canals in a single root [42]. The degree of union may be total or partial and often presents with a coronal incisure. Like gemination, fusion is more common in the primary dentition but can occur in the permanent dentition as well. It also typically involves the anterior region of the dental arch. While there is no gender predilection, incidence is higher among Native Americans and Asians [43].

Bolhari et al. (2016) defined dilaceration as an unusually abrupt angle between tooth crown and root or within the root itself, resulting from trauma during the development of the tooth. Dilaceration is a developmental disturbance in the shape of teeth. It refers to an angulation, or a sharp bend or curve, in the root or crown of a formed tooth. This disturbance is more likely to affect the maxillary incisors and occurs in permanent dentition [44]. Although this may seem more of an esthetics issue, an impacted maxillary incisor will cause issues related to occlusion, phonetics, mastication, and psychology in young patients. Dilaceration of the crown can diagnosed visually. Crown dilaceration will present itself as a tooth that is angled to face outward or inward [45]. It most commonly affects patients in their permanent dentition. Incidence is more common in females (six times) than males. There are reports of a 0.53% occurrence rate for the two anterior teeth of the upper jaw, but the most common teeth to experience dilaceration are the third molars of the lower jaw with an incidence of 24.1% [46].

Dineshshankar et al. (2014) stated that taurodontism (bull teeth) is an abnormality of tooth morphology that often occurs in multi-rooted teeth and is characterized by a short root and an enlarged crown containing an equally enlarged pulp chamber and no constriction at the level of the cemento–enamel junction. It is a condition that mostly occurs in the molar teeth [47], but it can also be seen in both the permanent and deciduous dentition, unilaterally or bilaterally. The distance from the roof of the pulp chamber to the root bifurcation is greatly increased [48]. Variants include hypotaurodontism, mesotaurodontism, and hypertaurodontism. These conditions may exist as an isolated trait (autosomal dominant) or as part of several syndromes including trichodentoosseous syndrome, otodental dysplasia, ectodermal dysplasia, tooth and nail syndrome, amelogenesis imperfecta, and others [49].

Concrescence is the joining of two completely formed teeth by the cementum along the root surface usually located in the posterior teeth and the maxillary arch, often involving a second molar tooth closely approximating the roots of an impacted third molar. The affected teeth may fail to erupt or only partially erupt. The prevalence rate is 0.04% [50]. Hypercementosis is the thick mantle of cementum that makes the root look fat, and it can be seen in Paget disease of the skeleton (osteitis deformans) [51].

Dens invaginatus is a developmental anomaly presenting with complex morphological variations. This anomaly is also referred to as "dens in dente." It has been described as deep infolding of the enamel organ into the dental papilla during tooth development. Starting from the foramen caecum or tip of the cusps, it can extend deep into the root, with or without pulp involvement, sometimes even resulting in a second apical foramen [52]. This anomaly frequently results in early pulp necrosis. The most affected teeth are maxillary lateral incisors, mostly bilateral. Clinically, the affected teeth may vary in presentation with an increased crown diameter, hypoplasia at the palatal pit, peg or conical morphology, bifid cingulum, a talon cusp, or a deep foramen caecum [53].

6.4 Anomalies in the position of teeth

Rotations are movements that occur around the long axis of teeth. Teeth can become rotated either before they emerge or after. Probably the most common cause of teeth that emerge rotated is trauma to the mouth during the development of teeth. Cysts, tumors, or supernumerary teeth can also cause teeth to rotate as they grow [54]. Also, crowding can force a tooth that grows later to grow narrower than it should in the dental arch. In addition, when there is too much space between teeth (e.g., after tooth extraction), the tooth may move toward the empty space [55]. A rotated tooth is highly visible and can cause malocclusion.

A variety of eruption problems can happen during the mixed dentition period. One such problem is ectopic eruption. Akbas et al. (2022) defined ectopic eruption as a condition in which the permanent teeth, because of deficiency of growth in the jaw or a segment of jaw, assume a path of eruption that intercepts a primary tooth, causes its premature loss, and produces a consequent malposition of the permanent tooth. Failure to treat ectopic eruption can result in loss of arch length, inadequate space for the permanent premolar, and malocclusion [56]. An example of this anomaly is ectopic eruption of first permanent molars that can be positioned too far mesial into the alveolar bone and become impacted below the second permanent molar. Ectopic eruption has a 3% incidence rate and occurs more frequently in males than in females [57].

Matsumoto and Stuani (2018) stated that tooth transposition is a unique and severe condition of ectopic eruption. It is defined as an interchange in the position of two permanent teeth located at the same quadrant in the dental arch [58]. Chattopadhyay and Srinivas (1996) explained that transposition can be complete (when the position of affected teeth is totally transposed) or incomplete (when only the crowns are transposed, while the roots remain in normal position). The incidence of transposition in the overall population is low (0.2%–0.38%) and it is most often found among women in the maxilla (76%) and mostly unilateral. The canines are affected in 90% of transposition cases, most often relative to the first premolar (71%) or maxillary lateral incisor (20%) [59]. Shapira and Kuftinec (1989) described that while it may be present both in the maxilla and mandible, transposition between the canine and maxillary first premolar is most common, followed by a lateral incisor transposed with a maxillary canine. There are no reports of transposition in deciduous dentition [60].

6.5 Other conditions that cause abnormalities of the teeth

Bruxism is a sleep-related movement disorder in which a person grinds, gnashes, or clenches their teeth. Teeth in persons with bruxism have a characteristic appearance; they are flattened, have fractures in the enamel of the tooth crown, and in extreme cases have exposed dentin, which can cause tooth sensitivity and headaches as well as affect the temporomandibular joint [61].

Attrition is the loss of tooth surface due to normal tooth function and occlusion. Some wearing is normal (physiologic), but accelerated wear beyond what is normal is pathologic and usually caused by bad habits [62] such as poor tooth brushing techniques, lip biting, nail biting, and so on.

Erosion is the chemical dissolution of tooth structure often attributed to regurgitation of gastric acid or excessive intake of acidic food or drinks. It is often seen on the palatal and lingual surfaces of the anterior teeth, which present with flattened surfaces of dental morphology.

Abrasion is excessive wear caused by mechanical forces. It is similar to attrition, but attrition is more often located on occlusal and incisal surfaces and abrasion is more common on the non-occluding surface [63].

Abfraction is the loss of tooth substance attributed to occlusal forces during biting, which cause the teeth to flex ever so slightly. Constant flexing causes enamel to break from the crown, usually on the buccal surface [64].

Dentinogenesis imperfecta is an autosomal dominant condition affecting both deciduous and permanent teeth. Affected teeth are gray to yellow-brown and have broad crowns with constriction of the cervical area, resulting in a tulip shape. Enamel is easily broken, leading to exposure of dentin that undergoes accelerated attrition. This dental condition is sometimes seen in patients with osteogenesis imperfecta. It can be classified into three types. Type I occurs in people who have osteogenesis imperfecta, and type II (progressive hearing loss) and type III usually occur in people without other inherited disorders [65].

There is a large group of inherited developmental defects in enamel collectively referred to as amelogenesis imperfecta. The condition is rare, affecting about 1 in 14,000 people. There are three general categories of this defect: (1) Hypoplastic type: inadequate formation of enamel matrix; both pitting and smooth types exist. Enamel may be reduced in quantity but is of normal hardness; (2) Hypomaturation type: a defect in the crystal structure of enamel leads to a mottled enamel with white to brown to yellow colors; and (3) Hypocalcified type: a defect not in the quantity but in the quality of enamel. It is poorly mineralized, soft, and chips and wears easily [66].

Syphilitic vasculitis around the time of birth can damage the developing tooth buds and lead to dental anomalies. Hutchinson teeth are abnormal permanent upper central incisors that are peg-shaped and notched, usually with obvious thinning and discoloration of enamel in the area of the notching. They are widely spaced and shorter than the lateral incisors; the width of the biting surface is less than that of the gingival margin. Mulberry molars (also known as Moon or Fournier molars) are multicuspid first molars in which the tooth's grinding surface, which is narrower than that at the gingival margin, has many small cusps instead of the usual four well-formed cusps [67]. The enamel itself tends to be poorly developed. Hutchinson's teeth is a sign of congenital syphilis. As well as having a triangle or peg-like shape, a Hutchinson tooth is smaller than the usual size, widely spaced, and has thin enamel and notches on the biting surface. The affected molar teeth (known as mulberry molars) have multiple cusps instead of four cusps. Congenital syphilis affects primary teeth especially in central incisors and molars. Hence you cannot notice Hutchinson teeth in a child until six years of age [68].

The cusp of Carabelli is an additional cusp located on the mesio-palatal surface of permanent maxillary molars. Bhavyaa et al. (2021) found that primary maxillary second molars had the highest prevalence (72f%) of this anomaly followed by permanent maxillary first, third, and second molars. It is a so-called fifth cusp on the molars and it is more likely to be present in permanent dentition and is considered a predilection spot for caries [69].

Discoloration of teeth can be caused by antibiotics such as tetracycline, which is used to treat many common infections in children and adults. Tetracycline can bind to the calcium needed for tooth, cartilage, and bone development, resulting in discoloration of both the primary and permanent dentitions. Minocycline hydrochloride, a derivate of tetracycline used in acne treatment, can cause discoloration of skin, nails, sclera, and teeth. This permanent discoloration varies from yellow or gray to brown depending on the dose of the medicine. The prevalence of tetracycline and minocycline staining is 3–6% [70].

7. Diagnostic imaging of abnormal teeth

Classic radiography of a single tooth or several teeth using retro-alveolar or retro-coronary X-rays is not ideal for diagnosing dental anomalies. They provide only 2D images of the teeth, and the retro-alveolar method produces distorted images while the retro-coronary method cannot show the apex of teeth, which is not helpful in cases of supernumerary or impacted teeth. Occlusal and axial X-ray methods can help visualize the position of supernumerary and abnormal teeth, but these methods are limited without 3D scans [71]. Orthopantomography is the most common method for radiologic examination of patients with dental anomalies. Compared to classic X-rays, orthopantomography uses less radiation and thus it is more suitable for younger patients. This method allows visualization of both jaws and the temporomandibular joint. In primary and mixed dentition, orthopantomography can visualize the embryonic tooth germs of permanent teeth and diagnose extra, missing, or impacted teeth [72]. Images of teeth scanned by this method can be stored and used multiple times. Although not the most detailed scan, orthopantomography provides a larger picture of the development of jaws and teeth, making it more useful than classic radiography methods. It can diagnose abnormalities, but treatment of abnormal teeth cannot be planned without more precise scans. For chirurgical and orthodontic treatments, 3D scans are a necessity. Cone-beam computed tomography (CBCT) is the best 3D imaging technique for planning treatment, extraction, and implantation [73]. CBCT offers excellent precision and an optimal algorithm for comprehensive treatment of dental patients. Nematolahi (2013) showed that CBCT imaged oral and maxillofacial structures with a high resolution of 0.001 mm3 voxels and that these 3D images can provide a better understanding of many anatomical structures as well as pathologic conditions, developmental anomalies, and traumatic injuries of teeth [74].

8. Conclusion

Knowing the normal anatomy of teeth is not only imperative in dentistry but also in other specialties because various medical conditions can negatively affect the appearance, structure, and position of teeth. With proper diagnosis, complications of abnormal teeth can be prevented. Complications can not only negatively affect smile esthetics but can also adversely affect speaking, biting, chewing, and so on as well as diminish a person's self-confidence and in some cases lead to depression and social isolation. The best option for diagnostic imaging of abnormal teeth as well as monitoring treatment progress is CBCT.

Conflict of interest

The authors declare no conflict of interest.

Author details

Milos Stepovic[1]*, Maja Vulovic[2], Ivona Bankovic[2], Miroslav Misic[3] and Radisa Vojinovic[4]

1 Faculty of Medical Sciences, University of Kragujevac, Serbia

2 Department of Anatomy, Faculty of Medical Sciences, University of Kragujevac, Serbia

3 College of Health Sciences, Academy of Applied Studies Belgrade, Serbia

4 Department of Radiology, Faculty of Medical Sciences, University of Kragujevac, Serbia

*Address all correspondence to: stepovicmilos@yahoo.com

References

[1] Noronha JC, Massara Mde L, Souki BQ, et al. First permanent molar: First indicator of dental caries activity in initial mixed dentition. Brazilian Dental Journal. 1999;**10**:99-104

[2] Kunzel W. Effect of an interruption in water fluoridation on the caries prevalence of the primary and secondary dentition. Caries Research. 1980;**14**:304-310

[3] dos Santos AP, Nadanovsky P, de Oliveira BH. A systematic review and meta-analysis of the effects of fluoride toothpastes on the prevention of dental caries in the primary dentition of preschool children. Community Dentistry and Oral Epidemiology. 2013;**41**(1):1-12. DOI: 10.1111/j.1600-0528.2012.00708.x

[4] Poulsen S, Holm AK. The relation between dental caries in the primary and permanent dentition of the same individual. Journal of Public Health Dentistry. 1980;**40**:17-25

[5] keie MS, Raadal M, Strand GV, et al. The relationship between caries in the primary dentition at 5 years of age and permanent dentition at 10 years of age – A longitudinal study. International Journal of Paediatric Dentistry. 2006;**16**:152-160

[6] Lynch RJM. The primary and mixed dentition, post-eruptive enamel maturation and dental caries: A review. International Dental Journal. 2013;**63**(2):3-13. DOI: 10.1111/idj.12076

[7] Memarpour M, Soltanimehr E, Eskandarian T. Signs and symptoms associated with primary tooth eruption: A clinical trial of nonpharmacological remedies. BMC Oral Health. 2015;**15**:88. DOI: 10.1186/s12903-015-0070-2

[8] Vittoba Setty J, Srinivasan I. Knowledge and awareness of primary teeth and their importance among parents in Bengaluru City, India. International Journal of Clinical Pediatric Dentistry. 2016;**9**(1):56-61. DOI: 10.5005/jp-journals-10005-1334

[9] Kjær I. Mechanism of human tooth eruption: Review article including a new theory for future studies on the eruption process. Scientifica (Cairo). 2014;**2014**:341905. DOI: 10.1155/2014/341905

[10] Frazier-Bowers SA, Puranik CP, Mahaney MC. The etiology of eruption disorders - further evidence of a 'genetic paradigm'. Seminars in Orthodontics. 2010;**16**(3):180-185. DOI: 10.1053/j.sodo.2010.05.003

[11] Arola DD, Gao S, Zhang H, Masri R. The tooth: Its structure and properties. Dental Clinics of North America. 2017;**61**(4):651-668. DOI: 10.1016/j.cden.2017.05.001

[12] Shadad O, Chaulagain R, Luukko K, Kettunen P. Establishment of tooth blood supply and innervation is developmentally regulated and takes place through differential patterning processes. Journal of Anatomy. 2019;**234**(4):465-479. DOI: 10.1111/joa.12950

[13] Ryakhovsky AN. 3D analysis of occlusal surfaces of teeth and their contacts. Part II. Assessment of the degree of wear of occlusal surfaces. Stomatologiia (Mosk). 2021;**100**(5):38-42. Russian. DOI: 10.17116/stomat202110005138

[14] Ryakhovsky AN. 3D analysis of occlusal surfaces of teeth and their contacts. Part I. development of a method for assessing the area of the occlusal surface, the severity of its relief and the histogram of contacts. Stomatologiia (Mosk). 2021;**100**(4):37-43. Russian. DOI: 10.17116/stomat202110004137

[15] Tafti A, Clark P. Anatomy, Head and Neck, Primary Dentition, In: StatPearls [Internet]. Treasure Island (FL): StatPearls Publishing; 2022. p. 2022

[16] Aruede G, Pepper T. Anatomy, permanent dentition. In: StatPearls [Internet]. Treasure Island (FL): StatPearls Publishing; 2022. p. 2022

[17] Nanci A. Chapter 7 - enamel: Composition, formation, and structure. Ten Cate's Oral Histology (Eighth Edition). Mosby. 2013:122-164

[18] Beniash E, Stifler CA, Sun CY, Jung GS, Qin Z, Buehler MJ, et al. The hidden structure of human enamel. Nature Communications. 2019;**10**(1):4383

[19] Goldberg M, Kulkarni AB, Young M, Boskey A. Dentin: Structure, composition and mineralization. Frontiers in Bioscience (Elite Edition). 2011;**3**(2):711-735

[20] Yamamoto T, Hasegawa T, Yamamoto T, Hongo H, Amizuka N. Histology of human cementum: Its structure, function, and development. Japanese Dental Science Review. 2016;**52**(3):63-74

[21] Ivancik J, Arola DD. The importance of microstructural variations on the fracture toughness of human dentin. Biomaterials. 2013;**34**(4):864-874

[22] Yumoto H, Hirao K, Hosokawa Y, Kuramoto H, Takegawa D, Nakanishi T, et al. The roles of odontoblasts in dental pulp innate immunity. Japanese Dental Science Review. 2018;**54**(3):105-117

[23] Gomes RR, Fonseca JAC, Paula LM, Acevedo AC, Mestrinho HD. Dental anomalies in primary dentition and their corresponding permanent teeth. Clinical Oral Investigations. 2014 May;**18**(4):1361-1367. DOI: 10.1007/s00784-013-1100-6

[24] Manuel T, Ceena D. Medication-related tooth discoloration: A review. Dental Update. 2014;**41**:440-2-440-44445. DOI: 10.12968/denu.2014.41.5.440

[25] Küchler EC, Costa AG, Costa Mde C, Vieira AR, Granjeiro JM. Supernumerary vary depending on gender. Brazilian Oral Research. 2011;**25**:76-79

[26] Rajab LD, Hamdan MAM. Supernumerary teeth: Review of the literature and a survey of 152 cases. International Journal of Paediatric Dentistry. 2002;**12**(4):244-254

[27] Bäckman B, Wahlin YB. Variations in number and morphology of permanent teeth in 7-year-old Swedish children. International Journal of Paediatric Dentistry. 2001;**11**(1):11-17

[28] King NM, Tongkoom S, Itthagarun A, Wong HM, Lee CK. A catalogue of anomalies and traits of the primary dentition of southern Chinese. Journal of Clinical Pediatric Dentistry. 2007;**32**(2):139-146

[29] Anthonappa RP, King NM, Rabie ABM. Prevalence of supernumerary teeth based on panoramic radiographs revisited. Pediatric Dentistry. 2013;**35**(3):257-261

[30] Al-Ani AH, Antoun JS, Thomson WM, Merriman TR, Farella M.

Hypodontia: An update on its etiology, classification, and clinical management. BioMed Research International. 2017;**2017**:9378325. DOI: 10.1155/2017/9378325

[31] Ritwik P, Patterson KK. Diagnosis of tooth agenesis in childhood and risk for neoplasms in adulthood. The Ochsner Journal. 2018 Winter;**18**(4):345-350. DOI: 10.31486/toj.18.0060

[32] Dinckan N, Du R, Petty LE, Coban-Akdemir Z, Jhangiani SN, Paine I, et al. Whole-exome sequencing identifies novel variants for tooth agenesis. Journal of Dental Research. 2018b;**97**:49-59

[33] Colin L, Peter M, Bikram T, Andrew F, David S. Hypodontia—A retrospective review of prevalence and etiology. Part I. Quintessence International (Berlin, Germany : 1985). 2005;**36**:263-270

[34] Chetty M, Beshtawi K, Roomaney I, Kabbashi S. MACRODONTIA: A brief overview and a case report of KBG syndrome. Radiology Case Reports. 2021;**16**(6):1305-1310. DOI: 10.1016/j.radcr.2021.02.068

[35] Hannah-Shmouni F, Trivellin G, Stratakis CA. Genetics of gigantism and acromegaly. Growth Hormone & IGF Research. 2016;**30-31**:37-41. DOI: 10.1016/j.ghir.2016.08.002

[36] Chen Y, Zhou F, Peng Y, Chen L, Wang Y. Non-syndromic occurrence of true generalized microdontia with hypodontia: A case report. Medicine (Baltimore). 2019;**98**(26):e16283. DOI: 10.1097/MD.0000000000016283

[37] Bargale SD, Kiran SD. Non-syndromic occurrence of true generalized microdontia with mandibular mesiodens - a rare case. Head & Face Medicine. 2011;**7**:19. DOI: 10.1186/1746-160X-7-19

[38] Shashirekha G, Jena A. Prevalence and incidence of Gemination and fusion in maxillary lateral incisors in Odisha population and related case report. Journal of Clinical and Diagnostic Research. 2013;**7**(10):2326-2329. DOI: 10.7860/JCDR/2013/5677.3516

[39] Rao P, Veena K, Chatra L, Shenai P. Twin tooth on either side: A case report of bilateral gemination. Annals of Medical and Health Sciences Research. 2013;**3**(2):271-273. DOI: 10.4103/2141-9248.113677

[40] Agnihotri A, Marwah N, Goel M. Geminated maxillary lateral incisor with Talon's cusp-a rare case report. Journal of Oral Health Community Dentistry. 2007;**1**:40-42

[41] Zhu M, Liu C, Ren S, Lin Z, Miao L, Sun W. Fusion of a supernumerary tooth to right mandibular second molar: A case report and literature review. International Journal of Clinical and Experimental Medicine. 2015;**8**(8):11890-11895

[42] Goh V, Tse OD. Management of Bilateral Mandibular Fused Teeth. Cureus. 2020;**12**(4):e7899. DOI: 10.7759/cureus.7899

[43] Khurana K, Khurana P. Esthetic and endodontic management of fused maxillary lateral incisor and supernumerary teeth with all ceramic restoration after trauma. Saudi Endodontic Journal. 2014;**4**:23-27

[44] Bolhari B, Pirmoazen S, Taftian E, Dehghan S. A case report of dilacerated crown of a permanent mandibular central incisor. Journal of Dentistry (Tehran). 2016;**13**(6):448-452

[45] Topouzelis N, Tsaousoglou P, Pisoka V, Zouloumis L. Dilaceration of maxillary central incisor: A

literature review. Dental Traumatology. 2010;**26**(5):427-433

[46] Calişkan MK, Tekin U. Surgical extrusion of a partially erupted and crown dilacerated incisor. Dental Traumatology. 2008;**24**(2):228-230

[47] Dineshshankar J, Sivakumar M, Balasubramanium AM, Kesavan G, Karthikeyan M, Prasad VS. Taurodontism. Journal of Pharmacy & Bioallied Sciences. 2014;**6**(Suppl 1):S13-S15. DOI: 10.4103/0975-7406.137252

[48] Jafarzadeh H, Azarpazhooh A, Mayhall JT. Taurodontism: A review of the condition and endodontic treatment challenges. International Endodontic Journal. 2008;**41**:375-388

[49] Marques Fernandes M, Ferreira Silva R, De Lucena BT, Ribeiro Tinoco RL, Fontanella V. Nogueira de Oliveira R. Taurodontism and its forensic value: a case report. The Journal of Forensic Odonto-Stomatology. 2018;**36**(2):40-43

[50] Consolaro A, Hadaya O, Miranda DAO, Consolaro RB. Concrescence: Can the teeth involved be moved or separated? Dental Press Journal of Orthodontics. 2020;**25**(1):20-25. DOI: 10.1590/2177-6709.25.1.020-026

[51] Rao VM, Karasick D. Hypercementosis--an important clue to Paget disease of the maxilla. Skeletal Radiology. 1982;**9**(2):126-128. DOI: 10.1007/BF00360497

[52] Thakur S, Thakur NS, Bramta M, Gupta M. Dens invagination: A review of literature and report of two cases. Journal of Natural Science, Biology and Medicine. 2014;**5**(1):218-221. DOI: 10.4103/0976-9668.127341

[53] Alkadi M, Almohareb R, Mansour S, et al. Assessment of dens invaginatus and its characteristics in maxillary anterior teeth using cone-beam computed tomography. Scientific Reports. 2021;**11**:19727. DOI: 10.1038/s41598-021-99258-0

[54] Păcurar M, Brezeanu L, Olteanu C, Dombi K, Sita D, Molnar C. The effect of rotation upon dental structure components following orthodontic fix appliance. Medicine and Pharmacy Reports. 2019;**92**(Suppl 3):S45-S49. DOI: 10.15386/mpr-1519

[55] Nayak G, Singh I. A variation in tooth position-180° rotated maxillary second premolar. Journal of Clinical and Diagnostic Research. 2013;**7**(8):1806-1807. DOI: 10.7860/JCDR/2013/5682.3285

[56] Akbas M, Karabıyık Z, Varol A. Ectopic tooth in Mandibular Canal, maxillary sinus, and mandibular condyle. Case Reports in Dentistry. 2022;**2022**:3118998. DOI: 10.1155/2022/3118998

[57] Sharma S, Chauhan SJ. Bilateral ectopic third molars in maxillary sinus associated with dentigerous cyst--a rare case report. International Journal of Surgery Case Reports. 2019;**61**:298-301. DOI: 10.1016/j.ijscr.2019.07.072

[58] Matsumoto MAN, Stuani MBS. Tooth transposition: A multidisciplinary approach. Dental Press Journal of Orthodontics. 2018;**23**(1):97-107. DOI: 10.1590/2177-6709.23.1.097-107.bbo

[59] Chattopadhyay A, Srinivas K. Transposition of teeth and genetic etiology. The Angle Orthodontist. 1996;**66**(2):147-152

[60] Shapira Y, Kuftinec MM. Tooth transpositions - a review of the literature and treatment considerations. The Angle Orthodontist. 1989;**59**:271-276

[61] Lobbezoo F, Ahlberg J, Raphael KG, Wetselaar P, Glaros AG, Kato T, et al. International consensus on the assessment of bruxism: Report of a work in progress. Journal of Oral Rehabilitation. 2018;**45**(11):837-844. DOI: 10.1111/joor.12663

[62] Sperber GH. Dental Wear: Attrition, erosion, and abrasion-a Palaeo-Odontological approach. Journal of Dentistry (Basel). 2017;**5**(2):19. DOI: 10.3390/dj5020019

[63] Skalsky Jarkander M, Grindefjord M, Carlstedt K. Dental erosion, prevalence and risk factors among a group of adolescents in Stockholm County. European Archives of Paediatric Dentistry. 2018;**19**(1):23-31. DOI: 10.1007/s40368-017-0317-5

[64] Nascimento MM, Dilbone DA, Pereira PN, Duarte WR, Geraldeli S, Delgado AJ. Abfraction lesions: Etiology, diagnosis, and treatment options. Clinical, Cosmetic and Investigational Dentistry. 2016;**8**:79-87. DOI: 10.2147/CCIDE.S63465

[65] Devaraju D, Devi BY, Vasudevan V, Manjunath V. Dentinogenesis imperfecta type I: A case report with literature review on nomenclature system. Journal of Oral and Maxillofacial Pathology. 2014;**18**(Suppl. 1):S131-S134. DOI: 10.4103/0973-029X.141363

[66] Chaudhary M, Dixit S, Singh A, Kunte S. Amelogenesis imperfecta: Report of a case and review of literature. Journal of Oral and Maxillofacial Pathology. 2009;**13**(2):70-77. DOI: 10.4103/0973-029X.57673

[67] Bernfeld WK. Hutchinson's teeth and early treatment of congenital syphilis. The British Journal of Venereal Diseases. 1971;**47**(1):54-56. DOI: 10.1136/sti.47.1.54

[68] Nissanka-Jayasuriya EH, Odell EW, Phillips C. Dental stigmata of congenital syphilis: A historic review with present day relevance. Head and Neck Pathology. 2016;**10**(3):327-331. DOI: 10.1007/s12105-016-0703-z

[69] Bhavyaa R, Sujitha P, Muthu MS, Nirmal L, Patil SS. Prevalence of the cusp of Carabelli: A systematic review and meta-analysis. Annals of Human Biology. 2021;**48**(7-8):572-584. DOI: 10.1080/03014460.2022.2032339

[70] Sánchez AR, Rogers RS 3rd, Sheridan PJ. Tetracycline and other tetracycline-derivative staining of the teeth and oral cavity. International Journal of Dermatology. 2004;**43**(10):709-715. DOI: 10.1111/j.1365-4632.2004.02108.x

[71] Ternovoĭ SK, Serova NS, Ivanova DV. Radiodiagnostic methods for dental anomalities. Vestnik Rentgenologii i Radiologii. 2012;(6):4-7 Russian

[72] Roslan AA, Rahman NA, Alam MK. Dental anomalies and their treatment modalities/planning in orthodontic patients. Journal of Orthodontic Science. 2018;**7**:16. DOI: 10.4103/jos.JOS_37_18

[73] Venkatesh E, Elluru SV. Cone beam computed tomography: Basics and applications in dentistry. Journal of Istanbul University Faculty of Dentistry. 2017;**51**(3 Suppl 1):S102-S121. DOI: 10.17096/jiufd.00289

[74] Nematolahi H, Abadi H, Mohammadzade Z, Soofiani GM. The use of cone beam computed tomography (CBCT) to determine supernumerary and impacted teeth position in pediatric patients: A case report. Journal of Dental Research, Dental Clinics, Dental Prospects. 2013 Winter;**7**(1):47-50. DOI: 10.5681/joddd.2013.008

Chapter 3

Frequency and Distribution of Pulpal Calcifications in Teeth Involved in Jaw Tumors

María Teresa Hernandez-Solis, Edith Lara-Carrillo, Víctor Hugo Toral-Rizo, Ronell Eduardo Bologna-Molina, José Edgar Garduño-Mejía, Norma Guadalupe Ibáñez-Mancera, Wael Hegazy-Hassan and Ana Miriam Santillán-Reyes

Abstract

Pulp calcifications are idiopathic mineralized masses associated with irritation, age, trauma, and systemic or genetic diseases. The objective of this work was to examine frequency and distribution of pulp calcifications in teeth involved in jaw tumors, analyzing their relationship with age, sex, location, size, and diagnosis of the lesion in a sample of 21 teeth associated with tumors of the jaws. Imaging analysis included CT scans, periapical X-rays, and orthopantomography of the clinical record; histological analysis included pulp tissue fixed in 10% buffered formalin for 24 h, with hematoxylin and eosin staining, and examined under light microscope. A chi-square test was applied to associate calcifications with all variables. The tumor lesions were from patients aged 17–66 years. Calcifications were observed in 38.1% of cases on image and histologically in 76.2%; 56% were nodular and 68.8% were distributed in chamber and root canal. The male sex presented a higher frequency of pulp calcifications, estimating a statistically significant difference with respect to women ($p = 0.004$); there was no statistical significance with the other variables. In conclusion, the pulp tissue of teeth affected by maxillary tumors presents a percentage of pulp calcifications similar to the tissue where the periradicular tissue is intact.

Keywords: pulpal calcifications, jaw tumors, pulpal histology, dental pulp, pulp stones

1. Introduction

Pulpal calcifications are masses of mineralized tissue that can develop freely in any area of the pulp tissue or be found adhered at the interface of the dentin and pulp of healthy or diseased, erupted, or unerupted teeth [1]. Morphologically, they can be nodular, oval, needle-like, or irregularly shaped and are predominantly composed of minerals such as hydroxyapatite, aluminum, copper, iron, potassium, lead, and zinc [2–4].

IntechOpen

Previously published studies have reported a prevalence of 8–90%, and they can be located both in the pulp chamber and in the root canal; they can be single or multiple, and in terms of size, they can be tiny below 200 μm or so large that they can obliterate the pulp chamber or root canal [5–7].

According to their topographic location, Satheeshkumar et al. [8] classified them as follows: single pulp calcification in the pulp chamber (Type I), multiple pulp calcifications present in the pulp chamber (Type IA), single pulp calcification present in the root canal (Type II), multiple pulp calcifications present in the root canal (Type IIA), multiple pulp calcifications present in pulp chamber and root canal (Type IIB), and continuous calcifications extending from the pulp chamber to the root canal (Type III).

The etiological factors of this process are still unknown; however, they are associated with different factors, such as trauma, pulp degeneration, orthodontic treatment, periodontal disease, caries, operative procedures, pulp inflammation, systemic diseases, or genetic conditions [9–13]. This process used to be associated with age; however, it has been observed that they are also found in young patients and in teeth that have not yet erupted [14].

Investigations regarding pulpal calcifications have been carried out over time with different methodologies based on imaging studies that include various techniques, such as periapical X-ray, panoramic X-ray, bitewing X-ray, cone beam tomography, or micro-CT and histological studies. These methodologies have been performed on teeth present in the mouth, extracted dental organs, or prehistoric skeletal remains [15–17].

Tumor lesions are abnormal tissue that grows in both the maxilla and mandible; they are relatively rare and affect soft and hard tissues and may extend to the facial region. Most of them are benign, but they tend to be aggressive and grow exponentially, displacing bone tables and thus the teeth [18].

The aim of this work was to show the frequency and distribution of pulp calcifications of teeth involved in tumor lesions of the jaws by means of imaging and histological analysis, analyzing whether there is any relationship between pulp calcifications and age, sex, location, size, and diagnosis of the lesion.

2. Materials and methods

For this purpose, an analytical, descriptive, and cross-sectional study was designed with nonprobabilistic sampling by convenience. The sample consisted of 21 caries-free teeth with intact enamel involved in eight tumor lesions in patients between 17 and 66 years old, which were removed by the therapeutic indication of the maxillofacial surgeon. All donors signed the informed consent, in which it was mentioned that the teeth would be used for scientific research and the tissue obtained from the surgical maneuver would be processed for histopathological diagnostic purposes. In addition, this project was approved and registered by the Research Ethics Committee of the Autonomous University of the State of Mexico (2021/P11).

Immediately after surgical excision, the tumor lesion was immersed in 10% buffered formalin for immersion fixation and transferred to the oral pathology laboratory at the School of Dentistry of the Autonomous University of the State of Mexico for histological processing. The teeth were separated from the tumor lesion with a scalpel blade no. 15 in soft tissue and flexible diamond disc (Plexoflex Fine Grain. DFS-Diamon, Riedenburg, Germany) when the alveolar bone was involved.

The teeth separated from the tumor lesion were kept in 10% buffered formalin while pulp tissue was obtained; this process was performed by marking with a flexible diamond disc (Plexoflex Fine Grain. DFS-Diamon. Riedenburg, Germany) and abundant irrigation along the longitudinal axis of the uniradicular teeth. For the multiradicular teeth, two marks were made, one in the crown and the other in the root zone, removing the hard tissues and obtaining the pulp tissue from the chamber and root canal.

Immediately after the pulp tissue was obtained, it was placed again in 10% buffered formalin for 24 h and then dehydrated in a series of alcohols to be included in a paraffin block and stained with hematoxylin and eosin. The slides were obtained with 5 μm thick tissue sections.

To perform the histological analysis and topographical location with the classification of Satheeshkumary et al. [8], the slides were observed using a Leica Microsystems DM750 microscope at a magnification of 10x. The analysis was performed by two examiners from the Endodontics and Oral Pathology areas. Imaging was performed with Motic, VM 3.0, Digital Slide Scanning System using the programs Phatomation PMA.start (Pathomation BV, Berchem, Belgium) and ImageJ-FIJI.

For the imaging analysis, panoramic radiographs, periapical radiographs, and tomographies were collected and included in the clinical records, which were reviewed to determine the absence or presence of calcifications.

Subsequently, the statistical analysis of the data was carried out using SPSS package version 26 (IBM Corp. Armonk, N.Y. USA). The descriptive statistical data of variables were obtained, and Pearson's chi-square test was performed to associate the variables (age, sex, diagnosis, location, and size of the lesion) with the presence of pulpal calcifications.

3. Results

The sample consisted of 21 teeth, seven incisors (33.3%), two canines (9.5%), five premolars (23.8%), and seven molars (33.3%). They were from eight tumor lesions with histopathological diagnosis: two ameloblastomas (25%), two keratocysts (25%), one central giant cell lesion (12.5%), one peripheral giant cell lesion (12.5%), one calcifying epithelial odontogenic tumor (12.5%), one peripheral ossifying fibroma (12.5%), one peripheral giant cell lesion (12.5%), one calcifying epithelial odontogenic tumor (12.5%), and one peripheral ossifying fibroma (12.5%).

The donor patients were four women (50%) and four men (50%), ranging in age from 17 to 66 years, with a mean age of 34.75 years and a standard deviation of 20.76 years.

The location of the tumor lesions was 50% in the maxilla and 50% in the mandible, with a minimum size of 2 cm and a maximum of 15 cm, a mean of 7.43 cm, and a standard deviation of 4.04 cm. All teeth were analyzed together with their imaging studies included in the clinical records; 10 teeth were reviewed with tomography (46.6%), eight with panoramic radiographs (38.1%), and three with periapical radiographs (14.3%), observing calcifications in the imaging analysis in 38.1% of the cases (**Figure 1**). Sagittal view of a CT scan showing pulp calcification in the coronal area (white arrow) of a developing third molar associated with a keratocyst). Histological examination with hematoxylin and eosin revealed that pulp calcifications were present in 76.2% of the analyzed teeth. With respect to shape, 56.25% were only

Figure 1.
Sagittal view of a CT scan showing pulp calcification in the coronal area (white arrow) of a developing third molar associated with a keratocyst.

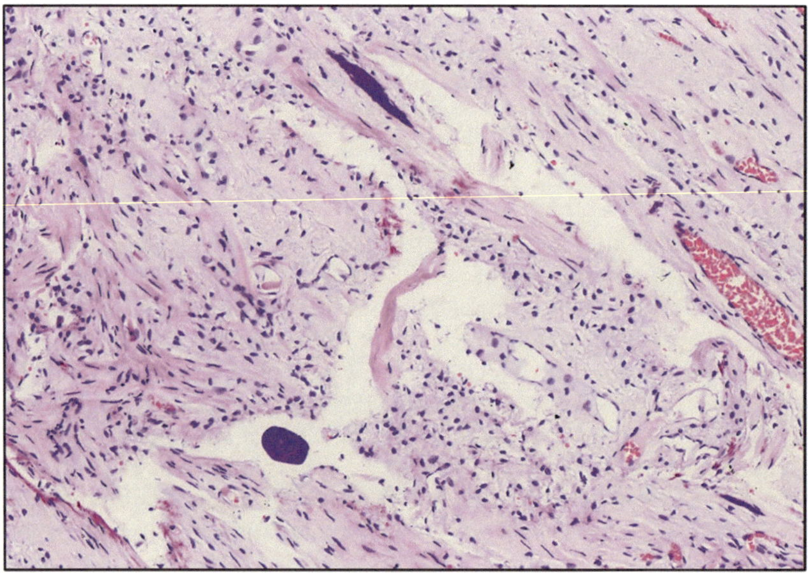

Figure 2.
Microphotograph of circular (nodule) and elongated pulp calcifications in the pulp tissue of a tooth associated with peripheral giant cell lesion (hematoxylin and eosin, 100× magnification, Motic VM 3.0. Digital Slide Scanning System).

nodules and 43.75% were nodules and needle-like in the pulp tissue itself (**Figure 2**). Microphotograph of circular (nodule) and elongated pulp calcifications in the pulp tissue of a tooth associated with peripheral giant cell lesion. Hematoxylin and eosin, 100× magnification, Motic VM 3.0. Digital Slide Scanning System).

Pulpal calcifications were distributed topographically according to the classification of Satheeshkumary et al. [8] as follows: 68.8% were in both the pulp chamber and root canal (Type IIB), 18.8% were multiple in the root canal (Type IIA) (**Figure 3**). Microphotograph of pulp tissue of a tooth involved in ameloblastoma, showing multiple calcifications of different sizes and shapes. Hematoxylin and eosin, 40× magnification, Motic VM 3.0. Digital Slide Scanning System), and 12.5% presented continuous calcifications from the crown to the root canal (Type III).

The presence of these calcifications was more frequent in males (87.5%) than in females (12.5%), showing a statistically significant difference with a value of $p = 0.004$.

In relation to age groups, more calcifications were observed in the 17- to 30-year-old group compared to the others, without showing statistically significant differences between groups ($p = 0.028$) (**Table 1**). It shows the frequency of pulpal calcifications in relation to age groups.

Pulp calcifications were found in 68.7% of mandibular teeth and 31.3% of maxillary teeth, but this difference was not statistically significant ($p = 0.717$).

Regarding the size of the tumor lesion and its relationship with calcifications, the 1 cm to 4 cm group had no calcified masses, while the 5 cm to 9 cm and 10 cm to 15 cm groups each had 8 teeth (50%); however, there was no significance between them ($p = 0.131$).

Regarding the histopathological diagnoses of the lesions and their relationship with calcifications, no significant association was found ($p = 0.484$), so the diagnosis

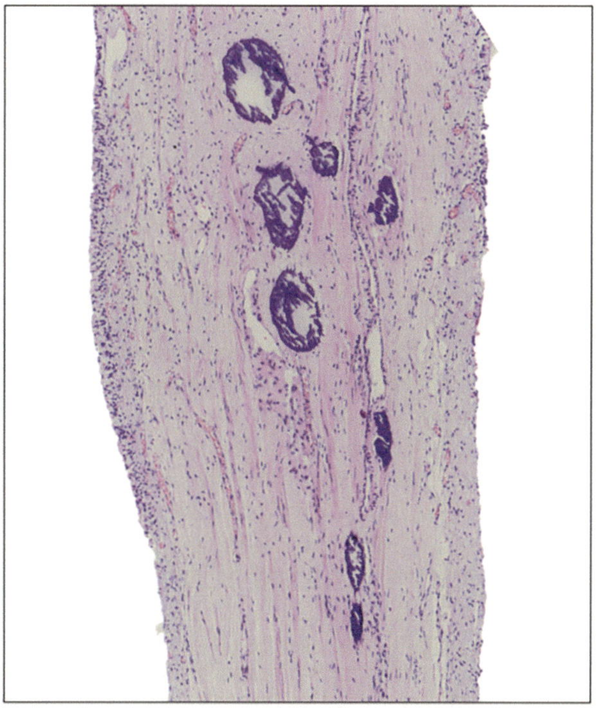

Figure 3.
Microphotograph of pulp tissue of a tooth involved in ameloblastoma, showing multiple calcifications of different sizes and shapes (hematoxylin and eosin, 40× magnification, Motic VM 3.0. Digital Slide Scanning System).

Age groups		17–30 years	31–59 years	60 + years	Total
Calcifications	Absent	2	2	1	5
	Present	12	0	4	16
Total		14	2	5	21

Table 1.
Frequency of pulp calcifications in relation to age groups.

Diagnostic	Peripheral ossifying fibroma	Ameloblastoma	Peripheral giant cell lesion	Central giant cell lesion	Keratocyst	Total
Calcification	1	3	0	0	1	5
	1	8	3	3	1	16
Total	2	11	3	3	2	21

Table 2.
Frequency of pulp calcifications in relation to histopathological diagnoses of tumor lesions.

does not determine the existence or absence of calcific precipitations (**Table 2**). It shows the frequency of pulp calcifications in relation to histopathological diagnoses of tumor lesions.

4. Discussion

The study of pulp calcifications has led to development research to determine the prevalence, distribution, and association with diseases such as caries, periodontal disease, or genetic conditions. However, in the literature review, no previous data were found on the study of pulp calcifications in teeth involved in maxillary tumors, so one of the objectives of this study was to publicize and disseminate the results, adding knowledge to this area.

Pulp calcifications can be detected through diagnostic means such as imaging and histological studies. Imaging studies are carried out by panoramic radiographs, periapical, bitewing, cone beam tomography, or micro-CT, which allow locating radiopaque nodules with a size greater than 200 μm since below this size they are not detectable by these means. For histological studies, they allow observing calcifications of different sizes and locations, since the lamellae are observed under the microscope with different magnifications, allowing the localization of these in any path of the pulp tissue.

Regarding the prevalence of pulp calcifications, previous studies have shown a range from 9.6 to 95% [3, 16]. These percentages vary due to the methodologies used for each investigation; usually, imaging studies show a lower frequency than histological studies since the lamellae are seen under the microscope, identifying any size of calcification. This finding agrees with our results and those of Huang et al. [19] since both studies performed imaging and histological analyses and observed similar results. Huang et al. [19] showed the presence of calcifications of 62% in the histological area and 30% in the radiological area, while our results were 76.2% in the histological area and 38.1% in the imaging area.

Histological studies allow a greater morphological appreciation of pulp calcifications, and we agree with Milcent et al. [2] that pulp calcifications are morphologically heterogeneous since they are different from each other in shape and size.

Another important factor is that the prevalence of calcifications in pulp tissue may be different depending on the ethnic group. For example, in Mexican patients, they have been observed in 84% [20], Peruvians 83.58% [21], Taiwanese 83.3% [22], Iranians 76% [23], Argentines 55% [24], Jordanians 51.4% [25], Turks 38% [26], and Yemenis 3.99% [13].

Sener et al. [24] and Ranjitkar et al. [1] performed the analysis of pulp calcifications with bitewing radiographs, finding a prevalence of 38% and 46.1%, respectively. In the present study, periapical, panoramic, and tomography radiographs were examined analyzing the coronal and radicular areas, and the percentage of calcifications was similar at 38.1%.

The frequency of pulp calcifications in males was higher than in females, which agrees with Hsieh et al. [23]. The possible cause is that men present greater dental attrition in relation to women, and the pressure exerted by the tumor in the apical area allows for premature contact points, which provides a local irritant in the pulp tissue.

Another interesting fact is that pulpal calcifications have been related to advanced age [9]. In contrast, in this study, the age group 17–30 years old was the one with the highest percentage of calcifications.

Mandibular teeth had a higher percentage of pulp calcifications, which agrees with Olivares et al. [20]. However, there are also studies where the maxillary dental organs are the predominant ones without a significant difference [24].

Regarding the size of the lesion and the histopathological diagnosis, no similar antecedents were found in the literature since these variables have not been investigated in pulp tissue under these conditions. Therefore, the present study generates knowledge that can be used as a basis for future lines of research.

However, one of the limitations of the study is the small size of the sample, since being a histological study, the samples of the tumor lesions should present dental organs, and these should be kept in an adequate fixation medium for the preservation of the pulp tissue, which is not frequent to find or to obtain them in the correct preservation medium.

5. Conclusion

In conclusion, the pulp tissue of teeth affected by maxillary tumors presents a similar percentage of pulp calcifications to studies where the periradicular tissue does not present any alteration.

Another important fact is that there was no statistically significant relationship between pulp calcifications and tumor size or histopathological diagnosis.

The results of this study may be useful as a basis for future lines of research since they broaden the knowledge of pulp calcifications and histological characteristics of pulp tissue of teeth in the presence of maxillary tumors. For future studies, it would be convenient to analyze this finding in a larger sample.

Conflict of interest

The authors declare no conflicts of interest.

Author details

María Teresa Hernandez-Solis[1], Edith Lara-Carrillo[2]*, Víctor Hugo Toral-Rizo[3], Ronell Eduardo Bologna-Molina[4], José Edgar Garduño-Mejía[5], Norma Guadalupe Ibáñez-Mancera[6], Wael Hegazy-Hassan[2] and Ana Miriam Santillán-Reyes[2]

1 Faculty of Dentistry, Doctoral Program in Health Sciences, Autonomous University of the State of Mexico, Toluca, Mexico

2 Center for Research and Advanced Studies in Dentistry "Dr. Keisaburo Miyata" Faculty of Dentistry, Autonomous University of the State of Mexico, Toluca, Mexico

3 Faculty of Dentistry, Orocentro Clinic, Autonomous University of the State of Mexico, Toluca, Mexico

4 Faculty of Dentistry, Center for Stomatological Molecular Pathology, Republic University, Montevideo, Uruguay

5 Department of Maxillofacial Surgery, Regional Hospital "Lic. Adolfo López Mateos", Toluca, Mexico

6 The Interdisciplinary Center of Health Sciences UST, Instituto Politécnico Nacional, Mexico City, Mexico

*Address all correspondence to: laracaedith@hotmail.com

References

[1] Ranjitkar S, Taylor JA, Townsend GC. A radiographic assessment of the prevalence of pulp stones in Australians. Australian Dental Journal. 2002;**47**(1): 36-40. DOI: 10.1111/j.1834-7819.2002.tb00301.x

[2] Milcent CPF, da Silva TG, Baika LM, Grassi MT, Carneiro E, Franco A, et al. Morphologic, structural, and chemical properties of pulp stones in extracted human teeth. Journal of Endodontics. 2019;**45**(12):1504-1512. DOI: 10.1016/j.joen.2019.09.009

[3] Kuzekanani M, Haghani J, Walsh J, L, Estabragh M. Pulp stones, prevalence, and distribution in an Iranian population. The Journal of Contemporary Dental Practice. 2018;**19**(1):60-65. DOI: 10.5005/jp-journals-10024-2212

[4] Berès F, Isaac J, Mouton L, Rouzière S, Berdal A, Simon S, et al. Comparative physicochemical analysis of pulp stone and dentin. Journal of Endodontics. 2016;**42**(3):432-438. DOI: 10.1016/j.joen.2015.11.007

[5] Bevelander G, Johnson PL. Histogenesis and histochemistry of pulpal calcification. Journal Dental Research. 1956;**35**(5):714-722. DOI: 10.1177/00220345560350050901

[6] Jannati R, Afshari M, Moosazadeh M, Allahgholipour SZ, Eidy M, Hajihoseini M. Prevalence of pulp stones: A systematic review and meta-analysis. Journal of Evidence-Based Medicine. 2019;**12**(2):133-139. DOI: 10.1111/jebm.12331

[7] Frank RM, Nalbandian J. Structure, and ultrastructure of the dental pulp. In: Berkovitz BKB, Boyde A, Frank RM, Höhling HJ, Moxham BJ, Nalbandian J, Tonge CH, editors. Teeth. Handbook of Microscopic Anatomy. 1st ed. Berlin Heidelberg: Springer Berlin; 1989. pp. 249-307. DOI: 10.1007/978-3-642-83496-7_5

[8] Satheeshkumar PS, Mohan M, Saji S, Sadanandan S, George G. Idiopathic dental pulp calcifications in a tertiary care setting in South India. Journal of Conservative Dentistry. 2013;**16**(1):50-55. DOI: 10.4103/0972-0707.105299

[9] Goga R, Chandler NP, Oginni AO. Pulp stones: A review. International Endodontic Journal. 2008;**41**(6):457-468. DOI: 10.1111/j.1365-2591.2008.01374.x

[10] Ravanshad S, Khayat S, Freidonpour N. The prevalence of pulp stones in adult patients of Shiraz Dental School, a radiographic assessment. Journal of Dentistry. 2015;**16**(4):356-361

[11] Ricucci D, Loghin S, Lin LM, Spångberg LSW, Tay FR. Is hard tissue formation in the dental pulp after the death of the primary odontoblasts a regenerative or a reparative process? Journal of Dentistry. 2014;**42**(9):1156-1170. DOI: 10.1016/j.jdent.2014.06.012

[12] Bastos JV, MIDS C. Pulp canal obliteration after traumatic injuries in permanent teeth—Scientific fact or fiction? Brazilian Oral Research. 2018;**32**(suppl 1):e75. DOI: 10.1590/1807-3107bor-2018.vol32.0075

[13] Sabeti M, Tayeed H, Kurtzman G, Mashhadi AF, Talebi AM. Histopathological investigation of dental pulp reactions related to periodontitis. European Endodontic Journal. 2021;**6**(2):164-169. DOI: 10.14744/eej.2021.96268

[14] Vibhute NA. Hard facts about stones: Pulpal calcifications: A review. Journal of Patient Care. 2017;**02**(01):1-4. DOI: 10.4172/2573-4598.1000105

[15] Sayegh FS, Reed AJ. Calcification in the dental pulp. Oral Surgery, Oral Medicine, Oral Pathology and Oral Radiology. 1968;**25**(6):873-882. DOI: 10.1016/0030-4220(68)90165-5

[16] Keleş A, Keskin C, Versiani MA. Micro-CT assessment of radicular pulp calcifications in extracted maxillary first molar teeth. Clinical Oral Investigations. 2022;**26**(2):1353-1360. DOI: 10.1007/s00784-021-04109-x

[17] Nicklisch N, Schierz O, Enzmann F, Knipper C, Held P, Vach W, et al. Dental pulp calcifications in prehistoric and historical skeletal remains. Annals of Anatomy. 2021;**235**:151675. DOI: 10.1016/j.aanat.2021.151675

[18] Neville BW, Damm DD, Allen CM, Chi AC. Oral and Maxillofacial Pathology. 4th ed. Missouri: W B Saunders; 2016. p. 928

[19] Huang L-G, Chen G. A histological and radiographic study of pulpal calcification in periodontally involved teeth in a Taiwanese population. Journal of Dental Sciences. 2016;**11**(4):405-410. DOI: 10.1016/j.jds.2016.05.001

[20] Olivares HML, Ovalle CJW. Prevalecence of pulp stones. Revista ADM. 2001;**58**(4):130-137. [In Spanish]

[21] Calero-HG T-LPL, García CR. Prevalence and distribution of pulp stones in a Group of Peruvian Adults: A cone beam computed tomography study. Odovtos. 2021;**23**(2):161-170. DOI: 10.15517/ijds.2021.44468

[22] Hsieh C-Y, Wu Y-C, Su C-C, Chung M-P, Huang R-Y, Ting P-Y, et al. The prevalence and distribution of radiopaque, calcified pulp stones: A cone-beam computed tomography study in a northern Taiwanese population. Journal of Dental Sciences. 2018;**13**(2):138-144. DOI: 10.1016/j.jds.2017.06.005

[23] Moshfeghi M, Tuyserkani F. Prevalence of pulp stones in an Iranian subpopulation: An assessment using cone beam computed tomography. General Dentistry. 2021;**69**(3):e1-e5

[24] Labarta AB, Mancipe Leguizamón S, Pineda Martínez CJ, Sierra LG. Radiographic evaluation of the frequency and distribution of pulpal nodules. Revista Facultad de Odontología (Buenos Aires). 2021;**20**:57-65. [In Spanish]

[25] Hamasha AA-H, Darwazeh A. Prevalence of pulp stones in Jordanian adults. Oral Surgery, Oral Medicine, Oral Pathology, Oral Radiology, and Endodontology. 1998;**86**(6):730-732. DOI: 10.1016/s1079-2104(98)90212-8

[26] Sener S, Cobankara FK, Akgünlü F. Calcifications of the pulp chamber: Prevalence and implicated factors. Clinical Oral Investigations. 2009;**13**(2):209-215. DOI: 10.1007/s00784-008-0212-x

Chapter 4

Aesthetic Smile Designing

Monisha P. Khatri, Shreya Kishore and Srujana Hemmanur

Abstract

'Smile is a universal language, shared by people all over the world'. It conveys a lot without saying anything at all. Our smile is of utmost importance and dental aesthetics play a vital role to create it. Smile designing (SD) consists of various components that aim to develop the overall personality of a patient. To create an aesthetic, smile is an organised and systematic approach that is required to evaluate, diagnose and resolve the problems associated with it. The ultimate goal of an aesthetic make-over is to ensure a stable masticatory system, where the oral hard and soft tissues remain in harmony. Hence this chapter aims to focus on various aspects of SD that includes its goals, components, procedure and future prospects.

Keywords: aesthetic, smile designing, golden proportion, gingival zenith, veneers

1. Introduction

The goal of smile designing is to create a harmonious balance between teeth, tissues, muscles, skeletal structures and joints while attending to the aesthetic requirement of the patient [1]. This requires complete understanding and knowledge of the symmetry, shape and proportions related to the hard and soft tissues of the oral cavity.

2. Components of smile designing

In dental treatment, aesthetics has traditionally been associated with profile enhancement. An organised and systematic approach is required to evaluate the aesthetic problems and diagnose and resolve them. The ultimate goal as clinicians is to achieve proportionate and pleasing composition in patients' smiles by creating an arrangement of various aesthetic elements. Harmonising an aesthetic smile requires an integration of facial and dental composition. The dental composition includes tooth components and gingival components, one of the vital elements of smile designing [1].

2.1 Tooth components

Tooth components consist of the following:

2.1.1 Dental midline

The dental midline is the vertical contact interface between the two maxillary central incisors. It is to be parallel to the midline of the face and perpendicular to the interpupillary line. Other landmarks that can help assess the dental midline are the midline of the nose, forehead and the philtrum of the upper lip. To evaluate the midline, one must consider the location and alignment. The dental midline should be parallel to the long axis of the face, perpendicular to the incisal plane, and should be over the papilla, that is the midline that should drop straight down from the papilla [1, 2].

2.1.2 Incisal lengths

The position of the maxillary incisal edge is the most important determinant in smile line as it acts as a reference point to decide the tooth proportions and the gingival levels. The parameters used to help establish the maxillary incisal edge positions are the amount of tooth display during smile, the phonetics and the patient input.

The amount of tooth display is determined when the mouth/muscles around the mouth are relaxed. Ideally, 3.5 mm of incisal third of the maxillary centrals are visible, which continues to reduce as the patient ages, due to loss in muscle tonicity.

Phonetics is a primary determinant, and the various phonetics used are (i) M sound, (ii) E sound, (iii) F and V sounds and (iv) S sound.

Patient input is an important guide to determine the smile line, as the patient's desire and satisfaction is the ultimate goal [1].

2.1.3 Tooth dimensions

Correct teeth proportions are essential in creating an aesthetically pleasing smile. Central dominance is a phenomenon that states that the centrals must be the dominant teeth in smile and should display pleasing proportions, as they are the key to an aesthetic smile. There are various guidelines for establishing correct proportions in an aesthetically pleasing smile, and they are as follows: [1, 3, 4].

i. *Golden proportion (by Lombardi* [1])

When viewed from the facial aspect, the width of each anterior is 60% of the width of the adjacent tooth. The mathematical ratio being 1.6:1:0.6 (**Figure 1**).

ii. *Recurring aesthetic dental proportions (by Ward* [1])

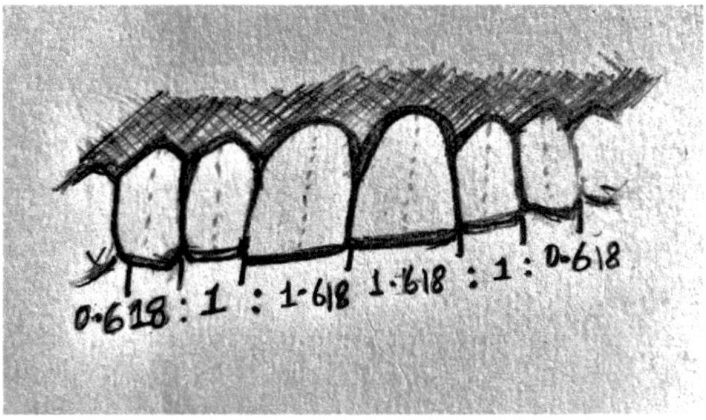

Figure 1.
Golden proportion based on apparent width from the frontal view.

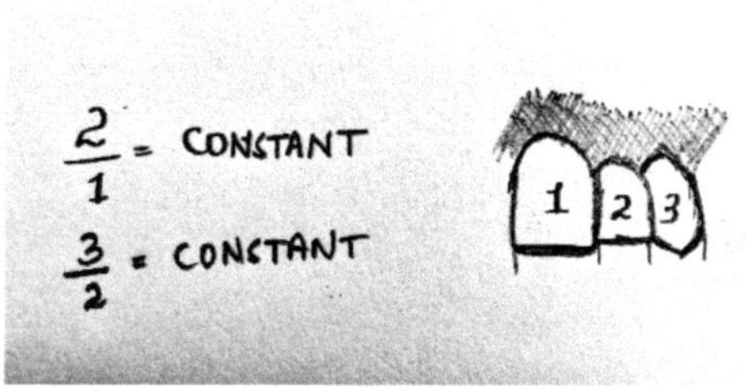

Figure 2.
Recurring aesthetic dental proportions.

The successive width proportion when viewed from the facial aspect should remain constant as we move posteriorly from the midline. This theory offers greater flexibility to match tooth proportions with facial proportions (**Figure 2**).

iii. *M proportions* (*by Methot* [1])

M proportions compare the tooth width with the facial width using software on a computer, due to which, involves more scientific calculations rather than artistic analysis.

iv. *Chu's aesthetic gauges* [1]

Dr. Chu's research refutes the golden proportions. A series of gauges were made available to make the intra-oral analysis easier. The gauges allow fast and simple diagnosis of tooth discrepancies and are used as a reference guide between the clinician and the lab technician, which in turn reduces the incidence of communication errors.

2.1.4 Zenith points

The most apical position of the cervical tooth margin where the gingival is most scalloped is known as the zenith point (**Figure 3**). It is located distal to the vertical axis of the tooth, except for lateral incisors where it is centrally located. Establishing the proper location of the zenith point is the critical step in altering the mesial and distal dimensions [4].

2.1.5 Axial inclinations

Also known as tooth inclinations, it compares the vertical alignment of maxillary teeth to central vertical midline. From the maxillary centrals to the maxillary canines,

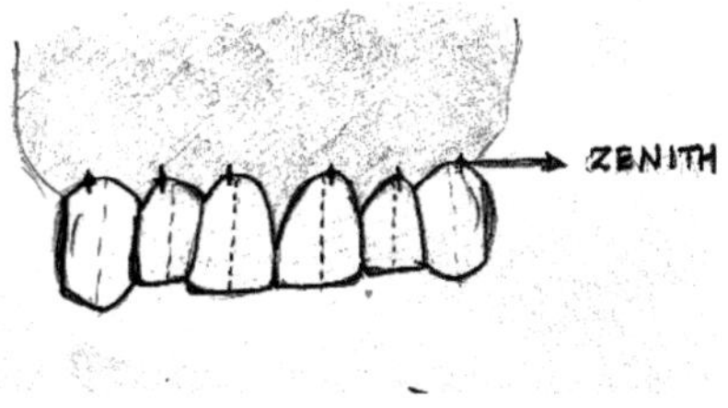

Figure 3.
Zenith points and their relation to the midline.

there should be a progressive increase in the mesial inclination of each subsequent anterior tooth, which looks natural. The evaluation of the axial inclination can be done in the frontal view photograph of the anterior teeth. The guide for labiolingual inclination is as follows: (i) maxillary central incisor is positioned vertically or slightly labial. (ii) maxillary lateral incisal edge is inclined slightly labially. (iii) the maxillary canine cusp is angulated lingually [1, 4].

2.1.6 Contact area and contact point

Interproximal contact area (ICA) is defined as the broad zone in which two adjacent teeth touch (**Figure 4**). An increase in ICA helps to create an illusion of longer teeth, which in turn helps in eliminating black triangles [1].

Interproximal contact point (ICP) is the most incisal aspect of the interproximal contact area.

2.1.7 Incisal embrasure

The incisal embrasures are supposed to display a natural, progressive increase in size from the centrals to the canines, which results in the interproximal contact points moving more apically as we move from the centrals to the canines to mimic the smile line. Failure in providing adequate incisal embrasure results in making the teeth appear too uniform and in loss of the incisor's individuality [4].

2.2 Gingival components

This composes of the soft tissue component of smile designing and they are as follows:

2.2.1 Gingival health

It is of paramount importance that the gingival tissues are in proper health prior to the start of the treatment. Healthy gingiva usually exhibits the following characteristic features: (i) pale pink, stippled, firm in consistency, (ii) located 3 mm above the alveolar crestal bone and (iii) located interdentally, 5 mm above the inter-crestal bone, should be pointed and should fill the gingival embrasure right up to the contact area [5].

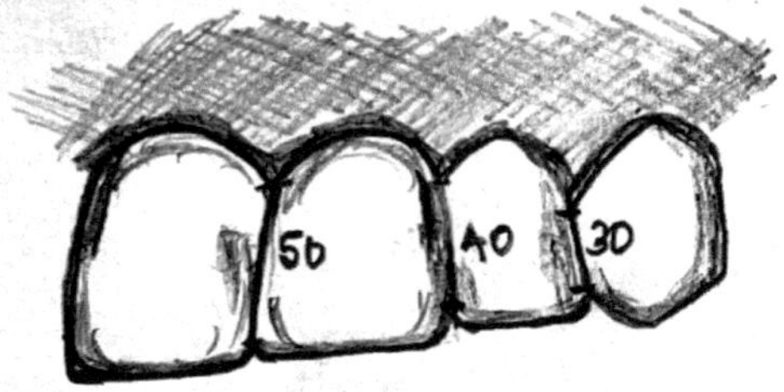

Figure 4.
Interproximal contact area.

2.2.2 Gingival levels and harmony

Establishing the ideal gingival levels for each individual tooth is the key to creating a harmonious smile. The cervical gingival height of the maxillary centrals should be symmetrical and can match that of the canines. The gingival margin of the lateral incisor can be 0.5–2.0 mm short of the central incisors [5, 6].

2.2.3 Smile line

Smile line refers to an imaginary line along the incisal edges of the maxillary anterior teeth, which should mimic the curvature of the superior border of the lower lip while smiling.

Keep in mind that the lip line should not be confused with the smile line. It refers to the position of the inferior border of the upper lip during smile formation and thereby determines the display of tooth or gingiva. Showing 3–4 mm or more of the gingiva often requires cosmetic periodontal recontouring to achieve ideal results, as in cases of gummy smiles [1, 5].

3. Treatment

Treatment modalities include veneers, direct or indirect and digital smile designing with help of software tools and CAD/CAM.

3.1 Veneers

Definition

A veneer is a layer of tooth-coloured material that is applied to a tooth to restore localised or generalised defects and intrinsic discolouration, a conservative aesthetic restoration of anterior teeth to mask discolouration, restore malformed teeth, close diastemas and correct minor tooth alignment [7].

History

Veneers were developed originally in the 1930s. California dentist Charles Pincus [8] developed thin facings of air-fired porcelain that could be fastened in place with adhesive denture powder. In 1955, Buonocore's [8] research into the acid etch technique provided a simple method of increasing the adhesion to enamel surfaces for acrylic filling materials. In the 1970s, Faunce [8] described a one-piece acrylic resin prefabricated veneer as an improved alternative to direct composite resin bonding. Veneer was attached both chemically and mechanically. With the advent of materials like glass-infiltrated ceramics and zirconia techniques, the ease and precision with which veneers can be prepared have increased multi-folds [8].

Classification

1. On the basis of extent—Partial/Full

   The partial veneers are the ones that partially cover tooth structure while full veneers encompass the full tooth structure with the veneer material.

2. On the basis of technique—Direct/Indirect

   Direct application of composite material to create direct veneers is done while lab preparation of veneers using either composite or porcelain creates indirect veneers.

3. On the basis of material used—Composite/Porcelain

   The use of different materials such as composite resin or porcelain are used to prepare veneers.

4. On the basis of mode of fabrication—Pre-fabricated/Custom made

   Prefabricated veneers are the ones that are modified to fit any tooth size or shape while custom-made veneers are fabricated for a specific patient and are customised built.

5. On the basis of tooth preparation—(**Figure 5**) [9]

   Class 0—no preparation: No preparation is done on any surface of the tooth for the luting of veneer.

   Class 1—window preparation: This involves preparing the veneer short of the incisal edge and retaining the enamel over the incisal edge. The disadvantage here is the difficulty of hiding the margin.

   Class 2—feather edge preparation: The preparation is taken or feathered to the incisal edge, with no reduction of the incisal edge length. The disadvantage of this preparation is that the margins can be subjected to shear forces in protrusion.

   Class 3—bevel preparation: A bevel is carried over the incisal edge from buccal to palatal, with 1–2 mm of incisal reduction. A tooth preparation that incorporates incisal overlap is preferable, as the veneer is stronger and provides a positive seat during cementation. This preparation design has the advantage of simple tooth preparation, and the aesthetic characteristics are easier to fabricate with the ceramist, as it is possible to develop incisal translucency. The margin should not be in a position where it will be subjected to protrusive forces during excursive

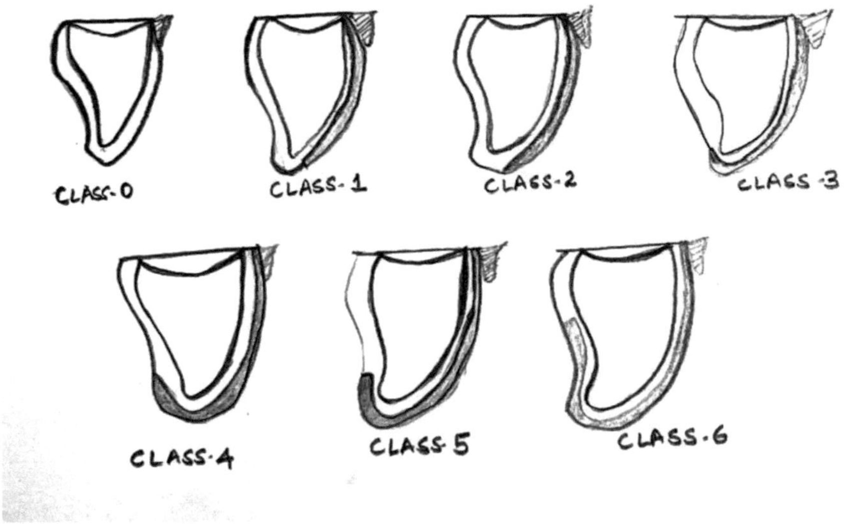

Figure 5.
Classification of veneers based on tooth preparation.

movements, therefore reducing the stress within the veneer while distributing the occlusal load over a wider surface area.

Class 4—overlapping preparation of incisal edge: The incisal edge is reduced with the preparation, then extended onto the palatal aspect. A positive seat is provided with this preparation, although there is a need to carefully evaluate the path of insertion to ensure that no undercuts are present.

Class 5—butt joint preparation: The butt joint preparation gives a 90-degree preparation all around the tooth that is to receive a veneer. The disadvantage is increased cutting of tooth structure.

Class 6—full veneer preparation: More or less like an anterior preparation for receiving a full veneer crown.

3.2 Indirect porcelain veneers

The indications, contraindications and advantages and disadvantages of indirect veneers are enumerated in **Tables 1** and **2**, respectively [8, 9].

3.3 Clinical procedure

3.3.1 Labial preparation

The preparation of the labial contour of anterior teeth must be done in three planes (3-plane reduction): incisal, middle third and cervical (**Figure 6**).

Careful depth reduction of the tooth is done to provide a minimum of 0.3 mm (feldspathic porcelain) or 0.6 mm (Empress, e max) preparation. The enamel thickness at the gingival third is 0.3–0.5 mm, up to 0.6–1 mm at the middle third and 1.0–2.1 mm at the incisal third.

The use of depth cutters or grooves and dimples has been recommended to control tooth preparation [9].

3.3.2 Incisal edge reduction

Four different preparation designs were possible (**Table 3**) (**Figure 7**) [9].

3.3.3 Proximal reduction

The proximal preparation can be done by stopping short of breaking the contact, or by preparing through the contact point.

- If contact points are left intact, it is preferable to leave the contact point with the margin ending approximately 0.25 mm or more labial to the contact region.

- The visibility of the tooth: porcelain interproximal interface may be viewed or be hidden by the use of an L-shaped preparation or elbow preparation to hide the margins interproximally.

- Breaking the contact is often used in changing the shape or position of teeth. With the additional space interproximally, adjustments in the contours and position of the teeth and width discrepancies can be done.

Indication	Contradictions
Moderate discolourations, such as tetracycline staining, fluorosis, devitalized teeth and teeth darkened by age, are not conducive to vital bleaching.	Patients with certain tooth to tooth habits such as bruxism or parafunctional habits
Teeth with generalised moderate facial discolouration from amalgam shine-through	Insufficient enamel for bonding and sealing of the peripheries.
Surface defects. Small cracks in the enamel caused by ageing, trauma or ice chewing	Class III and end-on malocclusions
Replacement of missing or fractured parts of the teeth	Deciduous teeth
Closing of diastemas, single or multiple spaces between the teeth and improving the appearance of rotated or malpositioned teeth. An aesthetic illusion can be created but the ideal treatment would be orthodontic management.	
Short teeth, teeth can be lengthened to a more aesthetic appropriate size	
Malocclusions or periodontally compromised teeth	
Agenesis of the lateral incisor	
Progressive wear pattern, but sufficient enamel must remain and the desired increase in length is not excessive.	
Functionally sound ceramo-metal or all-ceramic crowns with unsatisfactory hue.	

Table 1.
Indications and contraindications of indirect veneers.

Advantages	Disadvantages
Natural and stable colour with smooth lustrous surfaces.	Colour cannot be modified easily once bonded
Highly acceptable tensile bond strength.	Irreversibility of preparation vs. little or no preparation for direct composite resin bonding.
Long-lasting	Level of difficulty of fabrication and placement, time involved and expense.
Exceptional resistance to wear	Technical difficulties in avoiding overcontours and obtaining closely fitted porcelain/enamel margins. The margins can be especially brittle and difficult to finish.
Resistance to stain	Lower reparability compared to composite veneers.
Extremely good biocompatibility with gingival tissues	Susceptibility to pitting by certain topical fluoride treatments.
Much less absorption of fluidprevents the decrease in physical properties of ceramics.	
Surface lustre retention	
Lack of radiopacity. It appears as natural tooth structure on radiographs.	

Table 2.
Advantages and disadvantages of indirect veneers.

Figure 6.
Labial preparation.

Feather preparation	Window preparation	Bevel preparation	Incisal overlap
The preparation is taken to the incisal edge, without reducing the incisal edge.	The veneer is short of the incisal edge, retaining the enamel over the incisal edge.	A bevel is carried over the incisal edge from buccal to palatal, with 1–2 mm of incisal reduction. Preferred because the veneer is stronger and provides a positive seat during cementation.	The incisal edge is reduced with the preparation and then extended onto the palatal aspect. A positive seat is provided with this preparation. No undercuts must be present.
Disadvantage: The margins can be subjected to shear forces in protrusion.	Disadvantage: difficulty of hiding the margin.	Advantage: simple tooth preparation and aesthetic characteristics are easier to fabricate with the ceramist, as it is possible to develop incisal translucency. The proper seating of the veneer is also enabled with the positive seat that is provided. The margin should not be in a position where it will be subjected to protrusive forces during excursive movements, therefore reducing the stress within the veneer while distributing the occlusal load over a wider surface area	

Table 3.
Different preparation designs.

- Preparations may extend further proximally with the presence of caries and existing restorations. Never should a margin lie on restoration or caries and should always lie on sound enamel [9].

3.3.4 Cervical margin

- Chamfer with a maximum depth of 0.4 mm to ensure reproduction of natural tooth contours and prevents over-contouring.

- It also ensured simple seating minimising stresses and enhances future fracture resistance of the veneer.

- Supragingival or equigingival margins are preferred.

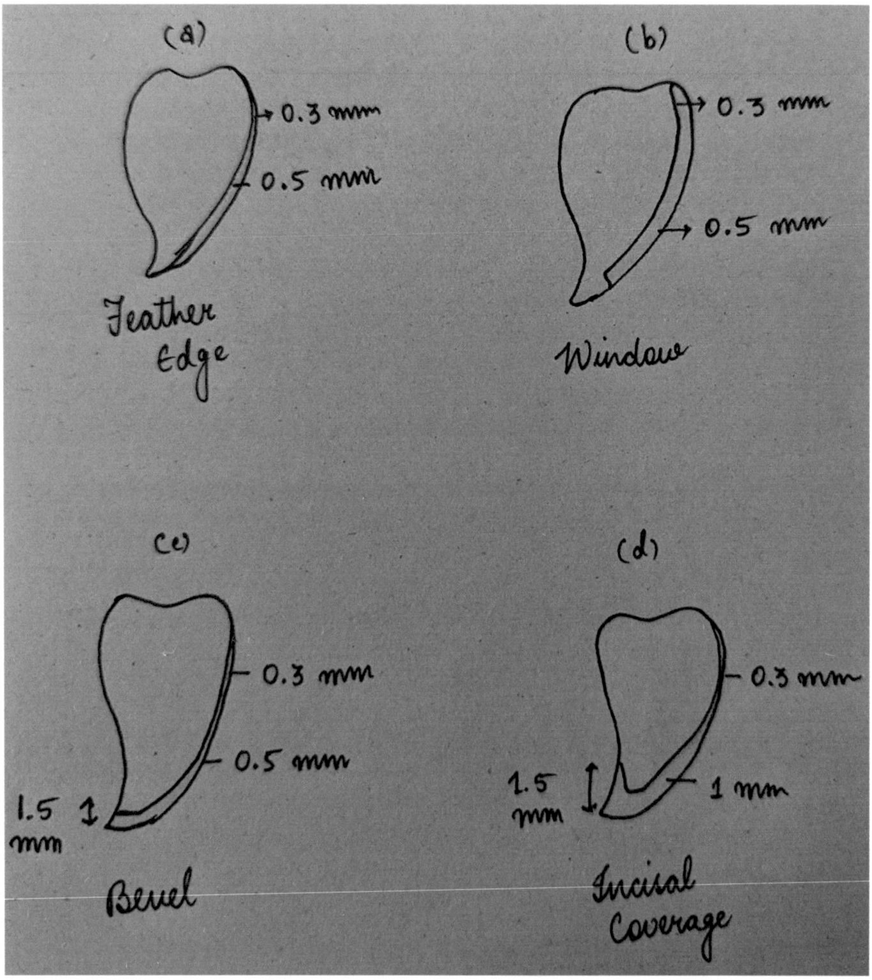

Figure 7.
Reduction according to type of veneer preparation.

The prepared tooth then must be finished using finishing burs and abrasives and polished using pumice such that smooth surfaces are obtained [10].

3.3.5 Impressions

After the tooth prep is done, retraction cords are placed such that all the gingival margins are exposed. An accurate impression material such as polyvinyl siloxane, polyether or reversible hydrocolloid (agar) is used and a final impression is made.

Either a single-step impression or a two-step putty wash impression is made.

With advancement in technology, optical impressions are also being made. These impressions integrated with CAD CAM can instantaneously produce the required veneer [10].

3.3.6 Temporisation

The provisional restorations must be placed to improve interim aesthetics and decrease sensitivity. They also provide essential diagnostic information like veneer colour, shape, length and incisal edge configuration.

If it is a single-tooth veneer, the safest option is a light cured composite, which is used to directly build the tooth up. Central spot etching is done. No overhanging resin composite at the margins should be seen (Direct technique).

When provisionals for multiple teeth are needed, it is preferable to use a clear matrix made on a preop diagnostic cast. The teeth are spot etched, washed and dried. The facial and incisal areas in the clear matrix are filled with resin composite and the matrix is placed on the teeth and cured after removal of the excess. Instead of the composite, bis-acrylic temporization material (e.g. Luxatemp, DMG, Germany) can also be used (Indirect technique) [8].

3.3.7 Try-in and luting

After the veneer is obtained, the fit and shade are all evaluated in the try-in procedure and luting is performed. The teeth are treated with either the total-etch or self-etch protocol and the intaglio surface of the veneer is treated with 10% hydrofluoric acid. Resin cement is used to lute the veneer and the excess cement is removed and finishing and polishing the finish line [9].

3.4 Indirect composite veneers

Aka semi-direct veneers

The clinical technique is similar to that of indirect porcelain veneer. Instead of porcelain, the resin composite is used to create veneers extra orally and these are cured and fired/autoclaved to provide extra resistance to fracture. The finishing and polishing of the composite veneers are also done extra orally. Resin cement is used to lute these semi-direct veneers after treating the tooth with self-etch or total-etch bonding protocol.

The advantages include better strength and fracture resistance, chair-side controlled contouring, single appointment completion and minor adjustments that can be done while carrying out the procedure.

The disadvantages include long chairside appointments and proper patient compliance and are technique sensitive.

3.4.1 Failures of indirect veneers

These are categorised as [8]:

1. Mechanical

   This includes (i) Fracture—poor positioning of the incisal margin, less incisal thickness and margin placed too sub-gingivally. (ii) Debonding—use of expired cement for luting and faulty veneer/tooth preparation.

2. Biological

Biological failures include (i) post-operative sensitivity (ii) improper curing of cement, (iii) exposed margins, (iv) poor marginal adaptation, (v) margins at DEJ and (vi) marginal microleakage due to poor fit and extension.

3. Aesthetic

Aesthetic complications include (i) improper shade selection and (ii) over contouring or improper margin placement in cases of gingival recession.

3.5 Direct veneers

A direct veneer is built up over the tooth surface by hand with composite material.

They are indicated in cases of aesthetically compromised anterior teeth, when one desires a younger and better smile, in cases of discoloured teeth, closure of diastemas and space closures, fractured teeth, anatomically malformed teeth, patient is a bruxer and clencher (Indirect veneers are contraindicated in such patients) [8, 11].

Whereas in patients with unacceptable occlusion, actively erupting tooth, presence of periodontal disease, insufficient coronal tooth structure, enamel irregularities or deficiencies, and extremely dark and stained tooth (underlying discolouration visible through the veneers) direct veneers are contraindicated [8, 11].

Advantages and disadvantages

The advantages associated with these veneers are that it can be done in a single visit, the dentists have chair side control of tooth anatomy, it can be repaired, patient compliance is not compromised, cheaper than the other alternative and there is minimal irreversible loss of tooth structure as the preparation is almost next to nil [8, 11].

On the contrary, they tend to discolour, there is increased chair side time, physical limitations of the material, they wear out more quickly, often require repair and have higher chances of replacement and marginal staining [8, 11].

3.5.1 Diagnostic considerations

Direct veneers can be done only when there is adequate amount of enamel present, there is tooth discolouration, in cases of orthodontic problems, in patients with habits like tobacco chewing or smoking, edge-to-edge occlusion, parafunctional habits and when one needs to meet patient's expectations immediately [9].

3.5.2 Treatment planning and shade selection

Two sets of models are made so that one can be used to study the patients' teeth and occlusion while another is used for a trial build-up (also known as an aesthetic mock up). A putty index is taken from it to replicate the mock-up [10].

Pre-operative photographs in natural lighting and monochrome must be taken to correctly evaluate the shade of the teeth. Shade tabs or colourimeters can also be used to determine the correct shade of the teeth. Another technique to achieve complex shades is the application of composite resin on non-bonded tooth surfaces and curing of the composite resin [8]. The first shade to be matched is for the dentin layer, followed by the shade for the enamel in the cervical and incisal areas [12].

3.5.3 Clinical steps

After shade matching, an intra-enamel tooth preparation should be done to a minimum depth of 1/4th to 3/4th thickness of enamel (depending on the severity of the case). Following this, the teeth are etched with 35% orthophosphoric acid, washed, dried and checked for frosty appearance. Next, the bonding agent is applied, air-dried and polymerised using light cure unit. After this, putty index is placed and placement of composite resin in layering technique is done to ensure shade matching. In case of dark discoloured teeth, an opaquer is used to mask the effect following which a dentin layer and enamel layer (cervical and incisal respectively) being applied and light cured. Finishing and polishing done 10 minutes after complete polymerisation are achieved. Burs, discs and wheels are used to reduce excess contours and final finishing and polishing are done using diamond discs and cups with polishing paste [9].

3.5.4 Failures of direct veneers

Composite veneers show around 60–70% of success rates in clinical studies. However, failures are seen to occur 2 to 5 years after placement what so ever the protocol or material is used.

Periodontal involvement: Due to the finish lines located near gingival margin and the presence of substances such as excess composite or bonding agents in the gingival crevices, which can act as irritants, gingival inflammation can occur.

Marginal breakdown: White lines may be seen as a result of finishing process. The composite tends to tear at the margins due to the effect of polymerisation shrinkage and the heat and friction produced during the finishing and polishing of the newly placed composite.

Occlusal-related chipping: This is seen especially in patients with parafunctional habits. Hence, the length and contouring of the veneer should be decided and perfected. Also, the occlusal table can be widened by adding extra material on the cuspid tip to resist shear forces.

Chipping at contact area: Seen due to occlusal trauma or fatigue caused by cyclic fatigue. A chamfer or rounded shoulder margin should be placed to avoid such chipping.

Layer separation: Occurs when composite is inadequately bonded to the tooth and other layers. Layering of composite under a completely sterile environment and carrying out proper polymerisation of each layer to prevent the presence of unpolymerized resin [9].

3.5.5 Strategies to repair failed direct veneers

Minor marginal breakdown can be treated with discing. Discing removes the damaged area and exposes the sound restoration underneath.

The repair of fractured composite such that new material is layered over existing composite should never be attempted. If a restoration needs repair, the old composite should be removed completely and the restoration must be repeated.

When a properly placed composite in a stress-bearing area fractures due to overloading stress, a better stress-bearing and stronger material should be used to restore the tooth instead of repeating the composite restoration [9].

4. Recent advances

4.1 Lumineers

Lumineers are made from a special patented Cerinate porcelain that is very strong but much thinner than traditional laboratory-fabricated veneers. The thickness is comparable to contact lenses [13]. Lumineers are a reversible procedure and it hardly requires removal of tooth structure. They will bond directly to the tooth and the longevity is up to 20 years.

Componeer (Coltene, Altstten, Switzerland) prefabricated veneers are thin composite resin shells (0.3 mm cervically and 0.6–1.0 mm to the incisal edge). These prefabricated veneers are made of a pre-polymerised hybrid composite resin and Synergy D6 (Coltene, Altstten, Switzerland) [14].

4.2 Digital smile designing (DSD)

With the new era of digitalization, use of softwares and computers allow for a repeatable and reproducible functional and aesthetic result. Components of DSD include a computer, DSD software, digital single-lens reflex (DSLR) camera or smartphone camera, intra-oral scanner, a printer and CAD/CAM system [15].

4.3 Photography and videography

Good quality photographs are a must as it forms the baseline to evaluate the changes as well as plan the treatment. It helps to form the facial and gingival reference lines such as the commissural lines, lip line and inter-pupillary line. The photographs and videographs should be taken accordingly (**Tables 4** and **5**) [15–17].

These images are uploaded in software that helps carry out the DSD. Various softwares are available depending on factors such as the dentist's ease of use, digital workflow, cost, time effectiveness and compatibility with other digital tools and systems, for example, Smile designer Pro, Tasty Tech ltd. Toronto, Ontario, Canada; Microsoft PowerPoint, Microsoft Office, Redmond, Washington, USA; Keynote, iWork, Apple, Cupertino, California, USA.

Given the history of humans and their inspiration by principles of nature, the digital tool just like a manual tool performs facial analysis [18]. The horizontal (inter-pupillary and inter-commissural lines) and vertical references (facial midline through the glabella, nose and the chin) are marked and drawn against each other to calculate the facial symmetry. The software then places the wide smile photograph of the patient behind these reference lines to measure the vertical midline

Frontal view-3	Proximal view-2	At 12 o'clock view	Intra occlusal view
1. Full face at rest. 2. Full face with teeth apart and wide smile. 3. Teeth apart retracted view of maxillary and mandibular arch.	1. Side view at rest. 2. Side view with full smile.	incisal edge of maxillary teeth visible and resting on lower lip with a wide smile.	second premolar to second premolar of maxillary arch.

Table 4.
Photographs.

Frontal video	Proximal video	At 12 o'clock video	Anterior occlusal video
Smiling with and without retractor	lips at rest and wide-smile	To allow visualisation of incisal edge	Keeping palatine raphe as a straight line from maxillary second premolar to second premolar

Table 5.
Videographs.

and horizontal plane for comparative evaluation between the teeth and face. Next, the dento-gingival analysis is done. Gingival display and smile curve is established and dental contouring is done based on the curvature of the lower lip and antero-posterior curvature of the teeth. Reference lines such as horizontal canine to canine tip, horizontal lines through incisal edges, vertical midline and gingival zenith are drawn to match the facial analysis. The dental length-to-width ratio is established by incorporating any one of the dental proportions such as the golden proportion, Pound's theory, recurring aesthetic dental proportion and Dentogenic theory or Visagism (using temperament in concept of smile design, introduced by Braulio Paolucci). All the above measurements are then transferred to a 3D cast. Wax-up is then done and reflected onto the cast. The software then uses tools to make necessary changes and the final image is presented to the patient, this final image can also be altered based on the patient's aesthetic needs and requirements. A 3D printer is used to process the digital wax-up, silicone index can be prepared out of the model and can be processed to create a provisional restoration. Mock-ups and alterations can be done on it. The final changes can be made and a scanner is used to feed it to the CAD/CAM machine [15–17]. The final approved 3D image of the preparation is obtained and final prosthesis is designed by the machine, this is called computer-aided designing. Appropriate size of the ceramic is chosen and the milling machine produces the veneers in accordance with the details fed into the computer [19]. An alternative method would be computer-aided designing/additive manufacturing using 3D printing to overcome the wastage of the material.

Various veneer materials are available and used by the CAD/CAM system such as feldspathic ceramics, mica-based ceramics, leucite-reinforced ceramics, lithium disilicate reinforced ceramics, glass infiltrated alumina and zirconia ceramics, polycrystalline alumina and zirconia, which includes yttria partially stabilised tetragonal zirconia polycrystals, magnesium partially stabilised zirconia and ceria stabilised zirconia/alumina nanocomposite. In 3D printing, the options available are light-cured resin, methacrylate-based resin, composites, hybrid composite resin, experimental PMMA-based nanocomposite, polymers and ceramic-filled hybrid resins such as experimental lithium disilicate-based ceramic-filled photo-sensitive resin [19, 20].

Advantages

Improved diagnosis and treatment planning.

Visualisation of treatment outcome.

Patients' involvement and satisfaction.

Better communication amongst dentists as well as the dentist and the patient.

High-quality pre- and post-treatment evaluation [15–17].

Limitations

Dependency on accurate photographs and videos.

Expensive.

Need to have updated knowledge and skill to work with technology [15–17].

5. Conclusion

Creating a smile requires thorough knowledge and skill to replicate the natural smile. One needs to consider the components of smile designing, which includes the tooth as well as the gingival counterparts. Consideration of various proportions to achieve the shape, symmetry, function and aesthetic is required. With availability of tooth-coloured materials, achieving a natural smile is now possible. Composite or ceramic veneers help achieve maximum beautification with minimal preparation. Moreover, popularisation of software tools and CAD/CAM systems will help achieve maximum precision and patient satisfaction.

Author details

Monisha P. Khatri[1*], Shreya Kishore[2] and Srujana Hemmanur[1]

1 Department of Conservative Dentistry and Endodontics, SRM Institute of Science and Technology, SRM Dental College, Ramapuram, Chennai, India

2 Department of Orthodontics, SRM Institute of Science and Technology, SRM Dental College, Ramapuram, Chennai, India

*Address all correspondence to: monishak@srmist.edu.in

References

[1] Bhuvaneshwaran M. Principles of smile design. Journal of Conservative Denstistry. 2010;**13**(4):225-232

[2] Nigam M et al. Esthetic design. Journal of Orofacial Research. 2015;**5**(4):130-133

[3] Calamia JR, Wolff MS. The components of smile design: New York University smile evaluation form revisited, update 2015. Dental Clinics. 1 Jul 2015;**59**(3):529-546

[4] Sabri R. The eight components of balanced smile. JCO. 2005;**36**(3):155-167

[5] Chakroborty G, Pal TK, Chakroborty A. A study on gingival component of smile. Journal of the International Clinical Dental Research Organization. 2015;**7**(2):111-115

[6] Calamia JR, Levine JB, Lipp M, Cisneros G, Wolff MS. Smile design and treatment planning with the help of a comprehensive esthetic evaluation form. Dental Clinics. 1 Apr 2011;**55**(2):187-209

[7] Lo ML. Book review: Mosby's dental dictionary. British Dental Journal. 2014;**217**(4):168

[8] Albers HF. Tooth colored restorations. In: Principle and Techniques. 9th ed. Hamilton. London: BC Decker Inc; 2002. p. 97

[9] Goldstein RE, Chu SJ, Lee EA. Stappert CF. In: Ronald E, editor. Goldstein's Esthetics in Dentistry. USA: John Wiley & Sons; 2018

[10] Tylman SD, Malone WF. Tylman's Theory and Practice of Fixed Prosthodontics. 7th ed. St. Louis: Mosby; 1978. pp. 96-134

[11] Robertson T, Heymann HO, Swift EJ Jr. Sturdevant's Art and Science of Operative Dentistry. Philadelphia, PA, USA: Elsevier Health Sciences; 2006

[12] Fahl N, Ritter AV. Composite veneers: The direct–indirect technique revisited. Journal of Esthetic and Restorative Dentistry. 2021;**33**:7-19. DOI: 10.1111/jerd.12696

[13] Zlatanovska K, Dimova C, Zarkova-Atanasova J. Minimally invasive aesthetic solutions - porcelain veneers and lumineers. DDF. 2017;**376**:111-120. DOI: 10.4028/www.scientific.net/ddf.376.111

[14] Gomes G, Perdigão J. Prefabricated composite resin veneers–A clinical review. Journal of Esthetic and Restorative Dentistry. 2014;**26**(5):302-313. DOI: 10.1111/jerd.12114

[15] Jafri Z, Ahmad N, Sawai M, Sultan N, Bhardwaj A. Digital smile design-an innovative tool in aesthetic dentistry. Journal of Oral Biology and Craniofacial Research. 2020;**10**(2):194-198

[16] Thomas PA, Krishnamoorthi D, Mohan J, Raju R, Rajajayam S, Venkatesan S. Digital smile design. Journal of Pharmacy and Bioallied Sciences. 2022;**14**(5):43

[17] Cervino G, Fiorillo L, Arzukanyan AV, Spagnuolo G, Cicciù M. Dental restorative digital workflow: Digital smile design from aesthetic to function. Dentistry Journal. 2019;**7**(2):30

[18] Basheer N, Madhubala MM, Mahalaxmi S. Future perspectives of biomimetics in restorative dentistry. Journal of Pharmaceutical Research International. 2020;**32**(25):19-28

[19] Liu PR, Isenberg BP, Leinfelder KF. Evaluating CAD-CAM generated ceramic veneers. Journal of the American Dental Association. 1993;**124**(4):59-63

[20] Dawood A, Marti BM, Sauret-Jackson V, Darwood A. 3D printing in dentistry. British Dental Journal. 2015;**219**(11):521-529

Section 2

Prevention and Oral Health

Chapter 5

Oral Health: Fundamentals, Importance, and Perspectives

Andréa Dolores Correia Miranda Valdivia, Maria de los Angeles Vázquez Sánchez, Diana Elizabeth Aguirre Cortés and Evangelina Gutiérrez Cortés

Abstract

Oral health is a fundamental and important part of the patient's general health. It is very common to find texts that focus on diseases rather than on health, the proposal of this chapter is to address the importance of oral health maintenance, its impact on people's lives, and the fundamental role of the dentist as a professional. Oral diseases are largely preventable or require only simple interventions if diagnosed and addressed at early stages. The aim of this chapter is to discuss the importance of the function of teeth and their supporting tissues for the health of the person, as well as the functions of esthetics, phonation, and mastication in the bio-psycho-social relationship of the human being, to demonstrate how fundamental, it is to maintain oral health.

Keywords: oral health, supporting tissues, quality of life, oral health concepts, diagnosis, oral

1. Introduction

Oral health is an integral part of general health since both share common causal pathways and affect each other in a bi-directional fashion [1]. The World Health Organization (WHO) defined oral health in 2011 as "a state of well-being, free from chronic orofacial pain, oral and throat cancer, oral ulcers, birth defects such as cleft lip and palate, periodontal or gum disease, dental caries, tooth loss and other diseases and disorders affecting the oral cavity" [2].

Oral health is essential for proper food intake. The condition of the teeth and supporting tissues have been shown to contribute to the maintenance of proper nutritional status, and consequently, to overall health [3]. The oral microbiome is comprised of over 600 prevalent taxa at the species level, with distinct subsets predominating in various oral habitats [4]. In the absence of effective oral hygiene, initial dental plaque formation on a clean tooth surface will occur within 48 hours and can be reflected in the oral environment [5]. Dental caries and periodontitis are the most common oral diseases and are major causes of tooth loss [6].

The area of oral health is very important because it is linked to the overall health of the patient, and much emphasis is placed on oral health problems rather than on oral health prevention. Prevention of oral diseases requires extensive knowledge of their causes such as socioeconomic inequalities, nutrition and dietary aspects, access to fluoride, and appropriate dental care, all of which may start early in life [7].

Many countries have established different policies and successful programs to reduce the prevalence and severity of oral diseases. Yet from a global perspective, much more needs to be done, positive action needs to be accelerated, and innovative solutions need to be evaluated and implemented at scale so that the vision of universal health coverage (UHC) for oral health becomes a realistic goal. This includes a renewed focus on integrating oral health care with primary health care (PHC) and integrating the promotion of prevention and oral health in settings outside specialist oral health facilities [1].

The aim of this chapter is to present the components and their function within oral health. In addition, show the impact of oral health on the quality of life, as well as comment on the role of the dentist in oral health.

2. Importance of teeth and supporting tissues

The oral cavity is mentioned by some authors as the entrance door of the external environment to the organism and has been related as the link to systemic health. It is formed by two zones: an external or vestibular one limited by the inner part of the lips and cheeks and the vestibular face of all the teeth. The other zone, the internal zone, is the space inside the mouth, limited by the inner side of the teeth, the upper part by the palate, and the lower part by the tongue and the floor of the tongue [8–10].

2.1 Function

It is in this place, the mouth, where the teeth are located, in which we differentiate between two zones, the visible portion called the crown, while the portion that is lodged in the dental alveolus is called the root. The tooth is a complex structure composed of different highly organized hard and soft tissues, including highly calcified enamel, dentin, and cementum, in addition to the soft tissues, which include the dental pulp and periodontal ligament [10–12].

Enamel, inorganic and calcified tissue, is the hard surface of the tooth and the hardest tissue in the body. It is directly exposed to the oral cavity and external stimuli such as mechanical stress. Dentin is the second layer of the tooth located below the enamel and cementum. It is a collagen mineralized structure formed of several microtubules that extend from the pulp wall to the dentin-pulp junction or cementum. Dentin has a mechanism in which its constituent cells called odontoblasts, are responsible for the secretion of the dentin matrix and the formation of secondary dentin which occurs in adult life during normal dentinogenesis. Once the tooth reaches adulthood, we consequentially find a decrease in the volume of the pulp cavity [13].

The pulp located in the central part of the tooth, the pulpar chamber and root canal, is the vital tissue that provides nutritional elements to the tooth. It also serves as a contributor of sensory function due to its composition of nerves and blood vessels that enter through the apical foramen. The pulp additionally fulfills its other formative and defensive functions through the formation of repair dentin, which it does as a response to harmful external stimuli [12, 13].

The portion of the maxillary and mandibular bone that surrounds the tooth is referred to as periodontal tissue and includes the gingival gingiva, the alveolar bone, and the periodontal ligament. The gingival gingiva is the mucosa that surrounds and protects the tooth acting as a barrier to mechanical trauma and infection specifically in the gingival sulcus where the gingiva inserts into the tooth. Its function within the gingiva is sensory, absorption of micronutrients and plays an important role in the innate immune response [14].

The periodontal ligament is a tissue composed of connective fibers containing specialized cells that surround the tooth. These fibers are responsible for keeping it attached to the alveolus in the maxillary bone, which contains a rich microvascular network within, thus performing the function of anchoring the tooth to the bone. A close relationship has also been demonstrated between the behavior of the periodontium and the support forces that act as "shock absorbers," since it is in continuous tension from constant chewing forces [11, 14, 15].

Humans develop two sets of teeth: primary and secondary, which vary in shape and number, among others. The primary dentition is formed by 8 incisors, 4 canines, and 8 molars, with a total of 20 teeth. These are currently recognized as important to preserve because of their relationship with the permanent teeth, as they are their precursors and are essential to contribute not only to the health of children but also to the welfare of adults [10, 11, 16].

Primary teeth begin their formation during gestation and begin to erupt in the mouth at approximately 6 months following a sequential eruption pattern, thus in the first year of life, a new function begins: chewing. This action can be affected by the premature loss of a dental piece since it not only causes a simple setback but also affects the masticatory function and results in the appearance of malocclusion problems by providing the indispensable space that will later be occupied by the secondary dentition [17, 18].

The crowns of primary teeth are smaller than those of secondary or permanent teeth, their precursors, and are formed by three thin layers: enamel (hard and external layer whose function is to protect the tooth from injuries and infections), dentin (layer formed by tubules responsible for providing nutrients to the tooth, unlike the enamel can regenerate) and pulp (formed by nerves and vessels important for the health of the tooth). Both enamel and dentin are found as thin layers compared to permanent teeth, which makes primary teeth more susceptible to dental caries [18].

Permanent teeth are larger than deciduous teeth, as well as their crowns, because as the jaws mature the permanent teeth gradually adapt in the arch. These teeth are formed by 8 incisors, 4 canines, 8 premolars, and 12 molars for a total of 32 teeth. The different occlusal and incisal morphologies that we observe in each of the groups of teeth are closely related not only to the function they perform and the position in which they are distributed in the structures involved in oral dynamics but are also linked to the mandibular movements that aid in mastication. For example, the anterior teeth are so called because of the position in which they are located, they are called incisors because their function is to incise or cut the food. The upper teeth are especially important, not only for their function but also from a phonetic and esthetic point of view, since they are directly related to the pronunciation of certain letters such as c, d, f, s, t, v, z. While the back teeth, the molars, located at the back of the jaws with their occlusal faces, are responsible for grinding food and facilitating chewing [10, 11, 18–20].

Teeth are not essential for life, however, they are crucial for the health of the individual, as well as his or her wellbeing since together with saliva they begin the processing of food, starting the digestive process in conjunction with hard and soft tissues and good salivation. The teeth crush the food during chewing and help the enzymatic action of the digestive system favoring the reabsorption of nutrients which is essential for growth and development in the infant stage. Good mastication is described as the act of grinding or crushing food through the teeth in a certain number of chewing cycles, which requires fine rhythmic motor coordination with the opening and closing of the mouth in synchrony with the lips and tongue. In this way, the anterior part of the tongue facilitates the formation of the food bolus and its transportation to a suitable position to continue grinding it together with the teeth in repeated cycles. Once sufficiently small particles are obtained, they are sent to the posterior part of the tongue to continue with the digestive process and in this way, the organism obtains the nutrients for the daily activities that the individual carries out [10, 21–23].

The deterioration of the occlusal surfaces or the absence of them has repercussions in the masticatory ability. This leads to changes in eating habits, food intake, as well as lack of nutrients, which results in alterations in nutrition. The rhythmic and repetitive function of mastication can be altered when form-related occlusal impairments occur. Most of the mastication is carried out by the premolars and molars, sometimes assisted by the canines, these movements are related to the anatomical and structural alignment of the teeth. Mastication includes a series of coordinated movements between teeth through occlusal contact with the tongue and jaws. Movements are guided by the dental anatomy and supporting structures, temporomandibular joints (TMJ), masticatory muscles, and central nervous system. This complex mechanism can be easily affected when any of the components such as teeth, periodontal articulation, and peripheral nervous system are altered. To maintain occlusal stability, it is necessary to maintain balance with the pressure of the lips and cheeks, occlusal and eruption forces, tongue pressure, periodontal support since they are intimately linked to the dental position. If any of them is altered in frequency or magnitude, the position of the tooth will change, modifying and altering the occlusion and consequently the function, which can lead to the loss of teeth, loss of partial structure, and deficiency in the support caused by trauma or periodontal disease. This will also affect occlusal stability, which is defined as the optimal functional state between teeth, articulation, arch, and muscles [18, 24] (**Figure 1**).

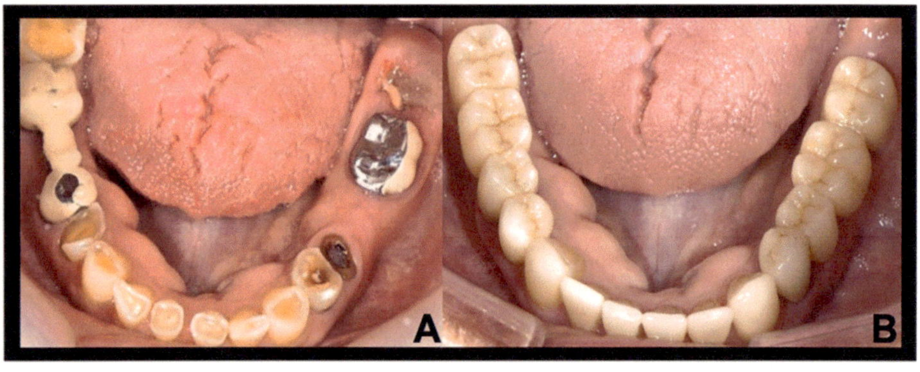

Figure 1.
Absence or deterioration of the occlusal surface has a repercussion in the masticatory ability which results in alterations in nutrition. (A) Patient before and (B) after oral rehabilitation (photos kindly provided by Dr. Francisco Javier Macías García).

Within the oral cavity and actively participating with the teeth in mastication is located the tongue which is a mass of muscle (intrinsic and extrinsic) that does not have a skeletal base and is essential for chewing, swallowing, taste, and speech, by combining a series of synchronous movements. In the dorsal part, of the anterior segment, we observe that it is covered by four types of papillae: filiform, fungiform, foliaceous, and circumvallate. The glands are involved in the sense of taste, helping to detect the primary flavors such as sweet, sour, and bitter, in the same way, the sense of taste is linked to the sense of smell, that is, the perception of taste is altered when the sense of smell is affected [19].

2.2 Esthetics

The face has an important esthetic significance where dental appearance is considered an essential component. Anthropomorphic studies have shown that the anterior teeth have a harmonious relationship with the facial structures. To achieve a pleasing appearance, intraoral and extraoral factors such as the periodontium, perioral muscles, skeletal components of the jaws as well as the relationship of the smile to adjacent tissues such as the nose and chin must be considered as they have an indisputable effect on the smile.

The smile is the facial display of emotion that is produced once the muscles of facial expression undergo contraction showing the teeth of the maxilla. This facial expression can influence the way in which the individual fits in and performs within society playing an important role linked to emotions, such as self-esteem. Although esthetics is completely subjective, it is influenced by various factors such as culture, social status, and level of education. For example, it is common for adolescents to form their esthetic ideals according to their own experience and based on cultural stereotypes [25, 26].

For a smile to be pleasing, harmony must be achieved with the size, shape, position, and color of the teeth as well as the proportion and relative symmetry between them plus the surrounding elements such as size and shape of the lip and the amount of visible gingiva. However, patient perception of an ideal smile corresponds to having white teeth that are oversized, while the literature points to dental alignment and color as the essential characteristics [27] (**Figure 2**).

In recent years, dentists and patients have turned their interest to dental esthetics and the facial appearance of treatments. The impact of having a pleasant dentition with a beautiful smile has even come to be considered a plus for employment,

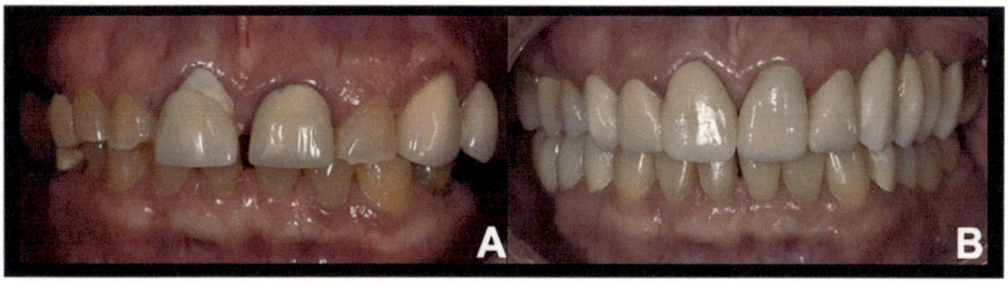

Figure 2.
Dental harmony between size, shape, position, and color. (A) Patient before and (B) after treatment (photos kindly provided by Dr. Francisco Javier Macías García).

interpersonal and social relationships, etc., increasing the demand for esthetic dental care every day [25, 28].

The oral cavity is involved in various functions among which are esthetics, speech, and mastication, the latter being where we observe the interrelationship with the intake and propulsion of food into the oropharynx, essential for the nutrition of the individual so that a decrease in masticatory function affects the overall health and welfare of the patient. Another function of the mouth is to serve as a second secondary airway to the nasal cavity.

As health professionals, dentists are committed to caring for the oral health of patients, keeping teeth in natural occlusion and function integrated in harmony with orofacial esthetics [13, 29–31].

3. Impact of oral health on people's quality of life

Oral health varies over the life course from early life to old age, is integral to general health and supports individuals in participating in society and achieving their potential [32]. Oral diseases are a major public health problem for countries and populations worldwide, although they often are not publicly recognized as such [1].

In May 2021, a historic resolution concerned with the lack of attention to oral diseases recognizing was made by the World Health Assembly. It stated that Oral health should be firmly embedded within the noncommunicable disease (NCD) agenda and that oral health care interventions should be included in universal health coverage (UHC) programs [33].

The combined estimated number of cases of oral diseases globally is about 1 billion higher than cases of all five main NCDs (mental disorders, cardiovascular disease, diabetes mellitus, chronic respiratory diseases, and cancers) combined [1, 34].

Oral diseases are a concept that encompasses everything from temporary and permanent tooth decay, periodontal disease, edentulism to oral cancer, all of which are currently recognized as public health problems for countries and populations around the world. Globally, these diseases affect almost 3.5 billion people and three out of four people affected live in middle-income countries [1].

Apart from the five main diseases (caries of deciduous and permanent teeth, severe periodontal disease, edentulism,lip and oral cavity cancer), many other diseases and conditions are relevant to oral health [1, 35].

Quality of life of the population is one of the generally accepted and important indicators that define the real opportunities of people that they need for a comfortable life. From the point of view of dental health, four types of factors were identified that are directly related to the quality of life of patients: Pathology of the dental-maxillary system, the state of the dental-maxillary system, influence of the condition of the dental-maxillary apparatus, and the condition of the dental-maxillary apparatus [36].

The circumstances in which people are born, grow up, live, work, and age [1], is the framework that influences, together with genetic, sociocultural, and environmental factors, the development of oral diseases as well as non-communicable diseases (NCDs).

3.1 Children

As is highlighted in the 1990 United Nations (UN) Convention on the Rights of the Child which has since been ratified in 2019, all children shall have the right to the best possible health and access to healthcare. This includes good oral health and access to dental care [37]. There is a clear connection between socio-economic factors and social vulnerability and oral health. Having parents with low educational level, single parents, unemployment in the family or parents from non-European countries increases the risk of the child having caries. For example, 95% of Swedish 3-year-old children and 73% of 6-year-old children are caries-free, but it has been shown that the worse the dental health of the parents, the higher the risk of caries in children [37]. The reality in the world population is that more than 2 billion people worldwide live with untreated dental caries. In primary teeth, untreated caries is the most common chronic childhood disease, affecting 514 million children worldwide [34, 38–41].

Traumatic dental injury is a widespread yet often overlooked condition. In March 2022, a revision of the International Classification of Diseases (ICD) was published, that now includes more detailed codes on dental trauma, allowing for better data collection and surveillance [1].

Nevertheless, oral mucosal conditions are underestimated and underdiagnosed not only by dentists but also by pediatricians, dermatologists, and other medical specialists. However, a retrospective study conducted in Poland in children aged 0-17 years found that the most frequent oral mucosal lesions were aphthae, mucocele, morsicatio buccarum, hairy tongue, fibroma, geographic tongue, papilloma, lip-tie, pyogenic granuloma, and traumatic erosions [38].

Seeing the mouth as part of the body and that children's dental health can affect both general health and future caries development is an important message to convey. The right of all children to good and equal health must not only be a vision but also an important goal to strive for it [37].

3.2 Adolescents

The WHO defined adolescents as "all persons between 10 and 19 years of age group and Youth as the 15–24 year age group. While Young People covers the age range 10–24 years." This transition from childhood to adulthood can negatively impact oral health. In adolescents, hormonal changes, diet, and inadequate hygiene habits added to other factors that modify the internal and external environment of the individual make them a risk group for poor oral health. Thus, anxiety, depression, low self-esteem, and psychosocial problems are some factors that negatively affect the oral health of both children and adolescents, resulting in pathologies such as bruxism and temporomandibular disorders (TMD), affecting their quality of life [42].

Eating is the most frequently affected dimension and toothache is the first cause of impact, showing a generally mild intensity and severity of impact. The impact on oral quality of life is greater in younger adolescents [43]. Significantly improving factors such as self-esteem, esthetics, social interaction, and self-perception in adolescents is important [42].

3.3 Adults and seniors

The dominant oral diseases are dental caries and periodontal diseases [44]. The dominant oral diseases in the adult population are dental caries and periodontal diseases. Tooth loss is often considered an inevitable result of aging and is socially accepted in many cultures. However, losing teeth and living with reduced or absent dentition can be psychologically traumatic, socially detrimental, and functionally limiting for the affected individual. A balanced diet can be difficult, especially when partial or full dentures to replace missing teeth are not accessible or affordable. For people over 60 years of age, the global average prevalence of edentulism is much higher, estimated at 22.7%: almost one in four people over 60 years of age lacks teeth [1]. However, several factors make older people particularly sensitive to these diseases, such as the influence of the aging process on immune system function, morbidities, and medication that can reduce salivary flow [44].

Good oral health is important for the general health, comfort, and basic dignity of older members of society. Because of the interaction between oral health and general health that has become increasingly established, especially in relation to many chronic diseases, such as cardiovascular disease, stroke, and diabetes, the oral health was redefined by World Dental Federation (FDI) as "multifaceted and includes the ability to speak, smile, smell, taste, touch, chew, swallow and convey a range of emotions through facial expressions with confidence and without pain, discomfort and discomfort of the Craniofacial Complex" [45].

Today, many older people in Sweden have many natural teeth remaining, and denture wearers are less frequent [46]. Similarly, in Japan, the goal is to ensure that people still have 20 of their teeth at age 80 so that they can maintain nutritional and social well-being, and campaigns are being conducted with a lifelong focus on preventing tooth loss, targeting all generations [1]. It is a reality that over the life course, oral diseases and conditions disproportionately affect poor and vulnerable members of society, which often include people with low incomes, people with disabilities, older people living alone or in residential care, refugees, prisoners, or people living in remote and rural communities, as well as people belonging to minorities or other socially marginalized groups [32]. Drawing on the experience of countries that have succeeded in turning this reality around, we must strive to raise awareness among these generations to establish preventive measures that will lead us to achieve the goal of better oral health.

Oral cancers are a disease group with high mortality and morbidity. The IARC Global Cancer Observatory (GLOBOCAN) estimated 377,713 new cases and 177,757 deaths from lip and oral cavity cancers worldwide in 2020 [1, 32, 46]. The main risk factors for lip, oral cavity, and oropharyngeal cancers are tobacco use, alcohol consumption, and betel quid use. Oral health plays a major role in well-being and self-esteem, while oral diseases heavily affect the quality of life, productivity, and ability to work as well as social participation [1].

4. Fundamental role of the dentist in oral health

Professional oral and dental supervision are critical components of patient health for quality life [47], oral functions such as eating, speaking, and relating to others are connected to prevention [48], and measures that have been incorporated into disease surveillance, facilitating comparisons of morbidity across health conditions, with the potential to guide large-scale priority setting [47]. The dentist has the role of the leader on the oral health team and, in this capacity; he/she is responsible for diagnosis, treatment planning, and the quality control of the oral treatment [49].

The clinical examination of the patient should start as the patient enters the clinic and is greeted by the clinician. The general examination includes:

1. Vital signs: Conscious state, pulse, blood pressure, and respiration.

2. Other signs: Weight, condition of the hands, skin lesions, and skin appendages.

3. Extraoral head and neck examination: Swellings, pallor, rash, erythema, palpation of all cervical, submental, and submandibular lymph nodes, salivary and thyroid glands, the temporomandibular joints (TMJ), muscles of mastication's and the cranial nerves should be examined.

4. Intraoral examination: Dental, periodontal, and mucosa examination [49, 50].

4.1 History of the present complaint

This should cover aspects relevant to the particular main complaint, such as:
Date of onset
Duration
Locations
Aggravating and relieving factors
Investigations thus far
Treatment already received.

The history and clinical examination are designed to put the clinician in a position to make a provisional diagnosis, or a differential diagnosis [50].

Clinical examination allows for a diagnosis, which often involves the patient's careful history (anamnesis). In fact, utilizing a patient's history provides a diagnosis in about 80% of cases [50].

4.2 Communicating the diagnosis

The dentist is competent at recognizing the presence of systemic disease and knowing how the disease and its treatment affect the delivery of dental care and recognizing the clinical features of oral mucosal diseases or disorders, including oral neoplasia, and identifying conditions that require management [49]. Follow their ethical and professional responsibility to refer the patient to another appropriate health care professional (HCP) if this is likely to be in the patient's best interests [50].

5. The role of oral immune system

Innate immunity plays a unique role in oral immunity, by triggering a crucial systemic response to protect the host and maintain homeostasis. Furthermore, the innate defense is pivotal in the activation and regulation of adaptive immunity [51].

Oral and periodontal health impact systemic health, and vice versa [52] the oral mucosa acts as a first line of defense and is like a barrier to protect from environmental exposures, physical and chemical damage, microbes, and toxins. Through its physical and immunological barrier functions [51], the role of the oral microbiome is crucial to homeostasis and immune responses in the mucosal tissues [48, 49].

In the oral cavity exists a highly dynamic microbial environment that harbors many distinct substrata and microenvironments housing diverse microbial communities [52]. Commensal microbiota is considered as the main driver of barrier immune function, shaping protective and homeostatic immune responses at the mucosa tissue [51]. The cells of the innate immune system are located strategically at the host-microbiome and are composed of three major compartments: the epithelial layer, lamina propria, and the mucosal-associated lymphoid tissue (MALT) [51].

The lining of the oral cavity is called the stratified flat epithelium and can be keratinized or non-keratinized. Its function is to protect the underlying tissue from mechanical damage, as well as to function as a primary barrier site and a portal for the entry of food and microorganisms. The epithelium is constantly replaced by cell division [51].

The oral cavity, often described as a community of commensal, symbiotic, and pathogenic microorganisms with a body space or other environment, has the second largest and diverse microbiota after the gut. It harbors over 700 species of microbial communities that are involved in promoting oral health and exist in a dynamic balance with the host [51]. It is also presented with near-constant environmental challenges, including host diet, salivary flow, masticatory forces, and the introduction of exogenous pathogens. The composition of the oral microbiome is shaped throughout life by factors that include host genetics and maternal transmission, as well as by environmental factors, such as dietary habits, oral hygiene practice, medications, and systemic factors [52].

Besides its role in mastication, saliva contains specific digestive enzymes to support food digestion such as amylase, lipase, and proteases [53] and it is involved in defending against external pathogens including bacteria, viruses, and fungi [54].

6. Conclusions and future perspectives

Some of the challenges in preventing and controlling oral diseases are related to the design and organization of (oral) health care systems, including oral health workforce models, individuals' capability for effective self-care, and access to and affordability of fluorides for oral health as well as data, surveillance, and research issues that limit the availability of basic information on oral health and disease [55–57].

More high-quality research is needed to understand fully the potential shared pathways between oral diseases, poor oral health, and other general diseases and conditions, the coexistence of multiple health conditions, as well as the impact of oral health interventions on general health [58]. It will require societal change addressing health care reform and social inequities to create better overall health and wellness, and more interest from many organizations in oral health heretofore [59].

Unprecedented advances in genomics, data science, and biotechnology are presented considering their contributions to a new era of health care in which interventions and treatment are increasingly tailored to individual patients [60]. Finally, mobile and remote technologies using smartphones and other digital approaches are expanding the scope of and approaches to health promotion and aspects of oral health care while challenging patient-provider relations and data protection [1].

Acknowledgements

The authors are indebted to financial support granted by Universidad Autónoma de Guadalajara and COECYTJAL.

Conflict of interest

The authors declare no conflict of interest.

DOI: http://dx.doi.org/10.5772/intechopen.111610

Author details

Andréa Dolores Correia Miranda Valdivia*, Maria de los Angeles Vázquez Sánchez, Diana Elizabeth Aguirre Cortés and Evangelina Gutiérrez Cortés
Faculty of Dentistry, Autonomous University of Guadalajara, Jalisco, Mexico

*Address all correspondence to: andrea.correia@edu.uag.mx

References

[1] Global Oral Health Status Report: Towards Universal Health Coverage for Oral Health by 2030. Geneva: World Health Organization; 2022. Licence: CC BY-NC-SA 3.0 IGO

[2] Bissett S, Preshaw P. Guide to providing mouth care for older people. Nursing Older People. 2011;**23**(10):14-21

[3] Jung YM, Shin DS. Oral health, nutrition, and oral health-related quality of life among Korean older adults. Journal of Gerontological Nursing. 2008;**34**(10):28-35

[4] Dewhirst FE, Chen T, Izard J, Paster BJ, Tanner ACR, Yu WH, et al. The human oral microbiome. Journal of Bacteriology. 2010;**192**(19):5002-5017

[5] Winning L, Lundy FT, Blackwood B, McAuley DF, El Karim I. Oral health care for the critically ill: A narrative review. Critical Care. 2021;**25**(1):353

[6] Kassebaum NJ, Bernabe E, Dahiya M, Bhandari B, Murray CJ, Marcenes W. Global burden of severe periodontitis in 1990-2010: A systematic review and meta-regression. Journal of Dental Research. 2014;**93**(11):1045-1053

[7] Peres KG, Nascimento GG, Gupta A, Singh A, Cassiano LS, Rugg-Gunn AJ. Scoping review of oral health-related birth cohort studies: Toward a global consortium. Journal of Dental Research. 2022;**101**(6):632-646

[8] Akintoye SO, Mupparapu M. Clinical evaluation and anatomic variation of the oral cavity. Dermatologic Clinics. 2020;**38**(4):399-411. DOI: 10.1016/j.det.2020.05.001

[9] Casamassimo PS, Flaitz CM, Hammersmith K, Sangvai S, Kumar A. Recognizing the relationship between disorders in the oral cavity and systemic disease. Pediatric Clinics of North America. 2018;**65**(5):1007-1032. DOI: 10.1016/j.pcl.2018.05.009

[10] Zhang W, Yelick PC. Tooth repair and regeneration: Potential of dental stem cells. Trends in Molecular Medicine. 2021;**27**(5):501-511. DOI: 10.1016/j.molmed.2021.02.005

[11] Volponi AA, Pang Y, Sharpe PT. Stem cell-based biological tooth repair and regeneration. Trends in Cell Biology. 2010;**20**(12):715-722. DOI: 10.1016/j.tcb.2010.09.012

[12] Husain MA. Dental anatomy and nomenclature for the radiologist. Radiologic Clinics of North America. 2018;**56**(1):1-11. DOI: 10.1016/j.rcl.2017.08.001

[13] Hodjat M, Khan F, Saadat KASM. Epigenetic alterations in aging tooth and the reprogramming potential. Ageing Research Reviews. 2020;**63**:101140. DOI: 10.1016/j.arr.2020.101140

[14] Famuyide A, Massoud TF, Moonis G. Oral cavity and salivary glands anatomy. Neuroimaging Clinics of North America. 2022;**32**(4):777-790. DOI: 10.1016/j.nic.2022.07.021

[15] Chang HH, Yeh CL, Wang YL, Huang YC, Tsai SJ, Li YT, et al. Differences in the biomechanical behaviors of natural teeth and dental implants. Dental Materials. 2021;**37**(4):682-689. DOI: 10.1016/j.dental.2021.01.003

[16] Stanton BF. Teeth: Vital to our Children's health. Pediatric Clinics of

North America. 2018;**65**(5):xv-xvi. DOI: 10.1016/j.pcl.2018.07.002

[17] Hammel JM, Fischel J. Dental emergencies. Emergency Medicine Clinics of North America. 2019;**37**(1):81-93. DOI: 10.1016/j.emc.2018.09.008

[18] Stanley J, Nelson Y, Ash MM. Wheeler: Anatomía, fisiología y oclusión dental. 9th ed. España: Elsevier; 2010. 343 p

[19] Sicher-DuBrul E. Anatomía Oral. Illinois: Ediciones Doyma; 1988

[20] Riojas Garza MT. Anatomía dental. 3th ed. México: Manual Moderno; 2014

[21] Souto-Souza D, Soares MEC, Primo-Miranda EF, Pereira LJ, Ramos-Jorge ML, Ramos-Jorge J. The influence of malocclusion, sucking habits and dental caries in the masticatory function of preschool children. Brazilian Oral Research. 2020;**34**:e059. DOI: 10.1590/1807-3107bor-2020.vol34.0059

[22] Takahara M, Shiraiwa T, Maeno Y, Yamamoto K, Shiraiwa Y, Yoshida Y, et al. Association of obesity, diabetes, and physical frailty with dental and tongue-lip motor dysfunctions in patients with metabolic disease. Obesity Research & Clinical Practice. 2021;**15**(3):243-248. DOI: 10.1016/j.orcp.2021.02.007

[23] Kim HE, Lee H. Factors affecting subjective and objective masticatory function in older adults: Importance of an integrated approach. Journal of Dentistry. 2021;**113**:103787. DOI: 10.1016/j.jdent.2021.103787

[24] Wanderley RL, Gomes Filho FN, Bonifácio MFR, Freire AR, Oliveira LFS, Medeiros MMD, et al. Masticatory function and nutritional status in Brazilian institutionalized elders: Influence of denture use. Pesquisa Brasileira em Odontopediatria e Clínica Integrada. 2021;**21**:e0244. DOI: 10.1590/pboci.2021.143

[25] Davis NC. Smile design. Dental Clinics of North America. 2007;**51**(2):299-318, vii. DOI: 10.1016/j.cden.2006.12.006

[26] Heravi F, Rashed R, Abachizadeh H. Esthetic preferences for the shape of anterior teeth in a posed smile. American Journal of Orthodontics and Dentofacial Orthopedics. 2011;**139**(6):806-814. DOI: 10.1016/j.ajodo.2009.07.029

[27] de Oliveira LV, Colussi PRG, Piardi CC, Rösing CK, Muniz FWMG. Self-perception of teeth alignment and colour in adolescents: A cross-sectional study. International Dental Journal. 2022;**72**(3):288-295. DOI: 10.1016/j.identj.2021.04.001

[28] Athayde GDS, Reis PPGD, Jorge RC, Americano GCA, Fidalgo TKDS, Soviero VM. Impact of masking hypomineralization opacities in anterior teeth on the esthetic perception of children and parents: A randomized controlled clinical trial. Journal of Dentistry. 2022;**123**:104168. DOI: 10.1016/j.jdent.2022.104168

[29] Riolina A, Hartini S, Suparyati S. Dental and oral health problems in elementary school children: A scoping review. Pediatric Dentistry. 2020;**30**(2):106-114. DOI: 10.1016/j.pdj.2020.04.001

[30] Cunha M, Santos E, Costa A, Pereira M, Varanda R, Loureiro S. Oral health, literacy and quality of life in the elderly people - systematic literature review. Revista de Enfermagem Referência. 2014;**serIV**:125-134. DOI: 10.12707/RIII12157

[31] Kantarci A, Stavropoulos A, Sculean A. Introduction: Vision of regenerative periodontology. Dental Clinics of North America. 2022;**66**(1):xi-xiii. DOI: 10.1016/j.cden.2021.08.002

[32] A75/10 Add.1. Draft global strategy on oral health. In: Seventy-fifth World Health Assembly, Geneva, 22-28 May 2022. Provisional agenda item 14.1. Geneva: World Health Organization; 2022. Available from: https://apps.who.int/gb/ebwha/pdf_files/WHA75/A75_10Add1-en.pdf [Accessed: 14 November 2022]

[33] WHA74.5. Oral health. In: Seventy-fourth World Health Assembly, Geneva, 24 May–1 June 2021. Summary and verbatim records. Geneva: World Health Organization; 2021 . WHA74/A74.R5. Available from: https://apps.who.int/gb/ebwha/pdf_files/WHA74/A74_R5-en.pdf [Accessed: 14 November 2022]

[34] Global Burden of Disease 2019. (GBD 2019) Results [Online Database]. Seattle: Institute of Health Metrics and Evaluation (IHME); 2020. Available from: https://vizhub.healthdata.org/gbd-results/ [Accessed: 14 November 2022]

[35] Rajendran R, Sivapathasundharam B, editors. Shafer's Textbook of Oral Pathology. Elsevier India: Noida; 2012

[36] Zhachko NI, Nespriadko-Monborgne TS, Skrypnyk IL, Zhachko MS. Improving Dental Health - Is Improving Quality of Life. Wiad Lek. 2021;**74**(3 p.II):722-725. DOI: 10.36740/WLek202103229

[37] Ridell K. Equal oral health for young children—A new approach? Acta Paediatrica. 2020;**110**(1):12-13

[38] Owczarek-Drabińska JE, Nowak P, Zimoląg-Dydak M, Radwan-Oczko M. The prevalence of oral mucosa lesions in pediatric patients. International Journal of Environmental Research and Public Health. 2022;**19**(18):1-16

[39] Jiang CM, Wong MC, Chu CH, Dai L, Lo EC. Changes in parental satisfaction and oral health-related quality of life of preschool children after receiving atraumatic restorative treatment with or without prior silver diamine fluoride application. Journal of Evidence-Based Dental Practice. 2022;**22**(3):1-10

[40] GBD 2017 Oral Disorders Collaborators, Bernabe E, Marcenes W, Hernandez CR, Bailey J, Abreu LG, et al. Global, regional, and national levels and trends in burden of oral conditions from 1990 to 2017: A systematic analysis for the Global Burden of Disease 2017 study. Journal of Dental Research. 2020;**99**(4):362-373. DOI: 10.1177/0022034520908533

[41] Kassebaum NJ, Bernabé E, Dahiya M, Bhandari B, Murray CJ, Marcenes W. Global burden of untreated caries: A systematic review and metaregression. Journal of Dental Research. 2015;**94**(5):650-658. DOI: 10.1177/0022034515573272

[42] Cerón Bastidas XA. Relación de Calidad de Vida y salud oral en la Población Adolescente. CES Odontología. 2018;**31**(1):38-46

[43] Alvarez-Azaustre MP, Greco R, Lleno C. Oral health-related quality of life in adolescents as measured with the child-OIDP questionnaire: A systematic review. International Journal of Environmental Research and Public Health. 2021;**9**:2-16

[44] Koistinen S, Olai L, Ståhlnacke K, Fält A, Ehrenberg A. Oral health-related quality of life and associated factors among older people in short-term care. International Journal of Dental Hygiene. 2019;**18**(2):163-172

[45] Gibney JM, Naganathan V, Lim MA. Oral health is essential to the well-being of older people. The American Journal of Geriatric Psychiatry. 2021;**29**(10):1053-1057

[46] Ferlay J, Ervik M, Lam F, Colombet M, Mery L, Piñeros M, et al. Global Cancer Observatory: Cancer Today. Lyon: International Agency for Research on Cancer; 2020. Available from: https://gco.iarc.fr/today [Accessed: 14 November 2022]

[47] Chaffee BW, Rodrigues PH, Kramer PF, Vítolo MR, Feldens CA. Oral health-related quality-of-life scores differ by socioeconomic status and caries experience. Community Dentistry and Oral Epidemiology. 2017;**45**(3):216-224. DOI: Doi/10.1111/cdoe.12279

[48] Fukai K, Dartevelle S, Jones J. Oral health for healthy ageing: A people-centred and function-focused approach. International Dental Journal. 2022;**72**(4):S2-S4

[49] Sanz M, Treasure E, van Dijk W, Feldman C, Groeneveld H, Kellett M, et al. Profile of the dentist in the oral healthcare team in countries with developed economies. European Journal of Dental Education. 2008;**12**(Suppl 1):101-110

[50] Scully C. Oral and Maxillofacial Medicine: The Basis of Diagnosis and Treatment. 3rd ed. Edinburgh; New York: Churchill Livingstone/Elsevier; 2013. 435 p

[51] Şenel S. An overview of physical, microbiological and immune barriers of oral mucosa. IJMS. 2021;**22**(15):7821

[52] Kapila YL. Oral health's inextricable connection to systemic health: Special populations bring to bear multimodal relationships and factors connecting periodontal disease to systemic diseases and conditions. Periodontology 2000. 2021;**87**(1):11-16

[53] Wu RQ, Zhang DF, Tu E, Chen QM, Chen W. The mucosal immune system in the oral cavity-an orchestra of T cell diversity. International Journal of Oral Science. 2014;**6**(3):125-132. DOI: 10.1038/ijos.2014.48. Epub 2014 Aug 8

[54] Ogawa H, McKenna G, Kettratad-Pruksapong M. Prevention of oral functional decline. International Dental Journal. 2022;**72**(4S):S21-S26

[55] Watt RG, Daly B, Allison P, Macpherson LMD, Venturelli R, Listl S, et al. Ending the neglect of global oral health: Time for radical action. Lancet. 2019;**394**(10194):261-272. DOI: 10.1016/S0140-6736(19)31133-X

[56] Vujicic M. Our dental care system is stuck: And here is what to do about it. Journal of the American Dental Association (1939). 2018;**149**(3):167-169. DOI: 10.1016/j.adaj.2018.01.006

[57] Cohen LC, Dahlen G, Escobar A, Fejerskov O, Johnson NW, Manji F. Dentistry in crisis: Time to change. La Cascada declaration. Australian Dental Journal. 2017;**62**(3):258-260. DOI: 10.1111/adj.12546

[58] Watt RG, Serban S. Multimorbidity: A challenge and opportunity for the dental profession. British Dental Journal. 2020;**229**(5):282-286. DOI: 10.1038/s41415-020-2056-y

[59] Silk H. The future of oral health care provided by physicians and allied professionals. Journal of Dental Education. 2017;**81**(8):eS171-eS179

[60] Ryan ME, Fox CH. Advances in precision oral health research: Opportunities for the future! Journal of Dental Research. 2019;**98**(12):1285-1286

Chapter 6

Oral Health: A Doorway to General Well-being

Lawrence O. Igbinosa, Richard Evbuomwan, Martins A. Okoromu and Uwa S. Osarenkhoe

Abstract

At every stage of life, oral health is crucial to overall health and well-being. Good oral health is beyond being free of pain. Maintaining good mental and social health also includes maintaining a healthy grin. The essential actions for observing proper oral hygiene are achieved by brushing, flossing, and consuming sugar-free, healthy foods, quitting smoking to enhance oral health and overall health, and scheduling a routine dental visit. The events of oral health range from Oral health education, primary prevention methods, and secondary prevention measures, and these forms the ingredients for oral health. Tooth decay, gum disease, and oral malignancies are the main illnesses that can have detrimental impacts on oral health. Most of these conditions can be prevented. Diabetes, heart disease, cognitive health, and nutritional deficiencies have all been linked to poor oral health, not necessarily as causative agents but as conditions that may worsen in poor oral health.

Keywords: oral health, general well-being, dental caries, periodontal diseases, diabetes, cognitive health

1. Introduction

The World Health Organization defined health in 1948 as a method that is still used today by world authorities. "Health is a state of complete physical, mental, and social well-being and not merely the absence of disease or infirmity." Additional refinements were added to the definition in 1986: "A resource for everyday life, not the objective of living. Health is a positive concept emphasizing social and personal resources, as well as physical capacities." This indicates that health is a resource to support a person's role in larger society rather than being a goal in and of itself. The ability to live a full life with meaning and purpose is made possible by a healthy lifestyle [1].

Speaking, smiling, smelling, tasting, touching, chewing, swallowing, and being able to communicate a variety of emotions through facial expressions with confidence and without suffering from pain, discomfort, or diseases of the craniofacial complex (head, face, and oral cavity) are all aspects of having good oral health [2]. Regardless of the age of a person, oral health is vital to general health and well-being [3].

The majority of oral health disorders can be treated when they are young and are mainly avoidable. Dental caries (tooth decay), periodontal disorders, tooth loss, and oral malignancies account for the majority of occurrences. Orofacial clefts, noma (a severe gangrenous disease that begins in the mouth and primarily affects young people), and oro-dental trauma are further oral disorders of public health significance.

According to the WHO Global Oral Health Status report from 2022, close to 3.5 billion people worldwide suffer from oral disorders, with three out of every four of these individuals residing in middle-income nations. 514 million children worldwide suffer from primary tooth decay, while 2 billion people are thought to have permanent tooth decay.

Oral diseases globally, have been on the rise. This has been due to increased urbanization and lifestyle changes. The combination of insufficient exposure to fluoride, widespread food with high sugar content, alcohol, tobacco, and reduction in community access to oral healthcare services has led to an increase in disease conditions [4].

2. Components of oral hygiene routine

Oral hygiene is the practice of keeping the mouth clean and disease-free. It involves brushing and flossing the teeth as well as visiting the dentist regularly for dental checkups.

The most popular oral hygiene technique is using a toothbrush to remove plaque. Brushing the teeth at least twice a day for two to three minutes with gentle pressure, preferably using the Bass technique or its modification, is advised as part of a healthy oral hygiene practice [5].

By flossing, one may get rid of food particles and other debris stuck in tight spaces between the teeth that the toothbrush can not access. This helps to prevent decay from spreading there [6].

Every six months, one should go to the dentist for a thorough examination and prophylactic cleaning. All the brushing in the world will not be able to get rid of calculus buildup. Frequent visiting will enable the dentist to spot and address emerging dental health issues before they worsen [7].

Healthy eating benefits the entire body, including the teeth. The teeth will remain strong and healthy far into old life if one eats a healthy diet high in calcium and other nutrients. This will boost health generally.

Additional dental care supplies might also improve oral health. Overall dental health can benefit from using products like mouthwash, interdental cleaners, and oral irrigators. These items are to complement brushing and flossing rather than replace them [8].

2.1 Elements in oral health programs

The majority of oral health regimens consist of three primary components. They are;

i. Oral health education/instruction

ii. Primary prevention methods (chair-side and non-chair-side).

iii. Secondary prevention measures [9].

2.1.1 Oral health education/instruction

Oral health education/instruction usually refers to oral hygiene instruction and/or oral health education. These educational exercises are designed to encourage good oral hygiene habits and to change people's perceptions of and attitudes regarding dental health. Not only do they target kids, but also their parents, teachers, and medical professionals. Schools and clinics frequently conduct oral health education and instructional programs for kids, parents, teachers, and healthcare professionals. The primary objective of oral hygiene training is generally acknowledged to be reinforcing and teaching toothbrushing. The most common methods used to spread oral health awareness are presentations, games, and printed materials. Schools are frequently chosen because they offer easy access to students, parents, teachers, and healthcare professionals [10].

2.1.2 Primary prevention measures

As a main preventive strategy, fluoride compounds are frequently employed. There is widespread agreement that the use of fluoride has significantly reduced dental cavities. Fluoride is most helpful at preventing dental cavities when a low level of fluoride is consistently maintained in the mouth, according to research. Fluoride can be provided in a variety of ways, including salt, mouthwash, toothpaste, and gels and varnishes that are professionally administered. Water fluoridation is thought to be the most economical method among these treatments for preventing caries. It has been hypothesized that the most significant impact of water fluoridation is not so much the prevention of new lesions but the remineralization of existing carious lesions, which slows or even stops their progression [11].

In poor nations, there are challenges in the implementation of water fluoridation. These challenges include a lack of a secure networked water delivery system and a lack of government backing or willingness. The decision to add fluoride to the water may be influenced in industrialized nations by anxiety over unfounded reports of negative effects and freedom of choice and autonomy arguments from "anti-fluoridationists" [12].

2.1.3 Secondary prevention measures

The typical approach utilized in early detection and treatment services is dental screening. Early detection and prevention of caries in primary dentition are crucial in maintaining high preventive levels. Dental procedures are frequently performed in industrialized nations either in permanent dental clinics or through "mobile dental clinics" located in vans that travel to different locations. In impoverished nations, where the price of basic instrument sets, dental supplies, and infection control items is too high and primary health workers are not adequately trained to perform basic oral care, these procedures are neither accessible nor economical. A novel approach to oral healthcare is required in these circumstances [13].

2.2 Risk assessment

A patient's susceptibility to infection, inadequate healing, bleeding, medication interactions, and physical and emotional capacity to tolerate dental treatment are all factors that must be taken into account in the patient's risk assessment. All these can be done during the preoperative, intraoperative, and postoperative phases. A patient's systemic disease's relative severity can be determined using common medical classification schemes (e.g., angina, heart failure, asthma, chronic kidney disease). Important diagnostic information is also provided by the quantity and kind of medications, test findings, and trips to the hospital or doctor. One approach to creating an international evaluation of the medical state is the American Society of Anesthesiologists physical classification system ASA.

ASA I—Normal healthy patient

ASA II—Patient with mild systemic disease. No significant impact on daily activity; unlikely to have an impact on anesthesia and surgery.

ASA III—Patient with significant or severe systemic disease that impacts daily activity; probable impact on anesthesia and surgery.

ASA IV—Patient with severe systemic disease that is a constant threat to life. Serious limitation on daily activity; major impact on anesthesia and surgery.

ASA V—A moribund patient who is not expected to survive without the operation.

ASA VI—A declared brain-dead patient whose organs are being removed for donor purposes [14].

3. How can oral health affect general well-being?

Oral conditions are frequently considered separate from other chronic conditions, but these are interrelated. Poor oral health is associated with other chronic diseases such as diabetes and heart disease. Oral disease also is associated with risk behaviors such as using tobacco and consuming sugary foods and beverages.

The link between oral health and general health is complex, multifaceted, and intertwined in such ways that the pathology of one may progress and influence the other. The changes in oral health can impact general health and well-being in various ways.

The presence of oral diseases such as dental caries and periodontitis when left untreated are associated with an increased probability of systemic conditions like cardiovascular disease, pneumonia, and gastritis. Conditions such as diabetes mellitus and uncontrolled hypertension are believed to have oral repercussions which may lead to periodontal disease and tooth loss.

Systemic conditions can present in the orofacial region as initial manifestations of an acute or chronic undiagnosed condition [15]. These manifestations may be the most severe feature of the systemic condition or a dominant cause in reducing the quality of life. They can also vary in frequency and mode of presentation.

The mouth acts as a window to our overall health in so many ways. The oral cavity houses millions of different bacteria. With the proper practice of good oral hygiene, most of the bacteria in the mouth remain harmless [16].

As much as they can cause inflammation in the oral cavity, they can also travel through the bloodstream and instigate inflammation in other parts of the body, contributing to a wide range of problems.

3.1 Diabetes

Oral infections weaken the body's ability to control blood sugar. Since there is a problem with the regular soon of blood glucose with diabetes, hence this increases its complications. There's a two-way link between oral health and diabetes: poor blood sugar control increases the risk of gingivitis and periodontitis, while inflammation from gum disease can worsen the regulation of blood sugar [17]. The periodontal disease becomes severe in patients with poorly controlled diabetes. This can be due to the problem of delayed healing associated with inadequate glycemic control. Conditions such as oral ulcers and fungal infections (oral candidiasis) are commonly seen in patients with poorly controlled diabetes [15].

3.2 Heart disease

Previously, correlations have been found to exist between poor oral health and cardiovascular disease. Periodontitis and Cardiovascular disease are understood to be multifactorial with a significant range of local and general risk factors [18]. Some of these risk factors are believed to be common to both diseases. Associated bacterial infections such as gum diseases as a result of poor oral hygiene can also spread to the heart, stimulating several inflammatory processes in the blood vessels [19]. This will further promote atherosclerosis, which is a hallmark of heart disease.

3.3 Cognitive health

Maintaining good dental health becomes even more crucial with age. This is due to the connection between having healthy teeth and gums and cognitive wellness. According to clinical evidence, maintaining good oral health may help to maintain brain function and prevent cognitive decline. Over time, brain cells may be harmed by molecules the body makes when it is in a chronic inflammatory condition [20].

3.4 Gastrointestinal system

More than just the smile might be lost to gum disease. The bacteria that cause gum disease can spread to the digestive tract. Once there, it may kill off helpful bacteria, upsetting the delicate balance required for gut health. It is even more important to take proper care of the mouth, teeth, and gums because of the special connection between the mouth and the digestive system. As the initial anatomic region of the gastrointestinal tract, the oral cavity may show systemic disorders involving the gut. In cases of gastroesophageal reflux disease, patients may present various complaints including halitosis, burning sensation of the tongue, dental erosion, and periodic sialorrhea [21].

3.5 Immune health

Failing to take care of the teeth and gums can even impact immune health. The inflammation caused by plaque buildup and gum disease can weaken the immune system and make it harder for the body to protect itself. This is bad news for everyone but is particularly harmful if one should have inflammatory or immune-related conditions like inflammatory bowel disease. Poor oral health may exacerbate symptoms and make conditions worse [22].

3.6 Stress response

One of the main causes of stress' negative effects on the body is the inflammatory process that occurs during the stress reaction. Health issues that heighten the body's stress response cause a chain of events to occur that lead to low-grade chronic inflammation [23]. Gum disease and tooth decay-related bacteria can start the body's stress response, which increases stress hormones and inflammation [24].

3.7 Nutritional deficiencies

Impaired or reduced nutritional intake or absorption may lead to different general conditions that can present in the oral cavity. Vitamin A deficiency has been associated with increased severity of the periodontal disease. Vitamin B deficiencies are associated with atrophic glossitis, cheilitis, and burning sensation of the tongue. Vitamin C (Scurvy) deficiency can manifest as gingival bleeding, ulceration, and periodontal disease. Vitamin K deficiency will lead to gingival bleeding. Iron deficiency can lead to glossitis, and can be a risk factor for candidiasis [25].

3.8 Psychological effects

The presentation, diagnosis, and subsequent management of oral cancers may lead to detrimental effects on the mental health of patients. Management options such as surgical jaw resection and subsequent radiotherapy may leave patients with significant disabilities in their oral cavity. Substitutes such as reconstruction plates, obturators, and/or dentures may be introduced to such patients to help to mitigate the effects of these treatments and reduce the perception of disability [26]. However, such mitigation may not go all the way to providing relief and happiness to these patients. The need to now live a life with a significant oral disability may no longer be meaningful to such patients [27]. Depending on the severity of their deformity, patients may have varying levels of psychological disturbances. From being alone, and lost in their thoughts to developing and cultivating suicidal tendencies.

3.9 Social effects

Tooth loss and halitosis can greatly affect the way people socialize and communicate, with many withdrawing from social gatherings as frequently as possible. This may be due to their inability to adequately communicate the way they normally would, or difficulty in being understood by their peers [28]. This can lead to problems in maintaining jobs that require adequate communication skills. Socializing among peers may also be affected, as the fear of being ridiculed may push patients to hide within shells of themselves [29]. Patients with halitosis may develop habits of excessive oral hygiene routines which in the long run can be detrimental to their oral tissues. The use of a hard toothbrush to brush away "all smell" with increased frequency may subject the patient to tooth wear lesions (abrasion) [30]. The excessive use of mouth rinses, especially antibiotic mouth rinses may lead to disturbances in their oral microbial balance [31].

3.10 Functional effects

Feeding is a vital part of daily activities as it provides the necessary nutrients required by the body. The ability to feed and replenish nutrients can be greatly influenced by the state of one's oral health. To adequately feed, one would require the teeth, tongue, and mouth for proper mastication of food [32]. In the case where this may be hindered, nutritional deficiencies will occur. Growth, development, and adequate functioning of the body and mind will be affected generally [33].

3.11 Economic effects

Oral diseases have overall expenses that include both direct and indirect costs. The direct costs are related to the care provided by dental experts including public and private expenditures [34]. The indirect costs are related to time lost from work, school, or other regular activities as a result of dental problems and treatment [35].

4. Conclusion

Oral health is an important aspect of overall health and well-being. The mouth is a gateway to the body, and good oral hygiene is essential for maintaining the health of the entire body. Poor oral health can lead to a variety of problems, including cavities, gum disease, and tooth loss, which can cause pain, difficulty eating, and reduced quality of life.

There is a strong link between oral health and general health. Poor oral hygiene can also contribute to other health problems such as diabetes and malnutrition. This highlights the importance of preventive care, including regular dental check-ups, brushing, and flossing, as well as avoiding behaviors that can harm oral health, like smoking and excessive alcohol consumption.

Maintaining good oral health is not only important for individual health but also has economic implications. The cost of dental treatment can be expensive, and untreated oral health problems can result in missed work and reduced productivity.

Thus, oral health is a key component of overall health and well-being. It is important to prioritize oral hygiene and to take steps to protect and maintain it, including regular dental visits and practicing good oral hygiene. By doing so, we can ensure the health of our bodies and improve our quality of life.

Author details

Lawrence O. Igbinosa[1,2]*, Richard Evbuomwan[2], Martins A. Okoromu[2] and Uwa S. Osarenkhoe[2]

1 University of Benin, Benin City, Nigeria

2 University of Benin Teaching Hospital, Benin City, Nigeria

*Address all correspondence to: surglaw@yahoo.com

References

[1] World Health Organization. The world Oral health report 2003 WHO global Oral health Programme. Community Dentistry and Oral Epidemiology. 2003;**31**:3-23. Available from: http://www.ncbi.nlm.nih.gov/pubmed/15015736

[2] Glick M, Williams DM, Kleinman DV, Vujicic M, Watt RG, Weyant RJ. A new definition for oral health developed by the FDI world dental federation opens the door to a universal definition of oral health. Journal of the American Dental Association. 2016;**147**:915-917. Available from: http://www.ncbi.nlm.nih.gov/pubmed/27886668

[3] Petersen PE, Bourgeois D, Ogawa H, Estupinan-Day S, Ndiaye C. The global burden of oral diseases and risks to oral health. Bulletin of the World Health Organization. 2005;**83**:661-669

[4] WHO. Oral health. Fact Sheets. 2022. Available from: https://www.ptonline.com/articles/how-to-get-better-mfi-results [Accessed: November 22, 2022]

[5] Hayasaki H, Saitoh I, Nakakura-Ohshima K, Hanasaki M, Nogami Y, Nakajima T, et al. Tooth brushing for oral prophylaxis. Jpn dent Sci rev. Japanese Association for Dental. Science. 2014;**50**:69-77. DOI: 10.1016/j.jdsr.2014.04.001

[6] Hujoel PP, Cunha-Cruz J, Banting DW, Loesche WJ. Dental flossing and interproximal caries: A systematic review. Journal of Dental Research. 2006;**85**:298-305. DOI: 10.1177/154405910608500404

[7] Langevin SM, Michaud DS, Eliot M, Peters ES, McClean MD, Kelsey KT. Regular dental visits are associated with earlier stage at diagnosis for oral and pharyngeal cancer. Cancer Causes & Control. 2012;**23**:1821-1829. DOI: 10.1007/s10552-012-0061-4

[8] Mitha S, ElNaem MH, Chandran J, Rajah NP, Fam TY, Babar MG, et al. Use of Oral cleaning devices and their perceived benefits among Malaysians in Kuala Lumpur and Johor Bahru: An exploratory structured approach. Journal of Pharmacy & Bioallied Sciences. 2018;**10**:216-225. Available from: http://www.ncbi.nlm.nih.gov/pubmed/30568379

[9] Lam A. Elements in oral health programs. The New York State Dental Journal. 2014;**80**:26-30. Available from: http://www.ncbi.nlm.nih.gov/pubmed/24851389

[10] Nakre PD, Harikiran AG. Effectiveness of oral health education programs: A systematic review. Journal of International Society of Preventive & Community Dentistry. 2013;**3**:103-115. Available from: http://www.ncbi.nlm.nih.gov/pubmed/24778989

[11] Iheozor-Ejiofor Z, Worthington HV, Walsh T, O'Malley L, Clarkson JE, Macey R, et al. Water fluoridation for the prevention of dental caries. Cochrane Database of Systematic Reviews. 2015;**2015**:CD010856. Available from: http://www.ncbi.nlm.nih.gov/pubmed/26092033

[12] Cheng KK, Chalmers I, Sheldon TA. Adding fluoride to water supplies. BMJ. 2007;**335**:699-702. Available from: http://www.ncbi.nlm.nih.gov/pubmed/17916854

[13] Chou R, Pappas M, Dana MT, Selph S, Hart E, Fu RF, et al. Evidence

Synthesis Number 210 Screening and Interventions to Prevent Dental Caries in Children Younger Than Age Five Years: A Systematic Review for the U.S. Preventive Services Task Force Acknowledgments. 2021; Available from: www.ahrq.govwww.ohsu.edu/epc

[14] Craig S, Miller DMD. Subject: Risk assessment. The American Academy of Oral Medicine. 2014;**1**:1-5. Available from: https://www.aaom.com/index.php?option=com_content&view=article&id=152:risk-1#:~:text=Risk

[15] Napeñas JJ, Brennan MT, Elad S. Oral manifestations of systemic diseases. Dermatologic Clinics. W.B. Saunders. 2020;**4**:495-505

[16] Attin T, Hornecker E. Tooth brushing and oral health: How frequently and when should tooth brushing be performed? Oral Health & Preventive Dentistry. 2005;**3**:135-140. Available from: http://www.ncbi.nlm.nih.gov/pubmed/16355646

[17] Midwood I, Hodge P. Diabetes and gum disease. Journal of Diabetes Nursing. 2018;**22**:1-4

[18] Jansson L, Lavstedt S, Frithiof L. Relationship between oral health and mortality rate. J Clin Periodontol. Nov 2002;**29**(11):1029-1034. DOI: 10.1034/j.1600-051x.2002.291108.x. PMID: 12472996

[19] Robert H. Shmerling. Gum disease and the connection to heart disease. Harvard Health Publishing. 2021. Available from: https://www.health.harvard.edu/diseases-and-conditions/gum-disease-and-the-connection-to-heart-disease

[20] Asher S, Stephen R, Mäntylä P, Suominen AL, Solomon A. Periodontal health, cognitive decline, and dementia: A systematic review and meta-analysis of longitudinal studies. Journal of the American Geriatrics Society. 2022;**70**:2695-2709

[21] Di Fede O, Di Liberto C, Occhipinti G, Vigneri S, Lo Russo L, Fedele S, et al. Oral manifestations in patients with gastro-oesophageal reflux disease: A single-center case-control study. Journal of Oral Pathology & Medicine. 2008;**37**:336-340. Available from: http://www.ncbi.nlm.nih.gov/pubmed/18284539

[22] Nijakowski K, Gruszczyński D, Surdacka A. Oral health status in patients with inflammatory bowel diseases: A systematic review. International Journal of Environmental Research and Public Health. 2021;**18**:18. Available from: http://www.ncbi.nlm.nih.gov/pubmed/34770034

[23] Liu Y-Z, Wang Y-X, Jiang C-L. Inflammation: The common pathway of stress-related diseases. Frontiers in Human Neuroscience. 2017;**11**:316. Available from: http://www.ncbi.nlm.nih.gov/pubmed/28676747

[24] Goyal S, Gupta G, Thomas B, Bhat KM, Bhat GS. Stress and periodontal disease: The link and logic!! Industrial Psychiatry Journal. 2013;**22**:4-11. Available from: http://www.ncbi.nlm.nih.gov/pubmed/24459366

[25] Sun A, Chen H-M, Cheng S-J, Wang Y-P, Chang JY-F, Wu Y-C, et al. Significant association of deficiencies of hemoglobin, iron, vitamin B12, and folic acid and high homocysteine level with recurrent aphthous stomatitis. Journal of Oral Pathology & Medicine. 2015;**44**:300-305. Available from: http://www.ncbi.nlm.nih.gov/pubmed/25048341

[26] Dreizen S. Oral complications of cancer therapies. Description and

incidence of oral complications. NCI Monographs. 1990;**9**:11-15. Available from: http://www.ncbi.nlm.nih.gov/pubmed/2188148

[27] Yuliana B, Serna H, Andrés J, Betancourt O, Patricia O, Soto L, et al. Trends of incidence, mortality, and disability- adjusted life years of oral cancer in Latin America Tendencia de la incidencia, mortalidad y años de Vida ajustados por discapacidad del cáncer oral en América Latina. 1980;**25**:1-8

[28] Khan SU, Ghani F, Nazir Z. The effect of some missing teeth on a subjects' oral health related quality of life. Pakistan Journal of Medical Sciences. 2018;**34**:1457-1462

[29] Deolia S, Ali M, Bhatia S, Sen S. Psychosocial effects of halitosis among young adults. Ann Indian Psychiatry. 2018;**2**:120

[30] WAJ S, Marchan S, Rafeek RN. The prevalence and severity of non-carious cervical lesions in a group of patients attending a university hospital in Trinidad. Journal of Oral Rehabilitation. 2007;**35**:070711234836004. DOI: 10.1111/j.1365-2842.2007.01763.x

[31] Veeresha KL, Bansal M, Bansal V. Halitosis: A frequently ignored social condition. Journal of International Society of Preventive & Community Dentistry. 2011;**1**:9-13. Available from: http://www.ncbi.nlm.nih.gov/pubmed/24478947

[32] Furuta M, Yamashita Y. Oral health and swallowing problems. Current Physical Medicine and Rehabilitation Reports. 2013;**1**:216-222

[33] Brownie S. Why are elderly individuals at risk of nutritional deficiency? International Journal of Nursing Practice. 2006;**12**:110-118. DOI: 10.1111/j.1440-172X.2006.00557.x

[34] Listl S, Galloway J, Mossey PA, Marcenes W. Global economic impact of dental diseases. Journal of Dental Research. 2015;**94**:1355-1361. Available from: https://pubmed.ncbi.nlm.nih.gov/26318590/

[35] Hayes A, Azarpazhooh A, Dempster L, Ravaghi V, Quiñonez C. Time loss due to dental problems and treatment in the Canadian population: Analysis of a nationwide cross-sectional survey. BMC Oral Health. 2013;**13**:17. Published 2013 Apr 15. DOI: 10.1186/1472-6831-13-17

Chapter 7

The Role of Sugar-Free Chewing Gum in Dental Caries Prevention

Marut Phuphaniat

Abstract

The key to avoiding dental caries is managing the microorganisms on the tooth responsible for demineralization. Chewing gum has physical adhesive capabilities. It also promotes saliva flow, assisting in eliminating food particles and reducing the formation of bacterial biofilms in the oral cavity. Saliva flow also helps to balance the pH balance in the oral cavity, thereby reducing the risk of dental caries. In addition, sugar-free gum contains various antibacterial and remineralization substances, such as xylitol and casein phosphopeptide-amorphous calcium phosphate (CPP-ACP). In conclusion, sugar-free gum can remove food particles and tartar, stimulate saliva, increase oral pH, inhibit demineralization, and increase remineralization. These qualities aid in preventing dental caries.

Keywords: chewing gum, sugar-free, dental caries prevention, xylitol, casein phosphopeptide-amorphous calcium phosphate (CPP-ACP)

1. Introduction

Oral health is an important problem for public health, especially dental caries, which have a relatively high incidence rate. Dental caries is not a severe disease but causes economic loss, pain, and suffering. In addition, the inability to chew food, bad breath, and lesions in the tooth are breeding grounds for bacteria that weaken the body and lead to other diseases. Despite the campaign to disseminate dental health knowledge to people of all ages to realize the problem and pay attention to oral health, dental caries is still a problem for people of all ages. Especially among children, the leading cause of dental health problems is improper dietary habits and improper oral health care [1]. In addition, parents have incorrect oral health care attitudes and a lack of knowledge about dental caries prevention, resulting in children not receiving proper treatment. Dental caries is caused by four factors: the tooth, food, bacteria, and the time of tooth surface acid exposure (**Figure 1**).

Food particles adhered proximally and in the groove of the tooth produce dental caries. It provides energy to the bacteria in the oral. The process by which bacteria break down food debris produces acid, leading to the demineralization of tooth. Therefore, when food remains in the oral for a long time or is frequently eaten, it often causes the demineralization of tooth, thus causing dental caries (**Figure 2**). School-age children are a group at high risk of developing dental caries. Many factors involve biological, social, behavioral, and psychological factors. Therefore, dental

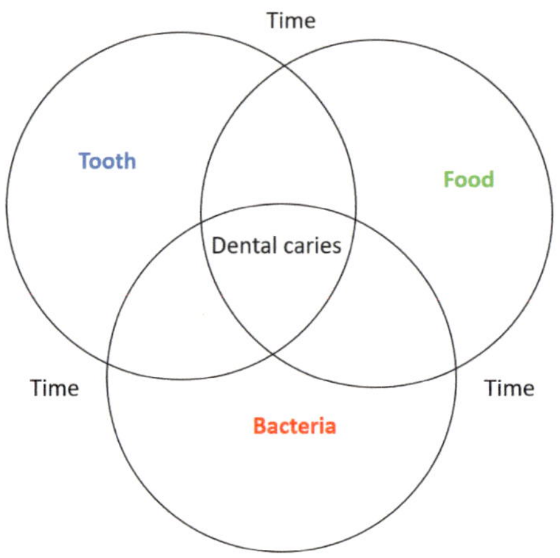

Figure 1.
Caries process.

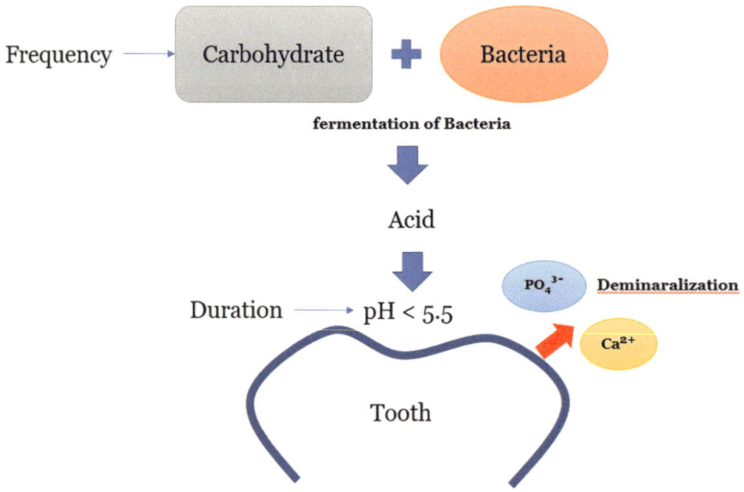

Figure 2.
Caries process.

caries is still a problem among school-age children. Especially during the age of 9 years, when the premolar and molar tooth are coming. Therefore, there is a chance of quickly accumulating food particles in the tooth' grooves [2].

There are many ways to prevent caries by controlling the causative factors of dental caries, including the tooth, food, bacteria, and time. The most commonly used methods are physical methods such as brushing tooth and chemical treatments such as fluoride. Brushing tooth with fluoride toothpaste is a common practice these days, but cavities still occur due to many factors, such as the frequency of overeating; as a result, brushing tooth three times a day is insufficient to stop dental caries progression. Therefore, it is necessary to control the risk of caries during the day. Sugar-free gum is a good choice because the adhesive gum's physical nature helps eliminate food residues after eating food by attaching the direct contact of the gum to the food scraps. In addition, chewing gum increases saliva flow reduces bacterial biofilm on the tooth surface, and maintains oral acid-base balance (pH > 5.5) [3, 4]. In addition, sugar-free gum contains a variety of substances that have antibacterial and remineralizing properties, such as xylitol and casein phosphopeptide-amorphous calcium phosphate (CPP-ACP), so sugar-free gum is more responsive to oral health care.

2. Sugar-free chewing gum and oral health

Sugar-free gum is chewing gum that uses a sweetener instead of sugar. Popular sweeteners such as xylitol, sorbitol, and mannitol [5]. The general composition of chewing gum includes sweetener (50–65%), gum base (18–30%), corn syrup (12–20%), color and flavor (1–2%), and softeners (0.3–3%). More than half of the composition is sweetener, which is responsible for enhancing the flavor and texture of the gum [6]. Sugar-free gum has significantly impacted oral health because of its ability to stimulate saliva and accelerate the removal of carbohydrates that accumulate from food intake. These actions may lead to healing, reduction of tooth decay, and other oral health benefits [7]. The benefits of chewing gum on oral health are as follows: (**Figure 3**)

- Remove food particles and dental plaque.
- Stimulate saliva.
- Increase plaque pH.
- Inhibit demineralization and enhance remineralization.

Nowadays, sugar-free gum has been produced for consumers to choose from in various styles. Chewing gum is another good way to clean tooth. Chewing gum may be used as auxiliary to dental brushing or when such method is impossible to be

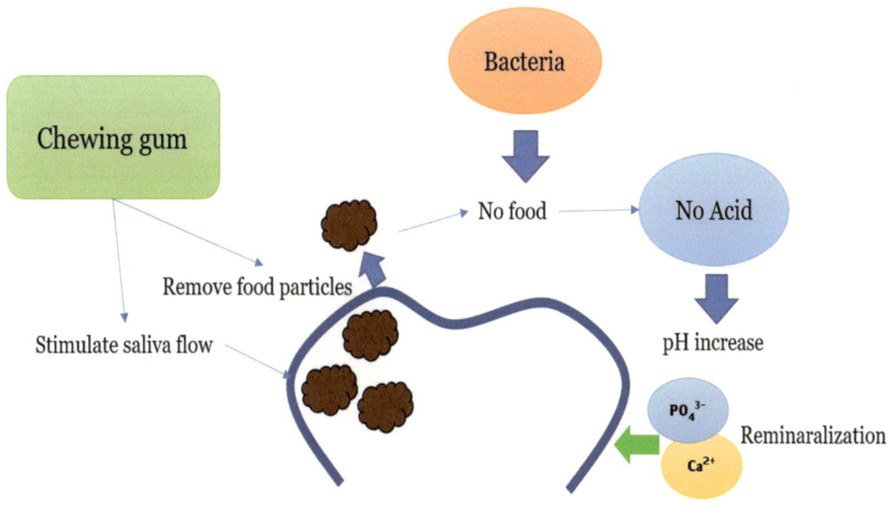

Figure 3.
The anti-cavity mechanism of chewing sugar-free gum.

performed, such as when you are outside, and it is not convenient to brush your tooth, you can chew gum to clean your tooth. Chewing sugar-free gum for 5–20 minutes after meals two to three times a day can help prevent tooth decay because gum helps to remove plaque and reduce gum inflammation. In addition, sugar-free gum stimulates saliva, which reduces acidity on the tooth surface and helps to self-cleansing can clean tooth [3, 8]. Several studies have concluded that sugar-free gum can inhibit the effects of oral bacteria [3, 4, 8]. This process is due to two main reasons: (1) the stimulation of saliva from chewing and (2) the use of sweeteners instead of sugar [9]. In addition, some substances promote healthy tooth, such as xylitol, andCPP-ACP, also known as Recaldent, which has been reported to enhance remineralization of the tooth surface [10–12]. A study on sugar-free gum chewing found that children who chewed sugar-free gum had fewer cavities than children in the control group who didn't chew gum [4, 13]. In addition, sugar-free gum was tested to prevent tooth decay in students. The students chewed 1 tablet of gum for 10 minutes after eating breakfast-afternoon every day. It was found that after 4 months of the experiment, the students had no more tooth decay [3, 8].

2.1 Compounds that affect oral health

2.1.1 Xylitol

Xylitol is a type of sweetener which is alcohol sugar. Xylitol was discovered in 1890 by German chemists Professors Emil Hermann Fischer and Rudolf Stahel by extracting the compound from beech [14] and later widely known in many countries such as the United States, Canada, Japan, etc. Xylitol has a molecular structure of 5 carbon atoms (**Figure 4**), chemical formula C5H12O5, and a molecular mass of 152.15 g/mol. Xylitol is a common sugar substitute. It is as sweet as sucrose. It can be used instead of sucrose in a ratio of 1:1, providing 2.43 cal/g of energy. In addition, it has the property of giving a cooling sensation while eating. Xylitol is a non-fermentable sugar that bacteria cannot use to synthesize for energy. Therefore, Xylitol is used as a glycemic control agent in Type 2 diabetes and as a product to prevent tooth decay [15].

Xylitol has been approved by the U.S. Food and Drug Administration (FDA) since 1960 as safe for children [16] and can be consumed in 30 g without causing harm. If consumed more than 30 g at a time can cause diarrhea. This may be because intestinal bacteria cannot metabolize large amounts of xylitol. In nature, it can be found in many natural plants, vegetables, and fruits, such as strawberries, carrots,

Figure 4.
Structure of xylitol.

corn cobs, nuts, etc. Marketing xylitol is available in liquid and solid form. Its applications are mainly made into chewing gum, followed by confectionery, medicine, and beauty products [17]. In addition, Xylitol is recognized by the American Academy of Pediatric Dentistry (AAPD) and serves as a guideline for tooth decay prevention [18]. Xylitol has anti-cavities properties for two reasons: promoting remineralization and inhibiting caries-causing bacteria (**Figure 5**).

2.1.2 Casein phosphopeptide-amorphous calcium phosphate

CPP-ACP, also known as "Recaldent", was discovered by Professor Eric Reynolds, School of Dental Science, University of Melbourne, Australia. CPP-ACP is divided into casein phosphopeptide (CPP) and amorphous calcium phosphate (ACP). CPP is a peptide derived from cow's milk by tryptic digestion of casein in cow's milk to produce a peptide containing phosphoseryl residue (Ser(P) – Ser(P) – Ser(P) – Glu – Glu –) a large amount of which binds calcium and phosphate in the ionized form to form a complex called ACP, which can cause calcium and phosphate ions. It is in a saturated state called metastable solution, i.e. it is saturated but does not form nucleation and does not precipitate calcium salts and phosphates, thus forming a nanocomplex between CPP and ACP known as CPP-ACP [19].

CPP-ACP has been proven to prevent tooth decay in laboratories, in vivo, and in humans; its efficacy is due to casein's ability to adapt to acidic environments. When the pH is acidic, the ACP is separated from the CPP, resulting in an increase in the levels of calcium and phosphate in the saliva, stabilizing the oral ACP [10] and increasing the calcium and phosphate content in the plaque. Furthermore, as an ion reserve, it inhibits demineralization [20] and stimulates the remineralization of tooth enamel by preventing the constant precipitation of calcium and phosphate in saliva [21]. It was also found that CPP directly affects bacterial apoptosis and impairs agglutination [22]. CPP-ACP can work with fluoride to form casein phosphopeptide-amorphous calcium fluoride phosphate (CPP-ACFP), readily strengthening enamel [20].

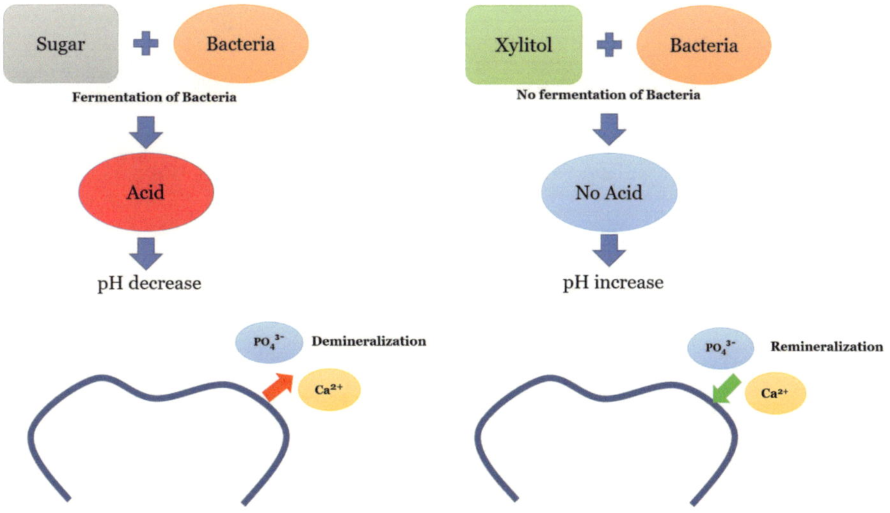

Figure 5.
The anti-cavity mechanism of xylitol.

The continued use of dental products containing CPP-ACP produces ion saturation in saliva and biofilm, making it available for subsequent precipitation in the form of ACP, favoring the dental remineralization process. The process of dental demineralization/remineralization: (1) oral pH < 5.5 causes demineralization of hydroxyapatite (HA), chemical formula $Ca_{10}(PO_4)_6(OH)_2$ (2) upon saturation of the oral environment and biofilm with Ca^{2+} and $PO4^{3-}$ ions, promoting the process of remineralization (**Figure 6**) [23].

In 1999, CPP-ACP was approved by the FDA that it is safe. CPP-ACP can be used in people with lactose intolerance because CPP-ACP does not contain lactose. However, CPP-ACP is not recommended for people with milk protein allergies [24]. CPP-ACP is used in various applications in food ingredients such as sugar-free candies or gum and in dental products such as toothpaste, mouthwash, dental implants, and glass ionomers. It is also available as a topical cream (paste) for ease of use [8].

3. Chewing gum and removing food particles on the tooth surface

Food debris is a factor in dental caries due to the accumulation of food particles in the mouth or on the tooth, especially starchy foods, which are the main energy source for bacteria. When the digestive bacteria are energized, they make acidic wastes that cause tooth demineralization. Therefore, removing food particles stuck in the tooth is necessary to prevent tooth decay. It is known that the gold standard for removing supragingival plaque is daily tooth brushing, flossing, and/or use of antimicrobial mouthwash. Still, in some cases, chewing gum may be used as auxiliary to dental brushing, or when such method is impossible to be performed. This led to the development of antimicrobial gum that would provide daily oral care and reduce plaque in the mouth [25]; due to the physical nature of gum as a binder, it

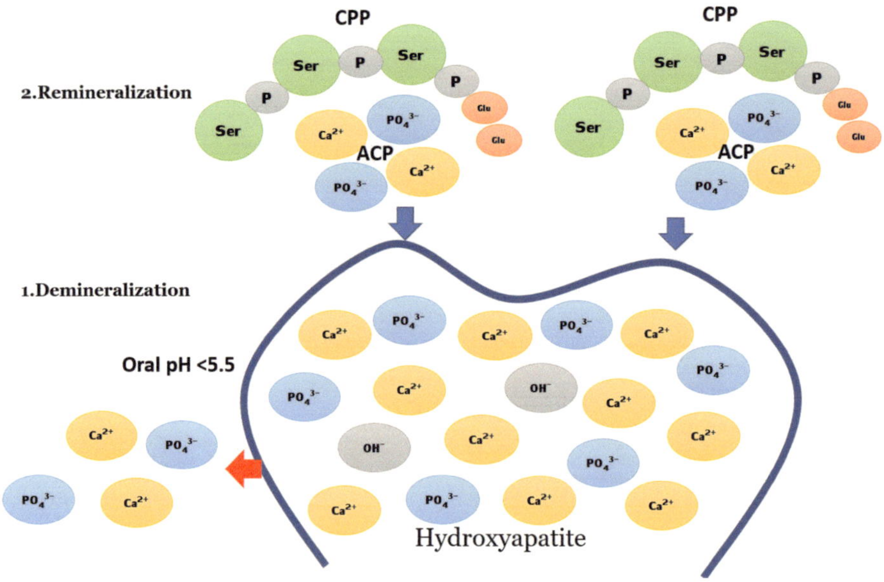

Figure 6.
Remineralization mechanism of CPP-ACP.

helps to eliminate food residues left after food consumption. Due to the physical nature of gum as a binding agent, it helps to eliminate food residue after consuming food. Disposal is partly due to the direct attachment of the gum to the food debris, along with the increased occlusion and saliva, which washes away the food particles [8]. In case we are outside, and it is not convenient to brush our tooth, it is possible to chew sugar-free gum to clean the tooth. In addition, a research report concludes that chewing gum may be used as an aid to flossing when brushing is impossible [26]. In addition to removing food particles, chewing sugar-free gum reduces plaque on the tooth caused by the accumulation of food particles on the tooth surface [27].

Phoophaniat and Joankrajang [8] examined the effectiveness of chewing gum containing CPP-ACP and xylitol in preventing dental caries in school-aged children. The sample group consisted of 90 students, divided into 3 groups by simple random sampling. A group of 30 students received one of the following interventions: experimental group 1 used chewing gums containing CPP-ACP, experimental group 2 used chewing gums containing xylitol, and control group. The experimental groups chew gums for 10 minutes, times times a day for 16 weeks. The dental plaque was assessed five times: before the trial, week 4th, week 8th, week 12th, and week 16th. Dental caries was assessed twice, before and after the trial. The data were analyzed by the friedman test, wilcoxon signed ranks test, kruskal-wallis test, and mann whitney U test. The result showed that there was a statistically significant decrease in dental plaque level in experimental groups, experimental group 1 had mean dental plaque of 1.10, 0.83, 0.40, 0.16, and 0.13 (χ^2 = 75.835, P-value < 0.001), and experimental group 2 had mean dental plaque of 1.00, 0.93, 0.51, 0.16, and 0.14 (χ^2 = 66.905, P-value < 0.001), and less dental plaque than the control group. Experimental group 1 ($Z = -1.000$, P-value = 0.500) and experimental group 2 ($Z = -1.414$, P-value = 0.250) dental caries did not increase compared to before the trial, and less tooth decay than control group (χ^2 = 7.164, P-value = 0.030). Suggestions that chewing gum containing CPP-ACP or xylitol, 1 tablet for 10 minutes, can reduce dental plaque. Consistent a study on chewing sugar-free gum in students found that students chewing one tablet of gum for 10 minutes after breakfast and afternoon daily was found to reduce the formation of dental plaque within 4 weeks and decreased dramatically at week 8 and then started to stabilize at week 12 because the amount of plaque on the tooth was already very low [3], consistent with research that found that most gum containing sugar substitutes, antimicrobial agents, and other minerals have an effect on reducing plaque [28]. The reduction of food particles and plaque on the tooth is one of the causative factors of tooth decay due to the physical nature of the gum and its beneficial compounds.

4. Chewing gum and oral bacteria reduction

Oral bacteria are endemic to everyone and are a common cause of oral health problems. The bacteria known as Streptococcus mutans are the main culprits that cause tooth decay. Controlling Streptococcus mutans can help reduce tooth decay. Dental plaque begins with a biofilm that binds the tooth surface between the saliva covering the tooth surface and bacteria. The physical nature of chewing gum helps to eliminate food debris, which is the energy source of bacteria, preventing them from growing [8]. In addition, sugar-free gum has other properties that eliminate bacteria, such as:

Xylitol is a sweetener instead of sugar. Xylitol molecules similar in shape to fructose are absorbed through the bacterial cell wall by the same mechanism that bacteria absorb sugar into cells. When xylitol gets into cells, bacterial enzymes cannot synthesize xylitol for energy, so bacteria cannot survive. Xylitol is a substance that helps reduce bacteria in the oral [8]. When compared to other oral bacteria, xylitol is more effective in inhibiting Streptococcus mutans [29]. Consistent with research findings, the samples had reduced plaque after 21 days of chewing gum containing xylitol [30]. Similarly, the research found that after chewing gum containing xylitol for 30 days, the samples had reduced plaque content. Thus confirming that chewing gum containing xylitol can reduce plaque [31].

CPP-ACP can increase calcium and phosphate in plaque. When there are many calcium ions in the extracellular fluid, the bacterial cell wall becomes porous, and the cell breaks down. This results in killing bacteria or inhibiting bacteria [12]. In addition, chewing gum can help stimulate the flow of saliva to produce more secretions than usual [3]. The secretion of saliva and the physical nature of the gum as a binding agent facilitates the removal of food residues left after ingestion. The elimination is partly due to the direct attachment of the gum to food particles. This, together with increased occlusion and saliva, thus prevented bacterial formation and reduced the appearance of oral bacterial biofilms, consistent with the results of a study by Phoophaniat et al. [4, 8] that after chewing CPP- ACP for 12 weeks, subjects had a statistically significant reduction in dental plaque content.

5. Chewing gum and oral pH balance

It is well known that oral pH is an integral part of the caries process. It comprises acidogenic and aciduric bacteria that are responsible for lowering the pH and subsequent destruction of HA matrix in enamel and dentine. Therefore, preventing tooth decay requires maintaining a pH balance that is not too low (<5.5). Chewing gum has a physical mechanism in which chewing stimulates saliva flow. It is commonly known that saliva is the pH balance system in the oral. Therefore, when there is a lot of saliva flowing, it helps to balance the pH of the oral cavity. In addition, the properties of compounds that have a positive effect on oral health also help to improve the pH balance in the oral cavity [8], as follows:

Xylitol has been reported to increase saliva flow and adjust oral pH [14, 32]. This may be due to its cooling effect on xylitol, which stimulates salivary flow. Together with the properties that help eliminate bacteria in the mouth, thus making the pH value not lower because bacteria cannot synthesize xylitol. It is not acidic, like digesting sugars in general [15]. Tests of various brands of xylitol gum were found to increase pH within 15 minutes. There was a statistically significant difference with the control group [33]. Consistent with the test in children 8–10 years, it was found that after chewing gum with xylitol, saliva pH immediately increased and increased dramatically in 15 minutes [34].

CPP-ACP has the property to increase the pH of saliva by casein modulating an acidic environment [10]. ACP refining cleaves from CPP, resulting in increased calcium and phosphate levels in saliva. Therefore, conditions within the oral cavity become basal according to the properties of calcium and phosphate. This is consistent with research findings that CPP-ACP can increase the pH of saliva [35]. Consistent with the results of tests in children 8–10 years of age, it was found that after chewing the CPP-ACP chewing gum, the saliva pH immediately increased and increased dramatically in 15 minutes [34].

6. Chewing gum and tooth strengthening

In this regard, the physical appearance of chewing gum has no effect, but with the properties of compounds that are beneficial to oral health, such as xylitol and CPP-ACP, chewing gum containing such compounds has outstanding features in promoting oral health, as follows:

Although xylitol is not very prominent in its remineralization, it does so as xylitol prevents enamel resorption by inhibiting the migration of calcium ions and phosphate ions from the lesions. Xylitol acts as a carrier of calcium ions to the tooth, especially in areas of demineralization. It accelerates the remineralization of the tooth. This mechanism increases the strength of the tooth [11].

CPP-ACP is the most outstanding tooth remineralization agent. CPPs consist of a multiphosphoseryl sequence, which can stabilize calcium phosphate in ACP solution within a nano complex solution. CPP-APP nanoscale complex molecules bind to the tooth surface. It prevents demineralization and can induce remineralization in enamel subsurface lesions. The hardness of the tooth will be more CPP-ACP binds with fluoride ions (F^-) to form amorphous calcium fluoride phosphate (ACFP) and stabilizes with CPP at the tooth surface, where these substances can diffuse into the lesion area. Calcium phosphate and fluoride solutions can help speed up the remineralization process with fluorapatite ($Ca_5(PO_4)_3F_2$), which resists acidic conditions and can penetrate deep into the lesion of the tooth (**Figure 7**). The advantage of being able to penetrate into the lesion is the ability to restore opaque enamel or white spot lesion [36]. It was also reported that CPP-ACP promoted the efficiency of fluoride use. This is caused by combining fluoride ions with ACP to make ACFP at the enamel surface. Based on these findings, CPP-ACP helps fluoride adhere to the tooth surface and enhances anti-demineralization efficiency [37], which has been shown to increase the effectiveness of fluoride in preventing tooth decay [38].

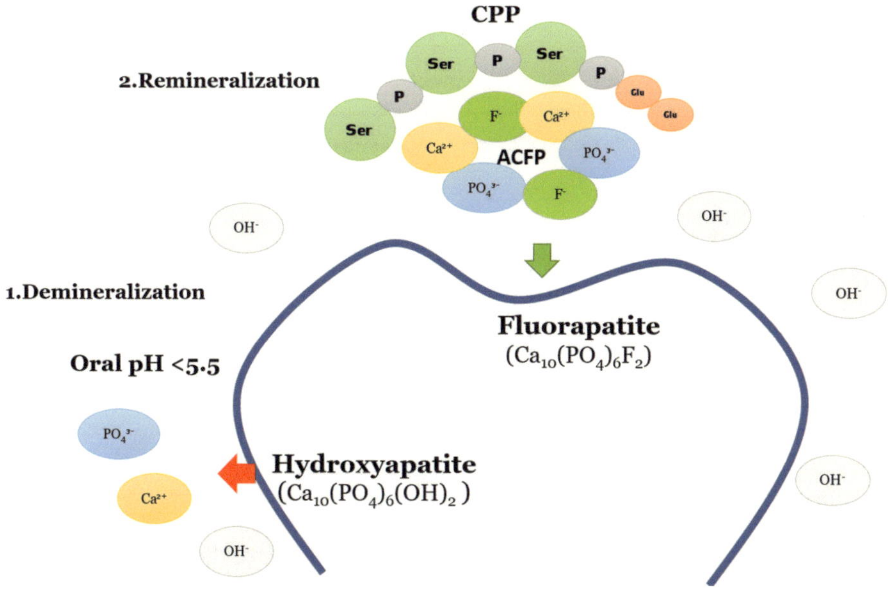

Figure 7.
Remineralization mechanism of CPP-ACP with fluoride.

7. Conclusion

Dental caries is generally caused by four factors: the tooth, diet, bacteria, and acid exposure time. Therefore, methods for preventing tooth decay must properly control those four factors. Sugar-free chewing gum has both physical and chemical properties to prevent tooth decay. The physical property of chewing gum is the direct attachment of the gum to food particles in conjunction with increased chewing and saliva. Increasing saliva can help wash away food particles and chemical properties from betel nut constituents such as xylitol and CPP-ACP. These compounds reduce oral bacteria, improve oral hygiene, and strengthen tooth enamel. The properties mentioned earlier of sugar-free chewing gum were able to control the four causative factors of caries incidence. Therefore, it was concluded that sugar-free chewing gum had caries-prevention properties. Additionally, sugar-free chewing gums are easily accessible, can be chewed in any situation, whether on-site or off-site, and are popular with school-age children, teens, and adults. Therefore, chewing gum, although not as effective as brushing or flossing, is another way to clean your mouth and control cavities.

Conflict of interest

The authors declare no conflict of interest.

Author details

Marut Phuphaniat
Faculty of Public Health and Allied Health Sciences, Sirindhorn College of Public Health Suphanburi, Praboromarajchanok Institute, Ministry of Public Health, Thailand

*Address all correspondence to: marut@phcsuphan.ac.th

References

[1] Bureau of Dental Health. The 8th National oral health survey, Thailand 2017. Nontaburi: Samcharoen Panich. 2018

[2] Trairatvorakul C. Preventive Dentistry in Children and Adolescents. 2nd ed. Bangkok: Bestbook; 2011. pp. 2-19

[3] Muralikrishnan K, Asokan S, Priya PR. Effect of different chewing gums on dental plaque pH, salivary pH, and buffering capacity in children: A randomized controlled trial. SRM Journal of Research in Dental Sciences. 2018;**9**:158-163

[4] Phoophaniat M, Joankrajang J, Kanpok T, Ninkamnoet S. Effectiveness of chewing gum containing casein phosphopeptide-amorphous calcium phosphate (CPP-ACP) on dental caries prevention in students. In: Proceedings of the 12th NPRU National Academic Conference; 9 July 2020; Thailand. pp. 1186-1194

[5] Soderling E, Pienihakkinen K, Gursoy UK. Effects of sugar-free polyol chewing gums on gingival infammation: A systematic review. Clinical Oral Investigations. DOI: 10.1007/s00784-022-04729-x

[6] Dhingra S, Marya CM, Jnaneswar A, Kumar H, Dahiya N, Dahiya V. Role of sugar free chewing gums in oral health. Journal of Oral Health and Community Dentistry. 2015;**9**:35-39

[7] American Dental Association. Oral health topics: Chewing gum [Internet]. 2021. Available from: https://www.ada.org/en/member-center/oral-health-topics/chewing-gum. [Accessed: 06 October 2021]

[8] Phoophaniat M, Joankrajang J. The effect of chewing gum containing casein phosphopeptide-amorphous calcium phosphate (CPP-ACP) and xylitol on dental caries prevention in students. CUTJ. 2022;**28**:87-100

[9] Edgar WM. Sugar substitutes, chewing gum and dental caries: A review. British Dental Journal. 1998;**184**:29-32

[10] Reynolds EC, Cai F, Shen P, Walker GD. Retention in plaque and remineralization of enamel lesions by various forms of calcium in a mouth rinse or sugar-free chewing gum. Journal of Dental Research. 2003;**82**:206-211

[11] Miake Y, Saeki Y, Takahashi M, Yanagisawa T. Remineralization effect of xylitol on demineralized enamel. Journal of Electron Microscopy. 2003;**52**:471-476

[12] Poureslami H, Hoseinifar R, Khazaeli P, Hoseinifar R, Sharifi H, Poureslami P. Changes in the concentration of ions in saliva and dental plaque after application of CPP-ACP with and without fluoride among 6-9 year old children. Dental Biomaterials. 2017;**4**:361-366

[13] Szoke J, Banoczy J, Proskin HM. Effect of after-meal sucrose-free gum chewing on clinical caries. Journal of Dental Research. 2011;**80**:1725-1729

[14] Makinan KK. The rocky road of xylitol to its clinical application. Journal of Dental Research. 2000;**2000**(79):1352-1355

[15] Burt BA. The use of sorbitol- and xylitol-sweetened chewing gum in caries control. Journal of the American Dental Association (1939). 2006;**137**:190-196

[16] Farster H, Quadbeck R, Gottstein U. Metabolic tolerance to high doses of oral xylitol in human volunteers not previously adapted to xylitol. Journal for Vitamin and Nutrition Research. 1982;**22**:67-88

[17] Salminen S, Salminen E, Marks V. The effect of xylitol on the secretion of insulin and gastric inhibitory polypeptide in man and rat. Diabetology. 1982;**22**:480-482

[18] The American Academy of Pediatric Dentistry. Guideline on xylitol use in caries prevention. American Academy of Pediatric Dentistry. 2011;**36**:175-178

[19] Divyapriya GK, Yavagal PC, Veereesh DJ. Casein phosphopeptide-amorphous calcium phosphate in dentistry: An update. International Journal of Dentistry and Oral Health. 2016;**6**:18-25

[20] Cross KJ, Huq NL, Palamara JE, Perich JW, Reynolds EC. Physiochemical characterization of casein phosphopeptide-amorphous calcium phosphate nanocomplexes. Journal of Biological Chemistry. 2005;**280**:15362-15369

[21] Morgan MV, Adams GG, Bailey DL, Tsao CE, Fischman SL, Reynolds EC. The anticariogenic effect of sugar-free gum containing CPP-ACP nanocomplexes on approximal caries determined using digital bitewing radiography. Caries Research. 2008;**42**:171-184

[22] Rose RK. Effects of an anticariogenic casein phosphopeptide on calcium diffusion in strepto-coccal model dental plaques. Archives of Oral Biology. 2000;**45**:569-575

[23] Madrid-Troconis CC, Perez-Puello SC. Casein phosphopeptide-amorphous calcium phosphate nanocomplex (CPP-ACP) in dentistry: State of the art. Revista Facultad de Odontología. 2019;**30**:248-262

[24] Azarpazhooh A, Limeback H. Clinical efficacy of casein derivatives. A systematic review of the literature. Journal of the American Dental Association (1939). 2008;**139**:915-924

[25] Kirkwood B, Miller M, Milleman J, Milleman K, Leung K. Four-day plaque regrowth evaluation of a peptide chewing gum in a double-blind randomized clinical trial. Clinical and Experimental Dental Research. 2020;**6**(3):318-327

[26] Moreira LV, Freitas Pontes M, de Almeida e Silva LD, Douglas-de-Oliveira DW, de Ávila e Silva Figueiredo A, Furtado Gonçalves P, et al. The effectiveness of apple, sugar-free chewing gum, Rolly Brush and tooth brushing in the reduction of plaque index: Crossover randomized clinical trial. Rev Estomatol. 2021;**29**:e10929. DOI: 10.25100/re.v29i1.11011

[27] Wessel S. Oral Health Benefits of Chewing Gum. Groningen: Rijksuniversiteit Groningen; 2016. pp. 43-55

[28] Matthews F. The use of sugar free chewing gum as a supplement in the prevention of dental caries. Journal of Oral Research. 2015;**4**:129-136

[29] Loesche WJ, Grossman NS, Earnest R, Corpron R. The effect of chewing xylitol gum on the plaque and saliva levels of Streptococcus mutans. Journal of the American Dental Association (1939). 1984;**108**(4):587-592

[30] Kaur K, Nekkanti S, Madiyal M, Choudhary P. Effect of chewing gums containing probiotics and xylitol

on oral health in children. Journal of International Oral Health. 2018;**10**:237-243

[31] Oza S, Patel K, Bhosale S, Mitra R, Gupta R, Choudhary D. To determine the effect of chewing gum containing xylitol and sorbitol on mutants streptococci and lactobacilli count in saliva, plaque, and gingival health and to compare the efficacy of chewing gums. Journal of International Society of Preventive & Community Dentistry. 2018;**8**:354-360

[32] Nayak PA, Nayak UA, Khandelwal V. The effect of xylitol on dental caries and oral flora. Clinical, Cosmetic and Investigational Dentistry. 2014;**6**:89-94

[33] Alshibani N, Shalabi M, AlMugbel K, AlSaqer E, AlFarraj N, Allam E. Xylitol content and acid production of chewing gums available in the markets of Saudi Arabia. Saudi Dental Journal. 2022;**34**(2):121-128. DOI: 10.1016/j.sdentj.2021.11.001

[34] Prathima GS, Narmatha M, Selvabalaji A, Adimoulame S, Ezhumalai G. Effects of xylitol and CPP-ACP chewing gum on salivary properties of children with molar incisor hypomineralization. International Journal of Clinical Pediatric Dentistry. 2021;**14**(3):412-415

[35] Hegde RJ, Thakkar JB. Comparative evaluation of the effects of casein phosphopeptide-amorphous calcium phosphate (CPP-ACP) and xylitol-containing chewing gum on salivary flow rate, pH and buffering capacity in children: An in vivo study. Journal of the Indian Society of Pedodontics and Preventive Dentistry. 2017;**35**:332-337

[36] Mount GJ. Defining, classifying, and placing incipient caries lesions in perspective. Dental Clinics of North America. 2005;**49**:701-723

[37] Al-Batayneh OB. The clinical applications of tooth mousse TM and other CPP-ACP products in caries prevention: Evidence-based recommendations. Smile Dental Journal. 2009;**4**:8-12

[38] Thakkar PJ, Badakar CM, Hugar SM, Hallikerimath S, Patel PM, Shah P. An in vitro comparison of casein phosphopeptide-amorphous calcium phosphate paste, casein phosphopeptide-amorphous calcium phosphate paste with fluoride and casein phosphopeptide-amorphous calcium phosphate varnish on the inhibition of demineralization and promotion of remineralization of enamel. Journal of the Indian Society of Pedodontics and Preventive Dentistry. 2017;**35**:312-318

Chapter 8

Periodontitis, Its Associations, and Prevention

Yasmine N. Alawaji

Abstract

The ultimate goal of studying associations with diseases is to plan, implement, and evaluate preventive strategies. Today, after reviewing the body of evidence, one needs to ask: What has not been learned yet regarding periodontitis associations and its prevention? Current recommendations to prevent periodontitis are mostly limited to individual patient care while population-based approaches are nearly absent. Current strategies are not only time-consuming and costly but can be also ineffective to combat disease burden in populations. To initiate and sustain successful outcomes, prevention needs to be applied at multiple levels. Interventions need to target unhealthy behaviors along with their associated social and physical environmental constraints. The chapter presents highlights from current research on associations with periodontitis, its limitations, and the need to understand pathways linking periodontitis with its exposures over the life course. Finally, a suggested multilevel strategy for periodontitis prevention was outlined.

Keywords: periodontitis, prevention, associations, pathways, life course, social determinants of health

1. Introduction

A common approach in our clinical care is to advice patients to adopt healthy behaviors including smoking cessation, oral self-care (e.g., tooth brushing and use of interdental aids), healthy dietary habits, and physical activity [1]. While plenty of theories have been suggested trying to explain or change health behaviors, such approach can be limited without considering social and physical environmental exposures [2, 3]. For example, using different strategies with patients to quit smoking including referral to smoking cessation clinics could have limited benefits [4–6]. In contrast, implementing sin tax and age restrictions on tobacco products by policy-makers can make these products less affordable and accessible. Consequently, more cost-effective and radical change may occur at population level; people may quit smoking or refrain from smoking in first place. The social and environmental exposures can be collectively called social determinants of health, which include socio-economic status, housing, residential neighborhood, environmental safety, social support network, social norms, structural racism, and discrimination [4, 5, 7–9]. Although targeting the social determinants of health can be effective in prevention of diseases along with their related behaviors, they could be more effective when applied as multilevel strategies including clinical care.

In Geoffrey Rose's seminal paper "*sick individuals and sick populations*", key concepts of studying disease etiology and its prevention were discussed [10]. Rose highlighted the importance of considering different ecological levels when studying the disease etiology. When an exposure is uniformly distributed within a population, it could hamper its detection, which necessitates studying its variations among populations. Besides Rose's remarks, lack of variations in distribution could be also indicative of methodological limitations in measuring, summarizing, or analyzing the disease and its exposures [11]. During periodontal indices' era, it was believed that gingivitis leads to periodontitis and consequently results in tooth loss in virtually everyone [12, 13]. These notions of lack of variations in susceptibility to periodontitis were attributed to several methodological limitations. Periodontal disease was measured using indices including Russell's Periodontal Index [14] which was based on the presence or absence of pockets without precise measurements. Thus, it did not properly differentiate between gingivitis and periodontitis [12]. The gingivitis and periodontitis were measured as continuous process using gradually increasing scores, summarized for as individual's mean index score, then summarized for a population as mean score based on individual's mean scores. Consequently, variations within and between individuals were lost [12, 13]. The periodontal indices, which measure and summarize periodontal disease as mean score, were abandoned by the end of 1980s and replaced by site-specific measurements [15, 16]. The improvement of periodontal measurement methods allowed detection of variations at multilevels (site-level, tooth-level, individual level, and between populations) [17–22]; consequently, it motivated research on host susceptibility and associations with periodontitis [23–25].

Since the periodontal indices were abandoned, periodontitis is summarized in population-based studies using prevalence, extent, and severity [12, 26]. Prevalence can be defined as the proportion of individuals with periodontitis, extent can be defined as mean proportion of sites or teeth with periodontitis, and severity can be defined as mean clinical attachment loss (CAL) or mean periodontal probing depth (PPD). Among these summaries, periodontitis severity is calculated as a mean score for population; such summary does not properly account for the multilevel nature of the measurements and its variations at site level, tooth level, and individual level before it is summarized for the population [11, 27]. The summary also averages diseased and non-diseased sites in individuals with or without periodontitis. Similar issues are encountered when periodontitis progression is calculated as mean annual score in longitudinal studies [28]. This summary does not provide proper insights regarding the variations at different periods of observation in addition to the limitations abovementioned for periodontitis severity. Therefore, it appears that methodological limitations of the twentieth century still apply to some current research approaches. Other limitations of studying associations with periodontitis were discussed in the following sections.

2. Associations with periodontitis

Among dominant notions in the twentieth century was dental plaque considered the cause of periodontal diseases; it was thought that unless plaque is prevented, gingivitis invariably leads to periodontitis and subsequently tooth loss [13, 29]. Russell, who was considered the father of periodontal disease epidemiology, analyzed two population-based data and concluded that plaque and age alone explain around

90.3% of variations in periodontal disease regardless of which population being studied [30]. Similarly, an experimental gingivitis study was used to support the evidence that plaque is the cause of periodontal disease [31]. Consequently, clinical practice of that era almost solely focused on plaque control [32–34]. Though, these findings reflect the abovementioned methodological limitations of periodontal and plaque indices. By 1980s, studies on untreated populations used site-specific measurements, unveiled variations in susceptibility to periodontitis despite accumulations of uniformly high levels of plaque and calculus [19, 22, 35–39]. Variations in susceptibility to periodontitis were observed among individuals, population groups, and teeth sites. Thus, it was no longer accepted that plaque is the only cause of periodontal disease. Several theories evolved trying to explain the role of dental plaque including nonspecific and specific plaque theories [40]. Currently, there is a general agreement that dysbiosis of oral biofilm may induce an inflammatory response in susceptible hosts to cause periodontitis [41]. Dysbiosis is a shift in the species within the oral microbiome that results in loss of homeostasis and may impact the human health; it can result due to an expansion of pathogenic microorganisms, altered metabolic capacity, decreased beneficial microorganisms, or reduced species diversity. Thus, the etiology of periodontitis can be attributed to "*sufficient cause*" where a set of components (exposures) can be sufficient for causing the disease that requires dysbiotic biofilm as a necessary component [42–44]. However, the biofilm alone is not sufficient cause of periodontitis. Hypothetical sets of sufficient causes for periodontitis are depicted in **Figure 1**.

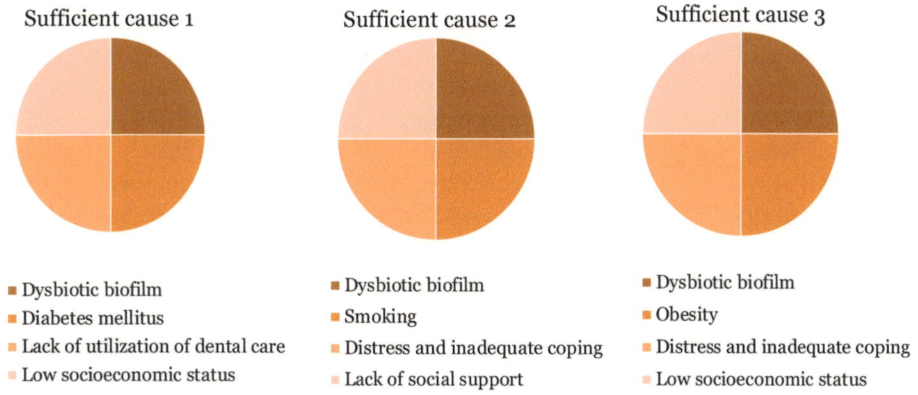

Figure 1.
Hypothetical sets of sufficient causes for periodontitis using sufficient cause component model (Rothman's pies) [45]. Dysbiotic biofilm is a necessary component but insufficient for causation [42, 44].

Periodontitis and other chronic diseases including atherosclerotic cardiovascular disease, obesity, and diabetes mellitus are associated with increased low-grade systemic inflammation and oxidative stress [46–49]. Evidence suggests that aggregation of chronic diseases within individuals is associated with higher level of inflammation and oxidative stress compared to having a single condition [46, 48, 49]. Chronic diseases also have common risk factors with periodontitis including underlying exposures and health behaviors [50, 51]. Underlying exposures can be either non-modifiable including age, sex, ethnicity, and genetics or modifiable such as the social determinants of health. Underlying exposures are mostly associated

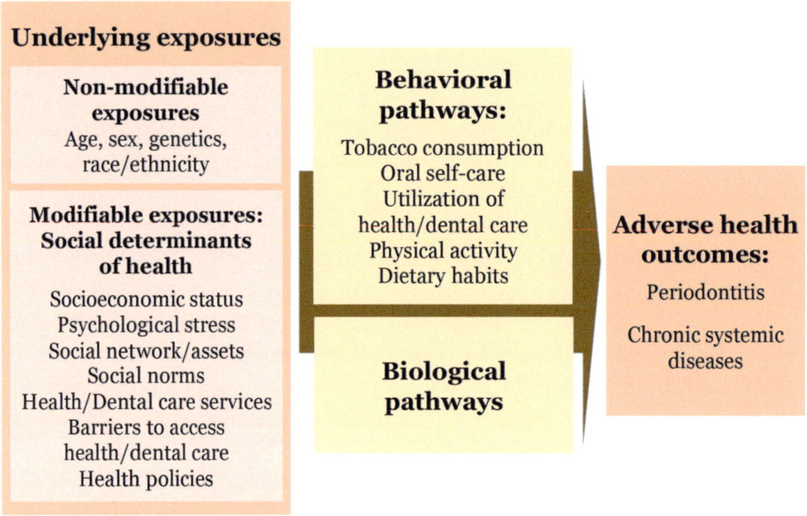

Figure 2.
Behavioral and biological pathways that could link the underlying exposures with the adverse health outcomes including periodontitis.

with periodontitis or other chronic diseases mainly through behavioral or biological pathways (**Figure 2**).

Several models were proposed to conceptualize disease causation including sufficient component cause model (**Figure 1**), web of causation, and structural equation models [44, 45, 52]. Nancy Krieger criticized the use of sufficient component cause model and web of causation for lacking accountability for unjust disparity in disease distribution when proposed her ecosocial theory [53, 54]. Krieger's theory is not a method of disease causation since it is concerned with individuals and their communities between the past and present. "*Shared observations of disparities in health do not necessarily translate to common understandings of cause; it is for this reason theory is key*" [53]. The term "*embodiment*" was introduced to refer to interactions between society and biology. The theory also focused on pathways of these interactions at multiple ecological levels over the life course while considering the accountability for injustice. Ecological levels can be conceptualized at multiple levels and domains including home, school, and work. In contrast to causal models that can be used for testing specific hypotheses, Krieger's theory can be useful for identifying knowledge gaps, generating research questions, articulating increments of evidence from separate studies to understand disease etiology, and taking the theory into action [54].

Two epidemiological concepts relevant to disease prevention that need to be distinguished are the life course approach and the natural course of disease. The life course epidemiology is concerned with understanding psychosocial and physical environmental exposures and when they can impact health behaviors or become biologically embedded over the life course (intrauterine life, childhood, adolescence, early adulthood, late adulthood, or across generations) [55–57]. While earlier approaches for life course focused on diseases as its end points, recent approaches shifted to study multiple aspects of physiological functions [55]. The exposures are studied for their impact independently, cumulatively, or for their interactions. In contrast, the natural course of disease is a disease specific concept and based on different stages of disease development (susceptibility, asymptomatic, symptomatic, and terminal stages) in the absence of treatment [11, 58, 59]. Recruitment of study population that is either never treated or all treated similarly for periodontal conditions can control for treatment status. However, disregarding impacts of treatment can lead to erroneous interpretations of study findings [11, 13]. This can be especially true when considering that periodontal treatment (non-surgical or surgical) has non-negligible impacts on the periodontitis severity, extent, and progression [11]. Unfortunately, studying the periodontitis associations while disregarding the treatment impacts is a common approach in periodontal research.

2.1 Impact of age on periodontitis: What has not been learned yet?

Based on cross-sectional surveys, periodontitis increased in severity, extent, and prevalence with age [11, 60–63]. When periodontitis defined using CAL, its association with age was more consistent, whereas PPD had less clear pattern [11, 15, 60, 64, 65]. When periodontitis included non-severe thresholds (CAL ≤4 mm, PPD ≤6 mm, or ≤ 1/3 bone loss), the prevalence steeply increased after 20 years through 30s until it reaches its peak between 35 and 45 years [11, 38, 60, 66, 67]. The prevalence before 20 years, although less frequent, widely varied within and among populations; estimates from different populations were between 1.0% and 69.2% [11, 60, 62, 66, 68–70]. Similar patterns were observed when periodontitis incidence was studied in longitudinal surveys [67, 71, 72]. In contrast, when periodontitis was defined at severe thresholds, majority of its variations, within populations, were mostly confined to adults [38, 64, 66, 73, 74]. Despite that clinical classifications consider molar-incisor pattern suggestive of localized aggressive periodontitis in young age groups [75–78], evidence from epidemiological surveys indicates that such pattern is common in all age groups [17, 18, 22, 29, 79]. Similar pattern was observed in a study on untreated population; most impacted teeth were upper molars and lower incisors of different age cohorts (**Figure 3**) [11].

There is general agreement that age association with periodontitis is due to accumulative effect of exposures and disease experience over time rather than having an increased susceptibility to periodontitis at specific age [15, 80]. However, it worth noting that such association remains robust after adjustment for different exposures including sex, social, behavioral, and medical exposures in cross-sectional or longitudinal studies [11, 60–62, 81, 82]. In a study in untreated Sri Lankan tea workers, periodontitis progression had an independent association with both age and follow-up time [83]. These age-related findings may suggest that effect of age on periodontitis cannot be completely explained as function of accumulation of exposures and disease experience over time. Among the difficulties of fully understanding the effect of age, in periodontal literature, is that majority of studies either focus on younger or older age groups excluding direct age-related comparisons [11]. Such approach could hide age-related patterns of periodontitis distribution and its associated exposures. Another limitation is that periodontitis progression is often studied as mean annual CAL or PPD for comparisons among age groups [28]. As previously stated, such summary does not properly account for variations in the rate of progression within or between individuals, subgroups, and at different periods.

An ideal study design to better understand the role of age on susceptibility to periodontitis would be a life course approach. Such approach can identify if there are sensitive periods in development, which can have behavioral or biological impacts [55]. Sensitive periods can be psychologically, socially, or physiologically defined [84]. Behaviors including smoking, unhealthy dietary habits, and inadequate coping with social adversity are often acquired during adolescence. This makes interventions

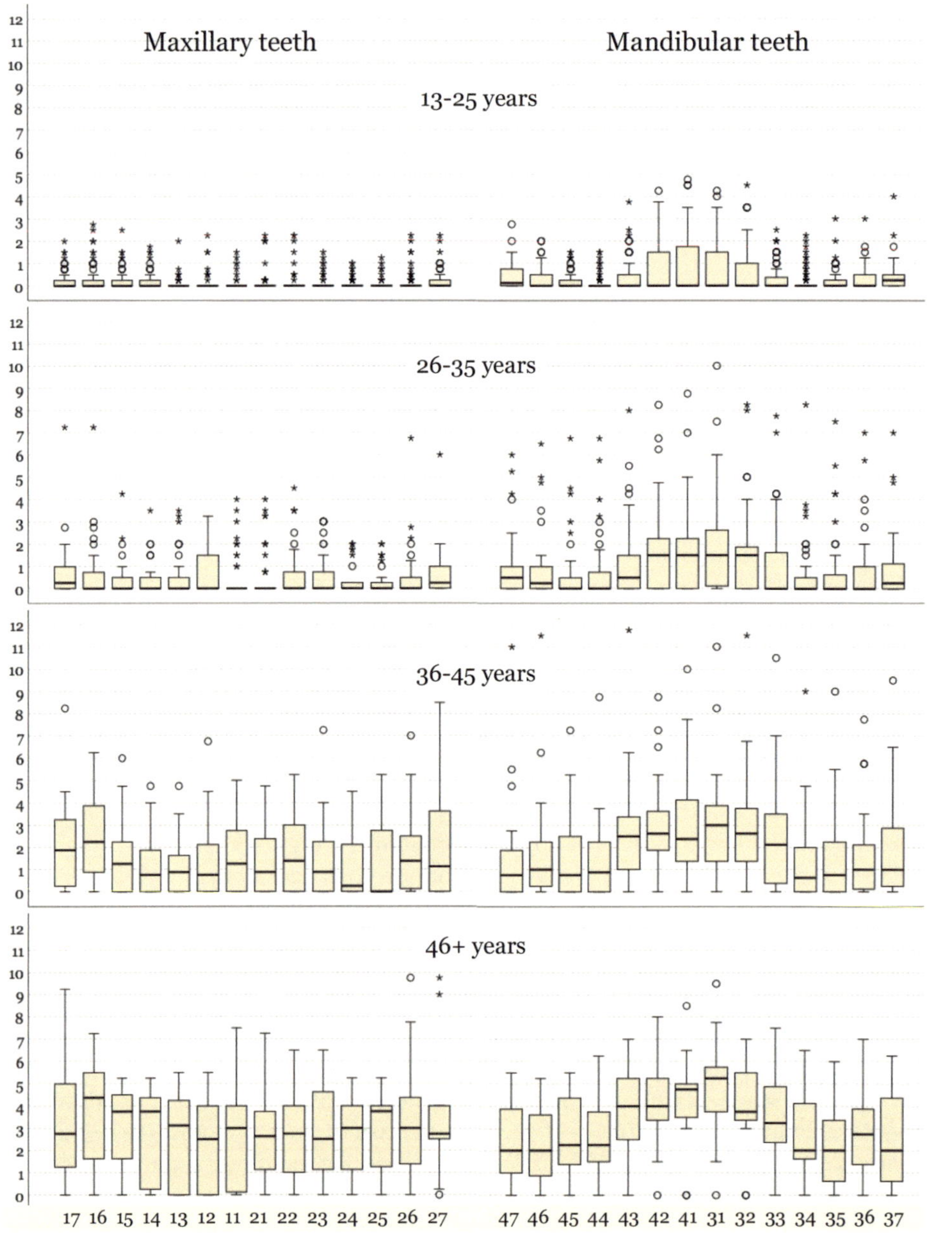

Figure 3.
Periodontitis severity for maxillary and mandibular teeth in an untreated population [11, 60]. The most impacted teeth were the lower incisors and upper molars across different age cohorts.

to combat certain behaviors such as smoking most effective at that stage [84]. At older age, compromised manual dexterity and physical disability could limit oral self-care and utilization of dental care. The life course study design is mostly limited due to its feasibility especially when considering the lengthy latency periods between exposures and periodontitis incidence.

2.2 Sex

Males almost consistently had higher periodontitis prevalence, extent, and progression compared with females in epidemiological studies [60, 61, 81, 85–87]. In contrast to suggestions that sex-related differences can be attributed to better oral health behaviors in females [80, 88], studies on populations with uniformly high levels of plaque and minimal access to dental care also found an increased periodontitis prevalence, extent, and progression in males [11, 60, 86, 87]. National surveys from Brazil, France, and the United States reported higher periodontitis prevalence in males despite the adjustment for behaviors including patterns of dental visits, smoking, and alcohol consumption [61, 81, 89, 90]. In contrast, a recent systematic review found almost no sex-related differences in terms of periodontitis progression rates [28]. This finding was based on a subgroup meta-analysis of two studies only. However, subgroup meta-analysis can be only conducted when there are minimally 10 available studies [91]. In addition, the sex-related differences were compared based on mean annual progression, which does not properly account for multilevel variations. In a study Brazilian population, males had an unadjusted risk ratio (RR) of 1.24 (95% confidence interval [CI]: 1.09, 1.42). After adjustment for smoking, males had an adjusted RR of 1.22 (95% CI: 1.08, 1.39). This may be interpreted as having small effect of smoking on sex. However, the study found an interaction between smoked packs per year and sex; females who smoked 30+ packs had higher RR of periodontitis progression after adjustments for age and level of education.

Sex dimorphism in immunological responses was suggested as plausible biological mechanism that may explain the differences in susceptibility to diseases [85]. Circulating levels of sex steroids could alter as a function of age mainly during puberty and menopause. However, such findings were mostly based on responses to acute infections. Although the use of life course approach would have been ideal to understand the sex dimorphism in susceptibility to periodontitis, cross-sectional surveys with wide age ranges may provide indirect evidence to identify these differences as function of age. This approach may narrow down the required observation time before conducting longitudinal studies. For example, in a study on untreated population, the sex-related differences in periodontitis prevalence and extent (**Figures 4** and **5**) were mainly observed after 20 years [26, 60]. Similarly, periodontitis had no association with sex in 15–19-year adolescents from several Latin American nations [70]. These findings may suggest the need to study the sex-related differences in exposures prior to 20 years.

2.3 Race/ethnicity

Studies from the United States almost consistently presented an association between race/ethnicity and adverse health outcomes including periodontitis [61, 94–96]. In the first National Health and Nutrition Examination Survey (NHANES), higher periodontitis prevalence was found in black Americans compared with whites with an adjusted odds ratio (OR) of 1.31 (95% CI: 0.78, 2.19) [94]. In NHANES III, the adjusted OR in Blacks increased to 2.09 (95% CI: 1.68, 2.60); the adjustments in both surveys were done for age, socioeconomic indicators, smoking, diabetes, and time elapsed since last dental visits. In NHNANES (2009–2012), both Hispanics and non-Hispanic blacks had higher periodontitis prevalence than

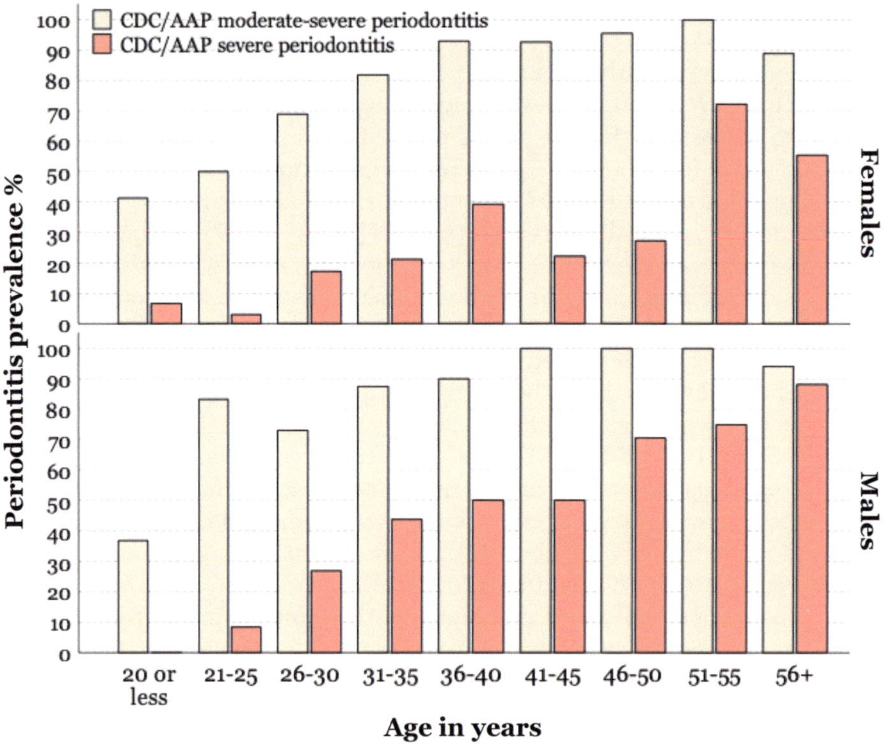

Figure 4.
Periodontitis prevalence per age and sex groups in an untreated population [11, 60]. Periodontitis defined based on criteria by the Centers for Disease Control and Prevention and American Academy of Periodontology (CDC/AAP) [92, 93].

whites [61]. Ethnicity-related associations were not confirmed in Brazilian population [81, 82]. Disparities in health outcomes based on race/ethnicity could not be explained by genetic differences but could rather reflect differences in social, cultural, and behavioral factors between the past and the present [53, 94, 96]. Thus, ethnicity is mainly considered as a social construct, and more information needs to be provided to explain its role within a broader context.

Borrell *et al.*, suggested that disparity in periodontitis prevalence between African Americans and Caucasians in the United States can be attributed to discrimination, racism, and residential segregation [94]. Residential segregation for example may impact health outcomes by potentially impacting exposures to environmental hazards, quality of education, employment opportunities, and influencing certain behaviors such as increased tobacco consumption is some neighborhoods. In a study that analyzed the association between ethnicity and periodontitis based on different socioeconomic status (SES), black Americans with high income had the highest periodontitis prevalence compared to both high-income whites and low-income blacks. This finding illustrates how increased income may not directly improve health outcomes. Possible explanations include increased income may occurred short term after development of periodontitis or was not accompanied by skills in managing resources. Social epidemiologists suggested that income increase in short term (e.g., winning lottery) can be associated with unhealthy behaviors such as increased smoking [84]. Another possible explanation is that high-income blacks may experience

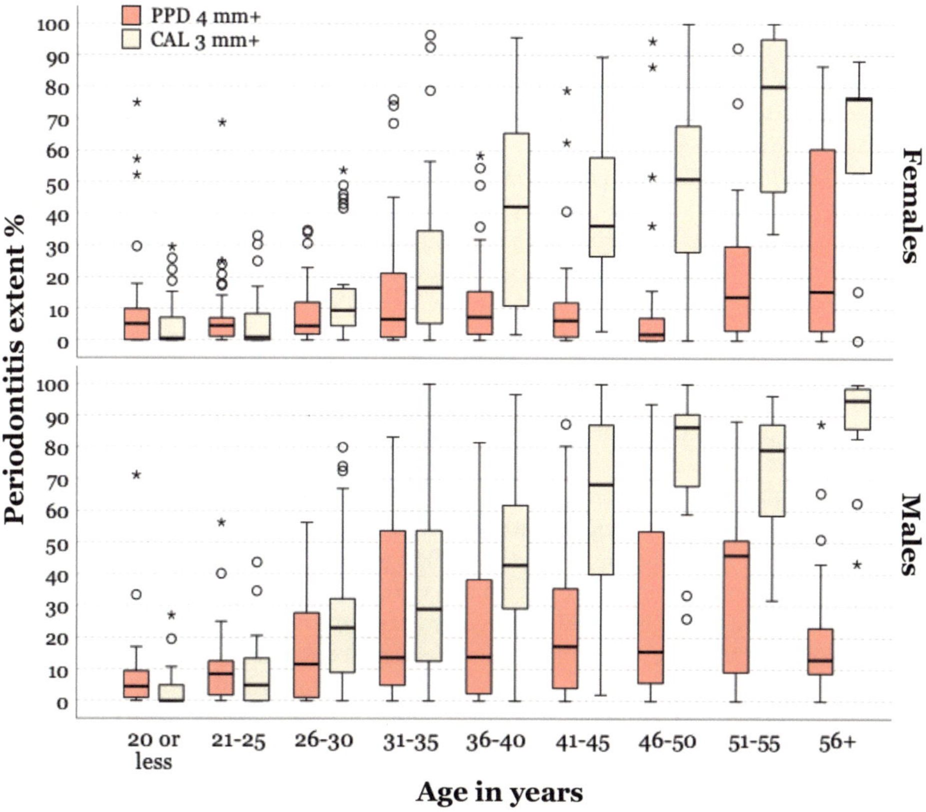

Figure 5.
Periodontitis extent per age and sex groups in an untreated population [11, 60]. PPD: Periodontal probing depth, CAL: Clinical attachment loss.

stressful conditions and increased social isolation, which adversely impact their health. Therefore, information about structural racism and residential segregation within a historical context provides better insights about the social and physical environmental constrains rather than "*blaming the victim*" for their behaviors and health conditions [53, 54, 97].

2.4 Socioeconomic status, its pathways, and embedment over the life course

SES is a term describes the social standing and power in society, and it is often measured using indicators such as level of education, income, and occupation. Social status may also be determined by the level of power "*In the context of workplace, there are those who occupy positions of supervisory authority over subordinates versus those who take orders from above*" [84]. Lower education was associated with periodontitis with a pooled OR of 1.86 (95% CI: 1.66, 2.10) from 18 studies [98]. The association between periodontitis and SES is almost consistently positive in the literature when using level of education and income as indicators [60, 61, 88, 95, 99]. However, such approach can be limited without studying this association within a broader context that could explain the link between the SES indicators and periodontitis. It has been suggested in the literature that income mainly acts as a mediator for the association between education and disparities in health outcomes; thus, some researcher would elect to report only one of these two indicators. Though, both education and income demonstrated independent association with periodontitis after adjustments in multivariate models [99]. The neighborhood SES also had an independent association with periodontitis after the adjustment for individual level SES, age, sex, marital status, ethnicity, smoking, and diabetes [100]. Low and medium neighborhood SES had an adjusted OR of 1.73 (95% CI: 1.29, 2.32) and 1.63 (95% CI: 1.23, 2.17), respectively, compared to high SES.

SES can be linked to disparities in health outcomes through interrelated pathways (**Figure 6**) where the association can be mediated by psychosocial exposures, material factors (resources), and health behaviors [9, 95, 99, 101]. In neo-material pathway, effect of education can be mediated by income, occupation's working conditions, related resources such as health insurance; consequently, it can improve housing, residential neighborhood, affordability of hygiene products, and access to health/dental care. In psychosocial pathway, higher educational attainment can be associated with having broader social network, which can provide social support, enhancement of control beliefs, problem-solving, and coping skills; consequently, improve response to stressors. In behavioral pathway, education can increase the chance of being surrounded by well-educated people, change the social norms, and increase health literacy, which can improve health behaviors.

SES needs to be conceptualized as a dynamic approach over the life course; income and occupation could fluctuate (increase or decrease) over time while education either remains stable or increases. A dynamic approach for SES also implies reciprocities where increased SES could improve health and vice versa. Some relevant questions to be asked regarding life course approach include when is the sensitive period in development where SES could impact behaviors or can be biologically embedded?

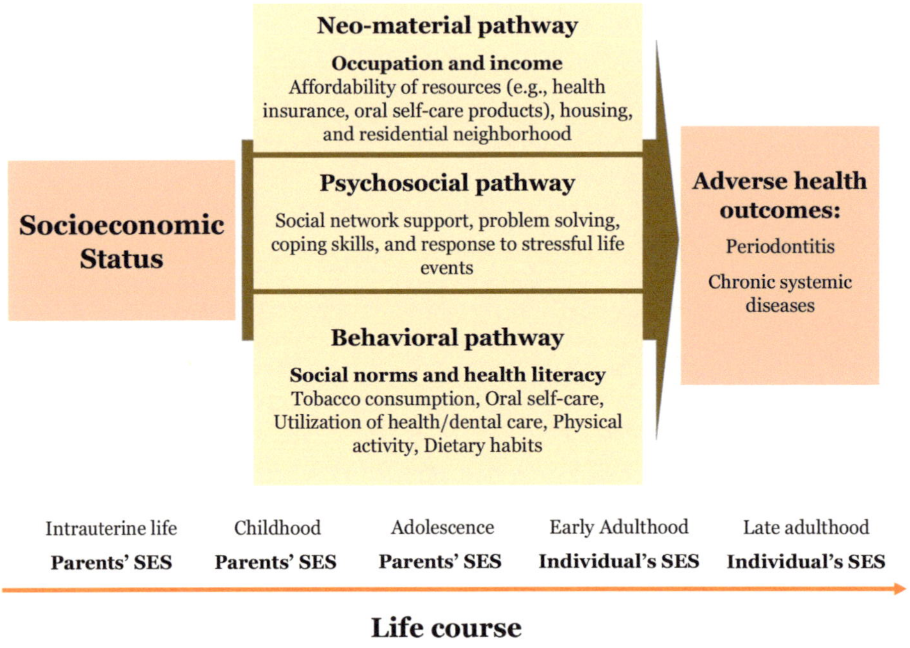

Figure 6.
Suggested pathways that could link socioeconomic status (SES) to periodontitis or other chronic systemic diseases over the life course.

What is the latency period between exposure to SES and having behavioral or biological impacts? Does the SES have accumulative impacts throughout life? Is the individual SES during adulthood or the parents' SES during childhood more relevant to periodontal health? Are childhood SES and adulthood SES correlated or independent? What is the impact of social mobility (transition to higher or lower SES)? It worth mentioning that in populations with rapid changes, social mobility needs to be assessed at smaller intervals over the life course rather than limiting the comparisons between childhood and adulthood.

Suggested models for dynamic approach for studying SES include [84]:

1. Critical (sensitive) period model assumes that Low SES could impact health at sensitive period in development; SES changes at later period cannot attenuate or reverse the impact already occurred.

2. Accumulation of risk model assumes that low SES's health impact increases incrementally.

3. Chain of risk (trajectory) model assumes that low SES in early life results in low SES later. However, impacts on health are only embedded later in life.

The importance of understanding the impact of SES on health and its related behaviors is to know when to intervene to achieve the best health outcomes. Based on a recent systematic review on seven longitudinal studies, early life SES had positive association with periodontitis [102]. However, the included studies had heterogeneous study designs and definitions of both SES and periodontitis; thus, made it difficult to arrive at a more specific conclusion. In one study, the parents' SES had no significant association with periodontal status in 15–16 years individuals; though, the parents' SES (income) was predictive for utilization of dental care [103]. In another study, early SES measured as parents' occupational status was associated with periodontitis in 26 years individuals, whereas the individual's SES in adulthood was not significant [104]. In contrast, the adult SES was more relevant to smoking and depression. The role of social mobility was also evaluated; highest periodontitis proportion was found in those with persistent low SES, followed by upward mobility group, then downward mobility group [104]. The lowest periodontitis proportion was found in those with persistent high SES. In another study that assessed the periodontal health at 15 years (baseline) up to 31 years, childhood SES was associated with periodontitis in adults, which was not mediated by adulthood SES or behaviors; RR was 1.85 (95% CI: 1.06, 3.24) [105]. So, studies that used the life course approach suggested that early SES during childhood can be more predictive of periodontitis rather than adulthood SES. These findings support the critical period model for impact of SES on periodontal health. However, the longest follow-up was up to 38 years, and the possible accumulative effect of SES cannot be excluded.

2.5 Psychosocial exposures, their pathways, and interactions

When negative life events occur, they undergo an appraisal process, if the stressors exceed the individual's coping ability, they can be perceived (appraised) as stressful [106, 107]. The physiological responses to stressors can be mediated by negative emotions including anxiety, depression, anger, or mixed emotions [107]. The "*allostatic load*" was a term introduced by McEwen to refer to wear and tear from chronic over- or underactivity of protective body systems against stressors including the autonomic nervous system, hypothalamic–pituitary–adrenal (HPA) axis, metabolic, immune, and cardiovascular systems [108]. The perceived stress can be associated with health conditions (including periodontitis) through interrelated behavioral and pathophysiological pathways (**Figure 7**) [109, 110].

Different psychological exposures had an association with periodontitis [112–114]. In one study, psychosocial stress was not associated with periodontitis but had positive association with oral health behaviors (smoking and oral self-care) in 65–74 years women [115]. Among different daily strains, only financial strain was associated with higher CAL and bone loss after adjustment for age, sex, and smoking [113]. Intrinsic mechanisms that could regulate and the stresses include coping [116]. Inadequate coping (high emotion-focused and low problem-focused coping) was associated with higher CAL and bone loss regardless of the level of financial strain [113]. Problem-focused coping includes reappraisal of stressful events as challenging events and applying strategies that focus on solving the problems [116]. In contrast, emotion-focused coping includes applying strategies that deals with the emotions rather than the problem; it can include positive strategies such as regulations of emotions, meditation, spirituality, distracting oneself, and seeking support from social network. However, it can be also negative when denying the event, avoiding dealing with it, or adopting negative health behaviors such as unhealthy dietary habits, smoking, consuming alcohol, and drugs. Thus, emotion focused coping can be an inadequate coping strategy with stresses. Other behaviors related to inadequate coping include negligence of oral self-care and lack of sleep [111, 117, 118]. In contrast to coping, social network support and positive life events could buffer the impact of stressors through external pathways [106, 119]. Social support can be received (actual support) or perceived, which refers to the subjective beliefs about the availability of support; the perceived social support can be more effective in buffering the impact of stressors (**Figure 7**) [106].

When impact of social network was studied for association with periodontitis, marital status had no significant association with periodontitis prevalence or progression rate [61, 82, 114, 120]. However, having fewer friends and being

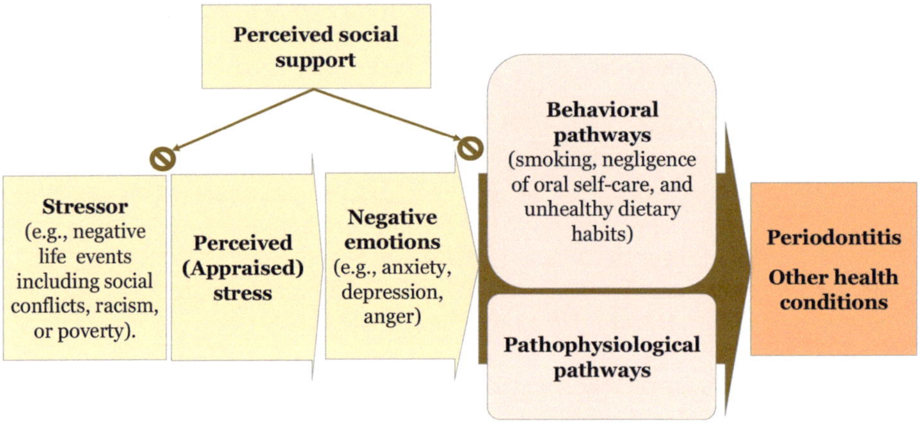

Figure 7.
Illustration of psychosocial interactions. Stressors can have adverse health impacts after being perceived (appraised) as stressful [106]. Negative emotions could be mediators of perceived stress and their impact on health could occur through interrelated pathophysiological pathways and behavioral pathways [109–111]. The perceived social support may buffer the impact of psychological stresses either before or after perceiving the event as stressful [106].

widowed were associated with higher periodontitis extent [120]. In a prospective study on healthcare professional men, having at least one friend was associated with a lower risk of periodontitis; RR was 0.70 (95% CI: 0.51, 0.96) after adjustments for age, marital status, smoking, body mass index (BMI), alcohol consumption, and diabetes mellitus [114]. Furthermore, participation in religious meetings had a lower risk of periodontitis and the adjusted RR was 0.73 (95% CI: 0.64, 0.83). In a cross-sectional study by Zini *et al.*, religiosity had protective association with periodontitis through spirituality (intrinsic pathway) and social support (extrinsic pathway) [121]. Other studies did not find an association between social support and periodontitis [60, 122].

Possible mechanisms of how social network positively impact health (**Figures 6** and 7) is by buffering impact of stressors, shaping norms for health behaviors, sharing information can increase health literacy and awareness, helping in problem-solving, sharing funds and resources, and improving overall wellbeing [119, 123, 124]. In a recent systematic review, social support had positive association with oral health literacy, quality of life, and behaviors including utilization of dental care in immigrants and ethnic minorities [125]. It is important to keep in mind, however, that impact of psychosocial exposures can be either positive or negative [11]. For example, perceived stress manifested as anxiety can stimulate sympathetic nervous system and HPA axis; consequently, help people to meet certain deadlines and get the job done [107]. The allostatic load, up to a certain limit, can be within the body's reparative capabilities before it can be associated with pathological changes. Similarly, social network could have positive influence on behaviors (e.g., oral self-care), negative influence (e.g., smoking), enable certain behaviors (e.g., utilization of dental care), or constrain such behaviors [123]. Social network could also be a source of conflicts, stressors, or transmission of diseases. Thus, measurement of social network in terms of its size and range (structural indicators) alone can be deficient without considering specific functions [106, 119] or social norms [11]. Among different social support functions, emotional and informational support was the most responsive to buffer wide range of stressors [106, 119]. However, the overall research on mechanisms, pathways, and interactions of psychosocial exposures is still limited in periodontal literature.

2.6 Oral health behaviors

Health behaviors can be largely shaped by social environment including social adversity, social norms, social support, and socioeconomic status [3, 101, 113, 120, 123]. Health behaviors appear to cluster within individuals. For example, those who are physically active can also follow healthy dietary habits, regularly utilize dental/health care, regularly practice oral self-care, and vice versa. Unhealthy behaviors often aggregate in individuals with low SES [84]. In addition, behaviors such as smoking, negligence of oral self-care, and lack of utilization of dental care could be indicative of inadequate coping with social adversity [111, 117, 118].

Impacts of smoking on periodontium were well documented, and evidence was considered adequate to conclude that smoking is a cause of periodontitis by the office of surgeon general in 2004 [126]. The frequency and duration of smoking had a dose response association with periodontitis prevalence and progression [82, 127]. Smoking also had positive association with periodontitis in adolescents (15–19 years) with an adjusted OR of 1.6 (p-value: 0.017) after adjustments for demographics, plaque scores, and bleeding on probing [70]. Number of years since quitting of smoking also had a gradient reduction in susceptibility to periodontitis. Based on a systematic review on 14 prospective studies, periodontitis incidence in smokers had an RR of

1.85 (95% CI: 1.5, 2.2) [128]. Smokers also had poor periodontal treatment outcomes (non-surgical and surgical) compared to never smokers [129–131].

Despite that oral self-care had clear association with gingivitis [31, 80], its association with periodontitis was not consistent. Based on systematic review on 14 studies, tooth brushing had an OR of 1.44 (95% CI: 1.21, 1.71) [132]. Infrequent oral self-care and presence of calculus had positive association with periodontitis in young populations (14–29 years) [62, 69]. Good oral self-care had protective association with overall periodontal treatment need in Portuguese population [133]. Other studies did not confirm the association of oral self-care with periodontitis [82, 134]. It worth noting here that self-reported oral self-care can be subject to social desirability bias, that is, study participants could report daily brushing and use of interdental aids because they are socially desirable behaviors [11]. Therefore, measurements of full mouth plaque scores can be used to validate such self-reported measures.

Several models were proposed to conceptualize the utilization of health/dental services [135–140]. Access is a term refers to potential access to health service based on enabling resources while utilization is the actual use of service [136]. Barriers to utilize health services can be structural such as accessibility (transportation), availability of resources, and accommodation (availability of appointments); financial barriers include affordability; personal barriers include acceptability (including trust in dental care providers), psychosocial factors, and awareness [135, 138, 140]. Populations with no (or minimal) access to dental care had an overall higher periodontitis prevalence, extent, severity, and progression compared to estimates from general populations [11, 17, 18, 22, 35, 37, 38, 60, 65]. Frequency of dental visits had positive associations with periodontitis in adults [81, 82] and young population (15 to 21 years) [69].

3. Periodontitis prevention: are we there yet?

The ultimate goals of studying the etiology of diseases are to plan, implement, and evaluate interventions to combat disease burden [141]. Clinical interventions, mostly based on behavioral theories, had positive periodontal outcomes for smoking cessation, diabetes control, dietary habits modifications, weight control, and increased physical activity [142]. The study periods ranged from few weeks to 24 months. Despite the positive outcomes for changing behaviors by clinicians, such strategies can be potentially challenging, costly, time-consuming, and have temporary or limited effectiveness if not implemented as multilevel strategies [2, 3, 10]. Economic burden of periodontitis in 2018 was estimated to be ≥150 billion dollars in the United States and ≥ 150 billion euros in Europe [143]. However, up to most recent recommendations to prevent periodontitis were mostly confined to individual clinical care [142, 144–146]. A recent review also indicated that preventive strategies for periodontitis at population level are almost non-existent [51].

Among several approaches of defining periodontitis cases that evolved over history, current classification system by the AAP and European Federation of Periodontology (EFP) seemingly has resolved several longstanding issues [75, 147]. Periodontitis is mainly defined under single category rather than relying on age. Such approach may help overcoming issues of discontinuity in reporting findings for younger and older individuals to better understand the effect of age on periodontitis [11]. Periodontitis staging follows the clinical stages of natural course of disease, and its practical implications were explicitly outlined [75]. Stage I and II were classified separately, though, their practical implications were similarly confined to non-surgical periodontal treatment. Stage III and IV mainly require surgical periodontal treatment and the latest additionally requires functional rehabilitations. The grading aims were to identify the progression rate, responsiveness to therapy, and possible impact on systemic health. The grading is assigned based on rate of progression and clinical phenotype then modified based on two risk factors: glycemic control and cigarette smoking. In contrast to staging, the practical implications of periodontitis grading were not explicitly suggested. Staging can be readily applied in population-based studies while grading's application can be limited due to relying on radiographic examinations, could fluctuate over time, and limited to two risk factors (diabetes mellitus and smoking) [11]. However, other exposures could also have direct impact on patient care including barriers to utilize dental care (e.g., anxiety, lack of trust, lack of awareness, lack of perceived need for care) which may impede patients from seeking clinical care or complying with supportive periodontal therapy. In addition, the staging/grading framework is limited to the latest stages of the natural course of disease; therefore, its implications are confined to individual patient care.

While most of the periodontitis exposures can have practical implications for different stakeholders, recommendations based on too many exposures related to periodontitis alone can be exhaustive and impractical [11]. The common risk factor approach to integrate prevention of several chronic diseases (e.g., cardiovascular disease, diabetes mellitus, and obesity) has been suggested [50, 51]. This approach simultaneously targets risk factors/indicators of several chronic diseases such as tobacco consumption, unhealthy diet, sedentary lifestyle, lack of glycemic control, and psychological stress when planning population-based strategies. The common risk factor approach might also be more manageable at the population level rather than suggesting disease specific preventive approaches. Also, duplication of efforts in combatting diseases can be avoided [50]. This can be especially important when considering that periodontitis is not a life-threatening disease and may not be perceived as a priority by policymakers compared with other chronic diseases (e.g., cardiovascular diseases). It is important to consider, however, that common risk factor approach can achieve its benefits the most when applied as a multilevel strategy. When considering common risk factor approach in clinical care, there is a possible need for integrated dental and medical care. Dentists may routinely refer their patients for medical care, but more help is required from medical care providers to refer their patients to dentists once they are diagnosed with chronic medical conditions such as diabetes mellitus [11]. Patients may prioritize medical over dental care either due to having more serious consequences, dental anxiety, or limited dental insurance coverage [11, 60].

3.1 Periodontitis prevention as a multilevel strategy

A framework suggested by the author for applying periodontitis prevention at multilevel was outlined in **Table 1** [11]. Periodontitis diagnoses integrated early stages (susceptibility stages) of the natural course of disease with the AAP/EFP staging criteria [75]. The diagnostic criteria follow the assumption that staging is irreversible; once a diagnosis is assigned, it can only be upgraded [75]. The common risk factor was a recommended approach for all levels of prevention along the natural course of disease. The framework may also enhance knowledge translation where evidence from clinical/epidemiological research can be applied in clinical, population, and public health practice and vice versa.

	Susceptibility stages (Pre-periodontitis)		Periodontitis stages		
	Stage of underlying exposures	**Stage 0 (Periodontitis risk factors/indicators)**	**Stage I + II**	**Stage III**	**Stage IV (Terminal stage)**
Staging diagnostic criteria	Underlying exposures (risk determinants). Non-modifiable: aging, males, and ethnicity. Modifiable: Social determinants of health (social and environmental constraints).	Underlying exposures. Periodontitis exposures/risk factors/risk indicators. Health behaviors: smoking, lack of utilization of dental care, unhealthy diet, and sedentary lifestyle. Medical exposures: diabetes mellitus, obesity, other chronic diseases.	Underlying exposures. Periodontitis exposures/ risk factors. CAL: 1–4 mm. No indications for surgical periodontal treatment.	Underlying exposures. Periodontitis exposures/ risk factors. CAL ≥5 mm. Indications for periodontal surgery: PPD ≥6 mm, osseous defects (Vertical bone loss, craters), class II, or III furcation involvements.	Underlying exposures. Periodontitis exposures/risk factors. CAL ≥5 mm. Indications for periodontal surgery: (PPD ≥6 mm, osseous defects (Vertical bone loss, craters), class II, or III furcation involvements. Loss of masticatory functional stability, excessive teeth mobility, drifting, and teeth loss due to periodontitis (≥5 teeth).
Level of prevention	Primordial	Primary	Secondary	Tertiary	
Stakeholders	Mainly policy makers and governments.	Mainly public health professionals but also includes dentists/ hygienists and medical professionals.	Dentists/ hygienists, and medical professionals.	Periodontists and medical professionals.	Periodontists, other dental specialities, and medical professionals.
Aims of prevention	Targeting the underlying exposures to prevent the development of periodontitis risk factors/ exposures.	Targeting the underlying and periodontitis risk factors/ exposures to prevent periodontitis incidence.	Targeting underlying and periodontitis risk factors/ exposures to prevent disease progression.	Targeting underlying and periodontitis risk factors/ exposures to prevent disease progression and complications.	Targeting underlying and periodontitis risk factors/ exposures to prevent further disease deteriorations.

	Susceptibility stages (Pre-periodontitis)		Periodontitis stages		
	Stage of underlying exposures	Stage 0 (Periodontitis risk factors/indicators)	Stage I + II	Stage III	Stage IV (Terminal stage)
Prevention strategies	Population-based prevention strategy: Use of common risk factors approach for combatting chronic diseases such as mass awareness campaigns regarding oral health and general health, imposing sin tax on tobacco and added sugar containing products, applying strategies for increasing resources for population with low SES such as facilitating dental and health insurance, improve access to dental and medical care, and increase outdoor/indoor spaces for physical activity.	High-risk prevention strategy: use of common risk factors for combatting chronic diseases for high risk individuals: Increasing awareness campaigns for promoting periodontal health and general health in susceptible individuals/ population groups including advocating for healthy diet, regular dental visits, regular medical visits, smoking cessations, increase physical activity, management of anxiety/ stress reduction protocols, and prophylaxis (removal of plaque, and calculus) by dental care provider.	Prevention by clinical care: Applying common risk factor approach for patients including advocating for healthy diet, regular dental visits, regular medical visits, smoking cessations, increase physical activity, and glycemic control by regularly taking medications, management of anxiety/ stress reduction protocols, promotion of oral-self-care, and prophylaxis (removal of plaque, and calculus) by dental care provider.	Prevention by clinical care: Applying common risk factor approach for patients including advocating for healthy diet, regular dental visits, regular medical visits, smoking cessations, increase physical activity, and glycemic control by regularly taking medications, management of anxiety/ stress reduction protocols, promotion of oral-self-care, prophylaxis (removal of plaque, and calculus) by dental care provider, and surgical periodontal treatment.	Prevention by clinical care: Applying common risk factor approach for patients including advocating for healthy diet, regular dental visits, regular medical visits, smoking cessations, increase physical activity, and glycemic control by regularly taking medications, management of anxiety/stress reduction protocols, promotion of oral-self-care, prophylaxis (removal of plaque, and calculus) by dental care provider, surgical periodontal treatment, and multidisciplinary functional rehabilitation.

Underlying exposures include non-modifiable exposures including age, sex, and genetics or modifiable exposures such as the social determinants of health. Periodontitis risk factors/exposures include medical and behavioral exposures that could have more direct association with periodontitis. CAL: Clinical Attachment Loss, PPD: Periodontal Probing Depth, SES: Socioeconomic status.

Table 1.
A multilevel prevention strategy for periodontitis with integrated use for individual clinical care and population/public health [11].

3.1.1 Susceptibility stages

In *stage of underlying exposures*, primordial prevention is applied to the population, and it aims to reduce development of disease specific risk factors including adopting unhealthy behaviors [11]. This level of prevention mainly benefits the population at an early stage in development, that is, during childhood and teenage [148, 149]. The strategies can be mainly applied by governments and policy makers to target the social determinants of health and take into consideration the non-modifiable exposures. Examples of preventive strategies include imposing sin tax on tobacco- and added sugar containing products, increase safe public places for indoor or outdoor physical activities, and reduce barriers to access dental and health care such as facilitating health and dental insurance for low profile jobs. The latest can be accompanied by conditional strategies that enforce regular attendance to dental/medical care.

In *stage 0*, primary level of prevention can be applied to individuals or population subgroups who developed periodontitis specific risk factors/indicators [11]. It aims to reduce periodontitis incidence and strategies can be mainly applied by public health practitioners. Examples include health awareness campaigns and use of behavioral theories to apply public health interventions for smoking cessation, glycemic control, modification of dietary habits, and weight loss.

3.1.2 Periodontitis stages

Periodontitis stages have similar diagnostic criteria proposed for periodontitis staging by the AAP/EFP [75]. Exposures of susceptibility stage still apply to periodontitis stages based on the current understanding that periodontitis develops in a susceptible host. The grading criteria were removed since their application may not be suitable for population-based studies due to the abovementioned limitations [11]. Periodontitis stages are mainly managed by clinicians following the secondary and tertiary levels of prevention. An additional strategy is to apply the common risk factor approach in clinical stages including the need for integrated dental and medical care. Medical professionals need to refer their patients once diagnosed with chronic systemic disease. Stage I and II were combined since they are managed similarly by non-surgical periodontal treatment. Diagnosis of stage I can be challenging when assessments are conducted by non-periodontists due to the use of low disease threshold. Population-based surveys commonly consider CAL $\geq$3 mm as minimum threshold to avoid misclassifying cases [92, 93, 150]. However, diagnosing periodontitis cases at lower thresholds allows early interventions, which can be particularly important in younger population [11].

3.2 Additional considerations for the multilevel prevention framework

Despite that gingivitis precedes periodontitis, it was not included in the framework because pristine gingival health may not exist even under optimal oral self-care [151–154]. Also, gingivitis can be reversible or self-limited. However, periodontitis staging is assigned with a purpose of being irreversible. Similarly, plaque per se is a universal finding even in individuals following stringent plaque control strategies [34, 155–157]. Plaque needs to be controlled in all members of the population. In addition, plaque is a reversible exposure, which makes it unsuitable for periodontitis staging.

When applying multilevel preventive strategies, it is recommended to consider outlining a specific plan for each exposure at a time [2]. For example, strategies for combatting tobacco consumption at population level could include imposing sin tax, age restrictions, and increase smoke-free spaces [6]; for smokers population subgroup, smoking cessation interventions and awareness campaigns can be implemented; for periodontitis patients, smoking cessation strategies including referral to smoking cessation clinics can be applied.

4. Concluding remarks

Key messages that need to be emphasized include:

- Variation in distribution is a key concept for studying diseases and their exposures. Therefore, using mean scores while disregarding the multilevel variations in periodontitis severity or annual progression rates makes these population-based summaries have questionable validity.

- Evidence suggests that majority of variations in periodontitis distribution in populations occur between 20 and 40 years. Thus, studies on associations with periodontitis need to take this age range into consideration.

- A major gap in knowledge regarding periodontitis and its associations is understanding the transition between adolescence and adulthood. Limited evidence suggests that exposures such as socioeconomic status may have its major impact during early life. However, the focus of current research on associations is mostly confined to adulthood.

- The use of life course approach is useful for understanding the pathways of associations with periodontitis, the nature of their impacts, their latency period, their impacts on health behaviors, and biology. However, since feasibility is a major limitation for such study designs, evidence from cross-sectional studies using wide age ranges may help in generating hypotheses and identifying the most appropriate assessment periods.

- The use of ecosocial theory by Krieger can be useful for generating hypotheses for mapping pathways of associations with periodontitis while considering the broader context over the life course [54].

- When studying the psychosocial exposures, it is important to consider both of their protective and harmful impacts.

- The periodontal treatment status needs to be considered in studies on associations with periodontitis. Consequently, the impacts and limitations of current periodontal treatment strategies can be evaluated at population-level.

- A multilevel periodontitis prevention framework based on common risk factor approach was suggested as an alternative to current approaches, which are limited to individual patient care.

Author details

Yasmine N. Alawaji[1,2,3]

1 Preventive Dental Science Department, College of Dentistry, King Saud bin Abdul-Aziz University for Health Sciences, Riyadh, Saudi Arabia

2 King Abdullah International Medical Research Center, Riyadh, Saudi Arabia

3 University of British Columbia, Vancouver, Canada

*Address all correspondence to: alawajiy@alumni.ubc.ca

References

[1] Suvan JE, Sabalic M, Araújo MR, Ramseier CA. Behavioral strategies for periodontal health. Periodontology. 2000;**90**(1):247-261

[2] Glanz K, Rimer BK, Viswanath K. In: Glanz K, Rimer BK, Viswanath K, editors. Health Behavior: Theory, Research, and Practice. 5th ed. San Fransisco (CA): John Wiley & Sons, Incorporated; 2015

[3] Okechukwu C, Davison K, Emmons K. Changing health behaviors in a social context. In: Berkman L, Kawachi I, Glamour MM, editors. Social Epidemiology. 2nd ed. New York (NY): Oxford University Press; 2014. pp. 365-395

[4] Adler KG. Screening for social determinants of health: An opportunity or unreasonable burden? Family Practice Management. 2018;**25**(3):3

[5] Magnan S. Social determinants of health 101 for health care: Five plus five. In: National Academy of Medicine Perspectives. Washington DC: National Academy of Medicine; 2017. pp. 1-9

[6] Borland R, Young D, Coghill K, Zhang JY. The tobacco use management system: Analyzing tobacco control from a systems perspective. American Journal of Public Health. 2010;**100**(7):1229-1236

[7] Ratcliff KS. Social determinants of health. In: The Social Determinants of Health: Looking Upstream. Cambridge: Polity Press; 2017. pp. 14-27

[8] World Health Organization. Social Determinants of Health. WHO; 2022

[9] Braveman P, Gottlieb L. The social determinants of health: It's time to consider the causes of the causes. Public Health Reports. 2014;**129**(Suppl. 2):19-31

[10] Rose G. Sick individuals and sick populations. International Journal of Epidemiology. 1985;**14**(1):32-38

[11] Alawaji YN. Periodontal Disease Epidemiology: Assessment of Untreated Periodontitis and Accuracy of Partial Mouth Recording Protocols. Vancouver (BC): The University of British Columbia; 2022

[12] Burt BA. The distribution of periodontal destruction in the populations of industrialized countries. In: Johnson NW, editor. Periodontal Diseases: Markers of Disease Susceptibility and Activity. 1st ed. New York (NY): Press Syndicate of the University of Cambridge; 1991. pp. 9-26

[13] Burt BA. The role of epidemiology in the study of periodontal diseases. Periodontology. 2000;**2**(1):26-33

[14] Russell AL. A system of classification and scoring for prevalence surveys of periodontal disease. Journal of Dental Research. 1956;**35**(3):350-359

[15] Papapanou P, Lindhe J. Epidemiology of periodontal diseases. In: Lang NP, Lindhe J, Berglundh T, Giannobile WV, Mariano VS, editors. Clinical Periodontology and Implant Dentistry. 6th ed. Chichester: Wiley-Blackwell; 2015. pp. 125-156

[16] Beltran-Aguilar ED, Eke PI, Thornton-Evans G, Petersen PE. Recording and surveillance systems for periodontal diseases. Periodontology 2000. 2000;**60**(1):40-53

[17] Loe H, Boysen H, Smith M. The natural history of periodontal disease in man. Study design and baseline data. Journal of Periodontal Research. 1978;**13**(6):550-562

[18] Timmerman MF, van der Weijden GA, Armand S, Abbas F, Winkel EG, van Winkelhoff AJ, et al. Untreated periodontal disease in Indonesian adolescents. Clinical and microbiological baseline data. Journal of Clinical Periodontology. 1998;**25**(3):215-224

[19] Haffajee AD, Socransky SS, Goodson JM. Periodontal disease activity. Journal of Periodontal Research. 1982;**17**(5):521-522

[20] Socransky SS, Haffajee AD, Goodson JM, Lindhe J. New concepts of destructive periodontal disease. Journal of Clinical Periodontology. 1984;**11**(1):21-32

[21] Goodson JM, Tanner AC, Haffajee AD, Sornberger GC, Socransky SS. Patterns of progression and regression of advanced destructive periodontal disease. Journal of Clinical Periodontology. 1982;**9**(6):472-481

[22] Cutress TW, Powell RN, Ball ME. Differing profiles of periodontal disease in two similar South Pacific island populations. Community Dentistry and Oral Epidemiology. 1982;**10**(4):193-203

[23] Johnson NW, Griffiths GS, Wilton JM, Maiden MF, Curtis MA, Gillett IR, et al. Detection of high-risk groups and individuals for periodontal diseases. Evidence for the existence of high-risk groups and individuals and approaches to their detection. Journal of Clinical Periodontology. 1988;**15**(5):276-282

[24] Beck J. Methods of assessing risk for periodontitis and developing multifactorial models. Journal of Periodontology. 1994;**65**(Suppl. 5):468-478

[25] Grossi SG, Zambon JJ, Ho AW, Koch G, Dunford RG, Machtei EE, et al. Assessment of risk for periodontal disease. I. Risk indicators for attachment loss. Journal of Periodontology. 1994;**65**(3):260-267

[26] Holtfreter B, Albandar JM, Dietrich T, Dye BA, Eaton KA, Eke PI, et al. Standards for reporting chronic periodontitis prevalence and severity in epidemiologic studies: Proposed standards from the Joint EU/USA Periodontal Epidemiology Working Group. Journal of Clinical Periodontology. 2015;**42**(5):407-412

[27] Alawaji Y, Alshammari A, Aleksejuniene J. Accuracy of estimating periodontitis and its risk association using partial-mouth recordings for surveillance studies: A systematic review and meta-analysis. International Journal of Dental. 2022;**17**:1-16

[28] Needleman I, Garcia R, Gkranias N, Kirkwood K, Kocher T, Iorio AD, et al. Mean annual attachment, bone level, and tooth loss: A systematic review. Journal of Periodontology. 2018;**89**(Suppl. 1): S120-S139

[29] Baelum V, Lopez R. Periodontal disease epidemiology - learned and unlearned? Periodontology. 2000;**62**(1):37-58

[30] Russell A. International nutrition surveys: A summary of preliminary dental findings. Journal of Dental Research. 1963;**42**(1):233-244

[31] Löe H, Theilade E, Jensen SB. Experimental gingivitis in man. Journal of Periodontology. 1965;**36**(3):177-187

[32] Hirschfeld L, Wasserman B. A long-term survey of tooth loss in 600 treated periodontal patients. Journal of Periodontology. 1978;**49**(5):225-237

[33] Lindhe J, Nyman S. The effect of plaque control and surgical pocket elimination on the establishment and maintenance of periodontal health. A longitudinal study of periodontal therapy in cases of advanced disease. Journal of Clinical Periodontology. 1975;**2**(2):67-79

[34] Rosling B, Nyman S, Lindhe J, Jern B. The healing potential of the periodontal tissues following different techniques of periodontal surgery in plaque-free dentitions. A 2-year clinical study. Journal of Clinical Periodontology. 1976;**3**(4):233-250

[35] Löe H, Anerud A, Boysen H, Morrison E. Natural history of periodontal disease in man. Rapid, moderate and no loss of attachment in Sri Lankan laborers 14 to 46 years of age. Journal of Clinical Periodontology. 1986;**13**(5):431-445

[36] Loe H, Anerud A, Boysen H, Smith M. The natural history of periodontal disease in man: Tooth mortality rates before 40 years of age. Journal of Periodontal Research. 1978;**13**(6):563-572

[37] Timmerman MF, van der Weijden GA, Abbas F, Arief EM, Armand S, Winkel EG, et al. Untreated periodontal disease in Indonesian adolescents. Longitudinal clinical data and prospective clinical and microbiological risk assessment. Journal of Clinical Periodontology. 2000;**27**(12):932-942

[38] Baelum V, Chen X, Manji F, Luan WM, Fejerskov O. Profiles of destructive periodontal disease in different populations. Journal of Periodontal Research. 1996;**31**(1):17-26

[39] Goodson J. Clinical measurements of periodontitis. Journal of Clinical Periodontology. 1986;**13**(5):446-455

[40] Bartold PM, van Dyke TE. An appraisal of the role of specific bacteria in the initial pathogenesis of periodontitis. Journal of Clinical Periodontology. 2019 Jan;**46**(1):6-11

[41] Kumar PS. Microbial dysbiosis: The root cause of periodontal disease. Journal of Periodontology. 2021;**92**(8):1079-1087

[42] Hujoel P, Kotsakis G. Fundamentals in the methods of periodontal disease epidemiology. In: Carranza F, Freire M, Jepsen S, Klokkevold PR, Newman MG, Preshaw P, et al., editors. Newman and Carranza's Clinical Periodontology. 13th ed. Philadelphia (PA): Elsevier; 2019. pp. 80-88

[43] Bouchard P, Carra MC, Boillot A, Mora F, Rangé H. Risk factors in periodontology: A conceptual framework. Journal of Clinical Periodontology. 2017;**44**(2):125-131

[44] Heaton B, Dietrich T. Causal theory and the etiology of periodontal diseases. Periodontology 2000. 2000;**58**(1):26-36

[45] Rothman KJ. Causes. American Journal of Epidemiology. 1976;**104**(6): 587-592

[46] Sanz M, Marco del CA, Jepsen S, et al. Periodontitis and cardiovascular diseases: Consensus report. Journal of Clinical Periodontology. 2020;**47**(3):268-288

[47] Graves DT, Ding Z, Yang Y. The impact of diabetes on periodontal diseases. Periodontology. 2000;**82**(1):214-224

[48] Ganesan SM, Vazana S, Stuhr S. Waistline to the gumline: Relationship between obesity and periodontal disease-biological and management considerations. Periodontology. 2000;**87**(1):299-314

[49] Sczepanik FS, Grossi ML, Casati M, Goldberg M, Glogauer M, Fine N, et al. Periodontitis is an inflammatory disease of oxidative stress: We should treat it that way. Periodontology. 2000;**84**(1):45-68

[50] Sheiham A, Watt RG. The common risk factor approach: A rational basis for promoting oral health. Community Dentistry and Oral Epidemiology. 2000;**28**(6):399-406

[51] Janakiram C, Dye BA. A public health approach for prevention of periodontal disease. Periodontology. 2000;**84**(1):202-214

[52] VanderWeele TJ, Rothman KJ. Formal causal models. In: Zinner S, editor. Modern Epidemiology. 4th ed. Philadelphia (PA): Wolters Kluwer; 2021. pp. 32-51

[53] Krieger N. Theories for social epidemiology in the 21st century: An ecosocial perspective. International Journal of Epidemiology. 2001;**30**(4):668-677

[54] Krieger N. Ecosocial Theory, Embodied Truths, and the People's Health. 1st ed. New York (NY): Oxford University Press; 2021

[55] Ben-Shlomo Y, Cooper R, Kuh D. The last two decades of life course epidemiology, and its relevance for research on ageing. International Journal of Epidemiology. 2016;**45**(4):973-988

[56] Jones NL, Gilman SE, Cheng TL, Drury SS, Hill CV, Geronimus AT. Life course approaches to the causes of health disparities. American Journal of Public Health. 2019;**109**(S1):S48-S55

[57] Baird J, Jacob C, Barker M, Fall C, Hanson M, Harvey N, et al. Developmental origins of health and disease: A lifecourse approach to the prevention of non-communicable diseases. Healthcare. 2017;**5**(1):14

[58] Jewell NP. Natural history of diseases: Statistical designs and issues. Clinical Pharmacology and Therapeutics. 2016;**100**(4):353-361

[59] U.S. Department of Health and Human Services. Natural history and spectrum of disease. In: Principles of Epidemiology in Public Health Practice. 3rd ed. Atlanta, GA: Centers for Disease Control and Prevention (CDC); 2012. pp. 1-59

[60] Alawaji Y, Alshammari A, Mostafa N, Carvalho RM, Aleksejuniene J. Periodontal disease prevalence, extent, and risk associations in untreated individuals. Clinical Experiment in Dental Research. 2022;**8**(1):380-394

[61] Eke PI, Wei L, Thornton-Evans GO, Borrell LN, Borgnakke WS, Dye B, et al. Risk indicators for periodontitis in US adults: NHANES 2009 to 2012. Journal of Periodontology. 2016;**87**(10):1174-1185

[62] Susin C, Haas AN, Valle PM, Oppermann RV, Albandar JM. Prevalence and risk indicators for chronic periodontitis in adolescents and young adults in South Brazil. Journal of Clinical Periodontology. 2011 Apr;**38**(4):326-333

[63] Susin C, Kingman A, Albandar JM. Effect of partial recording protocols on estimates of prevalence of periodontal disease. Journal of Periodontology. 2005;**76**(2):262-267

[64] Billings M, Holtfreter B, Papapanou PN, Mitnik GL, Kocher T, Dye BA. Age-dependent distribution of periodontitis in two countries: Findings from NHANES 2009 to 2014 and SHIP-TREND 2008 to 2012. Journal of Periodontology. 2018;**45**(Suppl. 20): S130-S148

[65] Dowsett SA, Eckert GJ, Kowolik MJ. Comparison of periodontal disease status of adults in two untreated indigenous populations of Guatemala, Central America. Journal of Clinical Periodontology. 2002;**29**(8):784-787

[66] Albandar JM. Periodontal diseases in North America. Periodontology 2000. 2000;**29**:31-69

[67] Kassebaum NJ, Bernabé E, Dahiya M, Bhandari B, Murray CJ, Marcenes W. Global burden of severe periodontitis in 1990-2010. Journal of Dental Research. 2014;**93**(11):1045-1053

[68] Hugoson A, Norderyd O, Slotte C, Thorstensson H. Distribution of periodontal disease in a Swedish adult population 1973, 1983 and 1993. Journal of Clinical Periodontology. 1998;**25**(7):542-548

[69] Lopez R, Fernandez O, Jara G, Baelum V. Epidemiology of clinical attachment loss in adolescents. Journal of Periodontology. 2001;**72**(12):1666-1674

[70] Morales A, Carvajal P, Romanelli H, Gómez M, Loha C, Esper ME, et al. Prevalence and predictors for clinical attachment loss in adolescents in Latin America: Cross-sectional study. Journal of Clinical Periodontology. 2015;**42**(10):900-907

[71] Thomson WM, Shearer DM, Broadbent JM, Foster Page LA, Poulton R. The natural history of periodontal attachment loss during the third and fourth decades of life. Journal of Clinical Periodontology. 2013;**40**(7):672-680

[72] Frencken JE, Sharma P, Stenhouse L, Green D, Laverty D, Dietrich T. Global epidemiology of dental caries and severe periodontitis - a comprehensive review. Journal of Clinical Periodontology. 2017;**44**(Suppl. 18):S94-S105

[73] Eke PI, Borgnakke WS, Genco RJ. Recent epidemiologic trends in periodontitis in the USA. Periodontology 2000. 2000;**82**(1):257-267

[74] Dye BA. Global periodontal disease epidemiology: Global periodontal disease epidemiology. Periodontology 2000. 2000;**58**(1):10-25

[75] Tonetti MS, Greenwell H, Kornman KS. Staging and grading of periodontitis: Framework and proposal of a new classification and case definition. Journal of Periodontology. 2018;**89**(Suppl. 1):S159-S172

[76] American Academy of Periodontology. American Academy of Periodontology task force report on the update to the 1999 classification of periodontal diseases and conditions. Journal of Periodontology. 2015;**86**(7):835-838

[77] Armitage GC. Development of a classification system for periodontal diseases and conditions. Annals of Periodontology. 1999;**4**(1):1-6

[78] Caton J. Periodontal diagnosis and diagnostic aids. In: Nevins M, Becker W, Kornman K, editors. Proceedings of the World Workshop in Clinical Periodontics July 23-27, 1989 Princeton, New Jersey. Chicago (IL): American Academy of Periodontology; 1989. pp. 1-32

[79] Löe H, Anerud A, Boysen H, Smith M. The natural history of periodontal disease in man. The rate of periodontal destruction before 40 years of age. Journal of Periodontology. 1978;**49**(12):607-620

[80] Burt B, Research S and TC of the AA of P. Position paper: Epidemiology of periodontal diseases. Journal of Periodontology. 2005;**76**(8):1406-1419

[81] Susin C, Dalla Vecchia CF, Oppermann RV, Haugejorden O, Albandar JM. Periodontal attachment loss in an urban population of Brazilian adults: Effect of demographic, behavioral, and environmental risk indicators. Journal of Periodontology. 2004;**75**(7):1033-1041

[82] Haas AN, Wagner MC, Oppermann RV, Rosing CK, Albandar JM,

Susin C. Risk factors for the progression of periodontal attachment loss: A 5-year population-based study in South Brazil. Journal of Clinical Periodontology. 2014;**41**(3):215-223

[83] Neely AL, Holford TR, Löe H, Ånerud A, Boysen H. The natural history of periodontal disease in man. Risk factors for progression of attachment loss in individuals receiving no oral health care. Journal of Periodontology. 2001 Aug;**72**(8):1006-1015

[84] Glymour MM, Avendano M, Kawachi I. Socioeconomic status and health. In: Berkman LF, Kawachi I, Glamour MM, editors. Social Epidemiology. 2nd ed. New York (NY): Oxford University Press; 2014. pp. 17-62

[85] Shiau HJ, Reynolds MA. Sex differences in destructive periodontal disease: Exploring the biologic basis. Journal of Periodontology. 2010;**81**(11):1505-1517

[86] Baelum V, Luan WM, Chen X, Fejerskov O. Predictors of destructive periodontal disease incidence and progression in adult and elderly Chinese. Community Dentistry and Oral Epidemiology. 1997;**25**(4):265-272

[87] van der Velden U, Abbas F, Armand S, Loos BG, Timmerman MF, van der Weijden GA, et al. Java project on periodontal diseases. The natural development of periodontitis: Risk factors, risk predictors and risk determinants. Journal of Clinical Periodontology. 2006;**33**(8):540-548

[88] Borrell LN, Papapanou PN. Analytical epidemiology of periodontitis. Journal of Clinical Periodontology. 2005;**32**(Suppl. 6):132-158

[89] Bouchard P, Boutouyrie P, Mattout C, Bourgeois D. Risk assessment for severe clinical attachment loss in an adult population. Journal of Periodontology. 2006;**77**(3):479-489

[90] Hyman JJ, Reid BC. Epidemiologic risk factors for periodontal attachment loss among adults in the United States. Journal of Clinical Periodontology. 2003;**30**(3):230-237

[91] Deeks JJ, Higgins JP, Altman DG, on behalf of the Cochrane Statistical Methods Group. Analysing data and undertaking meta-analyses. In: Higgins JP, Thomas J, Chandler J, Cumpston M, Li T, Page MJ, et al., editors. Cochrane Handbook for Systematic Reviews of Interventions. 2nd ed. Chichester (UK): John Wiley & Sons; 2021

[92] Page RC, Eke PI. Case definitions for use in population-based surveillance of periodontitis. Journal of Periodontology. 2007;**78**(Suppl. 7):1387-1399

[93] Eke PI, Page RC, Wei L, Thornton-Evans G, Genco RJ. Update of the case definitions for population-based surveillance of periodontitis. Journal of Periodontology. 2012;**83**(12):1449-1454

[94] Borrell LN, Burt BA, Gillespie BW, Lynch J, Neighbors H. Periodontitis in the United States: Beyond black and white. Journal of Public Health Dentistry. 2002;**62**(2):92-101

[95] Borrell LN, Crawford ND. Social disparities in periodontitis among United States adults 1999-2004. Community Dentistry and Oral Epidemiology. 2008;**36**(5):383-391

[96] Zambrana RE, Carter-Pokras O. Role of acculturation research in advancing science and practice in reducing health care disparities among Latinos. American Journal of Public Health. 2010;**100**(1):18-23

[97] Ryan W. Blaming the Victim. 1st ed. New York (NY): MIT Libraries; 1971

[98] Boillot A, Halabi BE, Batty GD, Rangé H, Czernichow S, Bouchard P. Education as a predictor of chronic periodontitis: A systematic review with meta-analysis population-based studies. PLoS One. 2011;**6**(7):e21508

[99] Borrell LN, Crawford ND. Socioeconomic position indicators and periodontitis: Examining the evidence. Periodontology. 2000;**58**(1):69-83

[100] Borrell LN, Burt BA, Warren RC, Neighbors HW. The role of individual and neighborhood social factors on periodontitis: The third National Health and Nutrition Examination Survey. Journal of Periodontology. 2006;**77**(3):444-453

[101] Stringhini S, Sabia S, Shipley M, Brunner E, Nabi H, Kivimaki M, et al. Association of socioeconomic position with health behaviors and mortality. Journal of the American Medical Association. 2010;**303**(12):1159-1166

[102] Schuch HS, Peres KG, Singh A, Peres MA, do LG. Socioeconomic position during life and periodontitis in adulthood: A systematic review. Community Dentistry and Oral Epidemiology. 2017;**45**(3):201-208

[103] Lu HX, Wong MCM, et al. Trends in oral health from childhood to early adulthood: A life course approach. Community Dentistry and Oral Epidemiology. 2011;**39**(4):352-360

[104] Poulton R, Caspi A, Milne BJ, Thomson WM, Taylor A, Sears MR, et al. Association between children's experience of socioeconomic disadvantage and adult health: A life-course study. The Lancet. 2002;**360**(9346):1640-1645

[105] Schuch HS, Nascimento GG, Peres KG, Mittinty MN, Demarco FF, Correa MB, et al. The controlled direct effect of early-life socioeconomic position on periodontitis in a birth cohort. American Journal of Epidemiology. 2019;**188**(6):1101-1108

[106] Cohen S, Hoberman HM. Positive events and social supports as buffers of life change stress. Journal of Applied Social Psychology. 1983;**13**(2):99-125

[107] Kubzansky LD, Winning A, Kawachi I. Affective states and health. In: Berkman LF, Kawachi I, Glymour MM, editors. Social Epidemiology. 2nd ed. New York (NY): Oxford University Press; 2014. pp. 320-364

[108] McEwen BS. Protective and damaging effects of stress mediators. New England Journal of Medicine. 1998;**338**(3):171-179

[109] Genco RJ, Grossi SG, Ho A, Nishimura F, Murayama Y. A proposed model linking inflammation to obesity, diabetes, and periodontal infections. Journal of Periodontology. 2005;**76**(Suppl. 11):2075-2084

[110] Kubzansky LD, Seeman TE, Glymour MM. Biological pathways linking social conditions and health. In: Berkman LF, Kawachi I, Glamour MM, editors. Social Epidemiology. 2nd ed. New York (NY): Oxford University Press; 2014. pp. 512-561

[111] da Silva AM, Newman HN, Oakley DA. Psychosocial factors in inflammatory periodontal diseases. A review. Journal of Clinical Periodontology. 1995;**22**(7):516-526

[112] Tanveer SA, Afaq A, Alqutub MN, Aldahiyan N, Almubarak AM, Shaikh AC, et al. Association of self-perceived psychological stress with the periodontal health of socially deprived women in shelter homes. International Journal

of Environmental Research and Public Health. 2021;**18**(10):2-8

[113] Genco RJ, Ho AW, Grossi SG, Dunford RG, Tedesco LA. Relationship of stress, distress, and inadequate coping behaviors to periodontal disease. Journal of Periodontology. 1999;**70**(7):711-723

[114] Merchant AT, Pitiphat W, Ahmed B, Kawachi I, Joshipura K. A prospective study of social support, anger expression and risk of periodontitis in men. The Journal of American Dental Association. 2003;**134**(12):1591-1596

[115] Aleksejunñiené J, Holst D, Eriksen HM, Gjermo P. Psychosocial stress, lifestyle and periodontal health. Journal of Clinical Periodontology. 2002;**29**(4):326-335

[116] Baker JP, Berenbaum H. Emotional approach and problem-focused coping: A comparison of potentially adaptive strategies. Cognitive Emotional. 2007;**21**(1):95-118

[117] Genco RJ, Ho AW, Kopman J, Grossi SG, Dunford RG, Tedesco LA. Models to evaluate the role of stress in periodontal disease. Annals of Periodontology. 1998;**3**(1):288-302

[118] Deinzer R, Hilpert D, Bach K, Schawacht M, Herforth A. Effects of academic stress on oral hygiene: A potential link between stress and plaque-associated disease? Journal of Clinical Periodontology. 2001;**28**(5):459-464

[119] Cohen S, Wills TA. Stress, social support, and the buffering hypothesis. Psychological Bulletin. 1985;**98**(2):310-357

[120] Sabbah W, Tsakos G, Chandola T, Newton T, Kawachi I, Sheiham A, et al. The relationship between social network, social support, and periodontal disease among older Americans. Journal of Clinical Periodontology. 2011;**38**(6):547-552

[121] Zini A, Sgan-Cohen HD, Marcenes W. Is religiosity related to periodontal health among the adult Jewish population in Jerusalem? Journal of Periodontal Research. 2012;**47**(4):418-425

[122] Vered Y, Soskolne V, Zini A, Livny A, Sgan-Cohen HD. Psychological distress and social support are determinants of changing oral health status among an immigrant population from Ethiopia. Community Dentistry and Oral Epidemiology. 2011;**39**(2):145-153

[123] Berkman LF, Krishna A. Social network epidemiology. In: Berkman LF, Kawachi I, Glymour MM, editors. Social Epidemiology. 2nd ed. New York (NY): Oxford University Press; 2014. pp. 234-289

[124] Sheiham A, Nicolau B. Evaluation of social and psychological factors in periodontal disease. Periodontology. 2000;**39**(1):118-131

[125] Dahlan R, Ghazal E, Saltaji H, Salami B, Amin M. Impact of social support on oral health among immigrants and ethnic minorities: A systematic review. PLoS One. 2019;**14**(6):1-21

[126] Office of the Surgeon General (US); Office on Smoking and Health (US). The Health Consequences of Smoking: A Report of the Surgeon General. Atlanta (GA): Centers for Disease Control and Prevention (US); 2004

[127] Tomar SL, Asma S. Smoking-attributable periodontitis in the United States: Findings from NHANES III. Journal of Periodontology. 2000;**71**(5):743-751

[128] Leite FR, Nascimento GG, Scheutz F, López R. Effect of smoking on periodontitis: A systematic review and meta-regression. American Journal of Preventive Medicine. 2018;**54**:831-841

[129] Grossi SG, Skrepcinski FB, DeCaro T, Robertson DC, Ho AW, Dunford RG, et al. Treatment of periodontal disease in diabetics reduces glycated hemoglobin. Journal of Periodontology. 1997;**68**(8):713-719

[130] Patel RA, Wilson RF, Palmer RM. The effect of smoking on periodontal bone regeneration: A systematic review and meta-analysis. Journal of Periodontology. 2012;**83**(2):143-155

[131] Javed F, Al-Rasheed A, Al-Hezaimi K, Almas K, Romanos GE. Effect of cigarette smoking on the clinical outcomes of periodontal surgical procedures. The American Journal of the Medical Sciences. 2012;**343**(1):78-84

[132] Zimmermann H, Zimmermann N, Hagenfeld D, Veile A, Kim T, Becher H. Is frequency of tooth brushing a risk factor for periodontitis? A systematic review and meta-analysis. Community Dentistry and Oral Epidemiology. 2015;**43**(2):116-127

[133] Machado V, Botelho J, Proença L, Alves R, Oliveira MJ, Amaro L, et al. Periodontal status, perceived stress, diabetes mellitus and oral hygiene care on quality of life: A structural equation modelling analysis. BMC Oral Health. 2020;**20**(1):229

[134] Merchant A, Pitiphat W, Douglass CW, Crohin C, Joshipura K. Oral hygiene practices and periodontitis in health care professionals. Journal of Periodontology. 2002;**73**(5):531-535

[135] Institute of Medicine (US) Committee on Monitoring Access to Personal Health Care Services. A model for monitoring access. In: Millman M, editor. Access to Health Care in America. Washington (DC): National Academies Press (US); 2022

[136] Aday LA, Andersen RM. Equity of access to medical care: A conceptual and empirical overview. Medical Care. 1981;**19**(12):4-27

[137] Andersen RM. Revisiting the behavioral model and access to medical care: Does it matter? Journal of Health and Social Behavior. 1995;**36**(1):1-10

[138] Saurman E. Improving access: Modifying Penchansky and Thomas's theory of access. Journal of Health Services Research & Policy. 2016;**21**(1):36-39

[139] McKernan S, Reynolds J, Mcquistan M. Access to dental care. In: Mascarenhas A, Okunseri C, Dye B, editors. Burt and Eklund's Dentistry, Dental Practice, and the Community. 7th ed. St. Louis (MO): Elsevier; 2020. pp. 95-124

[140] Penchansky R, Thomas JW. The concept of access. Medical Care. 1981;**19**(2):127-140

[141] Celentano DD, Szklo M. Introduction. In: Celentano DD, Szklo M, editors. Gordis Epidemiology. 6th ed. Philadelphia (PA): Elsevier; 2019

[142] Ramseier CA, Woelber JP, Kitzmann J, Detzen L, Carra MC, Bouchard P. Impact of risk factor control interventions for smoking cessation and promotion of healthy lifestyles in patients with periodontitis: A systematic review. Journal of Clinical Periodontology. 2020;**47**(Suppl. 22):90-106

[143] Botelho J, Machado V, Leira Y, Proença L, Chambrone L, Mendes JJ.

Economic burden of periodontitis in the United States and Europe: An updated estimation. Journal of Periodontology. 2022;**93**(3):373

[144] Preshaw PM. Detection and diagnosis of periodontal conditions amenable to prevention. BMC Oral Health. 2015;**15**(Suppl. 1):S1-S5

[145] Trombelli L, Franceschetti G, Farina R. Effect of professional mechanical plaque removal performed on a long-term, routine basis in the secondary prevention of periodontitis: A systematic review. Journal of Clinical Periodontology. 2015;**42**(Suppl. 16):S221-S236

[146] Tonetti MS, Eickholz P, Loos BG, Papapanou P, van der Velden U, Armitage G, et al. Principles in prevention of periodontal diseases: Consensus report of group 1 of the 11th European Workshop on Periodontology on effective prevention of periodontal and peri-implant diseases. Journal of Clinical Periodontology. 2015;**42**(Suppl. 16):S5-S11

[147] Caton JG, Armitage G, Berglundh T, Chapple IL, Jepsen S, Kornman KS, et al. A new classification scheme for periodontal and peri-implant diseases and conditions - introduction and key changes from the 1999 classification. Journal of Periodontology. 2018;**45**(Suppl. 20):S1-S8

[148] Kisling LA, das JM. Prevention Strategies [Internet]. StatPearls. 2022. Available from: https://www.ncbi.nlm.nih.gov/books/NBK537222/ [cited 2022 Apr 21]

[149] The Association of Faculties of Medicine of Canada. Basic concepts in prevention and health promotion. In: Donovan D, McDowell I, editors. AFMC Primer on Population Health [Internet]. 2nd ed. AFMC; 2018. Available from: https://phprimer.afmc.ca/en/ [cited 2022 Apr 21]

[150] Tonetti MS, Claffy N, European Workshop in Periodontology Group C. Advances in the progression of periodontitis and proposal of definitions of a periodontitis case and disease progression for use in risk factor research. Group C consensus report of the 5th European Workshop in Periodontology. Journal of Clinical Periodontology. 2005;**32**(Suppl. 6): 210-213

[151] Lang NP, Nyman S, Senn C, Joss A. Bleeding on probing as it relates to probing pressure and gingival health. Journal of Clinical Periodontology. 1991;**18**(4):257-261

[152] Lang NP, Bartold PM. Periodontal health. Journal of Periodontology. 2018;**89**(Suppl. 1):S9-S16

[153] Chapple IL, Mealey BL, van Dyke TE, Bartold PM, Dommisch H, Eickholz P, et al. Periodontal health and gingival diseases and conditions on an intact and a reduced periodontium: Consensus report of workgroup 1 of the 2017 world workshop on the classification of periodontal and peri-implant diseases and conditions. Journal of Periodontology. 2018;**89**(Suppl. 1):S74-S84

[154] Trombelli L, Farina R, Silva CO, Tatakis DN. Plaque-induced gingivitis: Case definition and diagnostic considerations. Journal of Periodontology. 2018;**45**(Suppl. 20): S44-S67

[155] Sreenivasan PK, Prasad KV, Javali SB. Oral health practices and prevalence of dental plaque and gingivitis among Indian adults. Clinical Expert in Dental Research. 2016;**2**(1):6-17

[156] Axelsson P, Nystrom B, Lindhe J. The long-term effect of a plaque control program on tooth mortality, caries and periodontal disease in adults. Journal of Clinical Periodontology. 2004;**31**(9):749-457

[157] Ramfjord SP, Morrison EC, Burgett FG, Nissle RR, Shick RA, Zann GJ, et al. Oral hygiene and maintenance of periodontal support. Journal of Periodontology. 1982;**53**(1):26-30

Section 3

Prosthetics and Implantology

Chapter 9

Development of Gallium Silicon Titanium Alloys for Dental Implants

Tharanikumar Sivakumar, Chandrasekaran Krithika and Nandikha Tharanikumar

Abstract

The aim of this study is to synthesize a new metallic aluminum and vanadium-free titanium alloy biomaterial for better osseointegration and implantation in the physiological system. The *in vitro* and *in vivo* methods were used to examine their biological compatibility, evaluated quantitatively and qualitatively. Results of Ga-Si-Ti alloy showed a higher ultimate tensile strength, yield strength and a higher percentage of elongation and more or less equal to Young's modulus when compared with the Ti and Ti-Ga alloy. *In vivo* study, a PA view of whole-body radiography all groups exhibited a substantial difference in the linear bone density of newly formed bone. Ga-Si-Ti group showed the highest bone mineral density than Ti and Ti-Ga group in the micro CT *ex vivo* study. The study exhibited a significant difference between the groups and the proportion of cortical bone volume to trabecular bone volume BV/TV in percentage. This is related to the anti-resorptive action of gallium and osteoblastic property of silicon, in addition to the benefits of commercial pure-Ti alloy.

Keywords: gallium, silicon titanium, osteoporosis, aluminum, vanadium, toxicity of Ti-Al-V micro CT, Histomorphometry

1. Introduction

1.1 Background

While the pride of humankind exists in a manner that helps different lives, having scientifically achieved the capability to replace the lost body parts along with their functions is considered to be the upfront of modern scientific civilization. The impact of the missing body parts and their dependence results in physical and psychosocial distress. The best assistance one can offer is to diminish, if not eliminate such an inability. In this direction, loss of teeth due to injury or any other pathology may lead to partial or complete edentulousness, that affects the psychosocial status and the functional mastication for which dental implants are preferred.

Study by the AAOMS reveals a 69% incidence of one tooth edentulous in the age group of 35–44 years [1]. Also, when people reach the age of 75 years, a minimum of one-fourth of the adult population would be completely edentulous. Rise, but the quality of life is not much improved. For instance, among the overall population, the percentage belonging to 65 years and above is on the rise. The total count of this category population in the year 2000 was 282 million and is expected to rise by 49%, rise to 420 million by 2050. Overall, the impact of an increase in population and a higher chance of that population being considered to be more than the age 65 leads to a considerable rise in the actual patient count, considering that 35 million people were older than age 65 in the year 2003. This number is required to increment by 87% by 2025. Hence, the increased need for dental implant treatments is due to the failures of a fixed prosthesis and the consequences, also the life longevity of the aging population and their dental needs.

Oral implantology has evolved and improved to be a centre of the art and technology of modern dentistry. The field demands its practitioners to have a distinguished knowledge of details in significant areas, including scientific updates, updating knowledge about different types of radiographs including regular radiographs, computed tomographic, Micro-CT, robust working knowledge about the anatomy, surgical procedures, prosthetic requirement and follow-up care. Every dentist who places or restores implants ought to be aware of the possible problems and should be able to know how to control them.

Oral implantology is a challenging field to master and the impact of failure may be a disastrous one. The dentist should know the material science and have awareness of how to prevent the failure of implants. Among various biomaterials, metallic biomaterials are used mainly and in various forms. The metals like Gold, Palladium and other noble metal alloys have the longest history of utilization. However, their high cost limits their applications. At present, the usable metallic materials in the medical science are Co-Cr-Mo, 316 L SS, Titanium-based alloys and miscellaneous alloys like amalgam and Gold. Titanium material resists corrosion, has strength to weight ratio, weight less and has good mechanical properties. Its biocompatibility, non-toxic, durable, easily available and cost-efficient makes Titanium, the choice of material for many uses in dental application

The mechanical strength and biocompatibility of Titanium make it the metal of choice for dental implants. Titanium with the mesoporous layer of Gallium and Silicon can act as a drug delivery system for enhancing osseointegration thereby expanding the clinical application of Ti alloy. Si is a potent anabolic element that promotes bone formation and Gallium inhibits bone osteolysis, review of literatures suggest verifying animal studies have shown that the synergistic impact of Ga and Si on promoting osseointegration can be applied to clinical trials.

The aim of this study is To Synthesize a new metallic Aluminum and Vanadium free Titanium alloy biomaterial for better osseointegration and implantation in the physiological system.

To Synthesize Gallium Silicon Titanium Alloy biomaterial.

To Characterize the mechanical and structural properties for use in dental implant application.

To Characterize the alloy for biocompatibility in the physiological system.

Advantage of gallium silicon titanium alloy implants to the population

- Gallium metal has the antiresorptive property and prevents the osteoclastic activity when implants placed in bone

- Silicon metal has the anabolic property and promotes the osteoblastic activity when implants placed in bone

- Titanium metal has high inert to the tissue and the metal density which is to replace missing hard tissue.
- Synergistic effect of Gallium silicon Titanium alloy would be having the inert tissue response with osteoclastic and promotes osteoblastic activity in bone promotes osseointegration of implant to bone in mankind with estrogen deficient females in old age and osteoporosis patient

2. Materials and methods

2.1 Synthesization of titanium, titanium gallium and gallium silicon titanium alloys

Code-263265-Gallium: 99.99% with molecular weight 69.72G/MOL, Code-343250-Silicon: 99.95% with molecular weight 28.09G/MOL and Code-GF96834493-1EA-Titanium: 99.6% were purchased from Aldrich and used as such with no further purification.

Titanium metal pieces are used as such.

Titanium Gallium alloy metal pieces were weighed as per the alloy composition described by [2] presented in **Table 1**.

Gallium Silicon Titanium alloy metal pieces are weighed and prepared as per the alloy composition described by [3] presented in **Table 2.**

The arc melting technique was used for melting metals and making alloys using a vacuum arc furnace as shown in **Figure 1**. The heating is done by striking an arc between a tungsten electrode and the metal pieces Ti, Ga and Si, for alloy preparation were shown in **Figure 2** that was placed in a water-cooled copper crucible. The melting was carried out in an evacuated chamber which was backfilled with inert Argon gas exactly before melting.

The working principle of a laboratory arc melting unit is similar to the standard TIG welding unit. The heat required for melting the metal pieces in the crucible was generated by the electric arc between electrode and metal. Argon gas provides an inert atmosphere and prevents oxidation of the sample. Melting was repeated 3–4 times so that a homogeneous alloy is obtained. The typical process of flow of melting metals followed in this study is shown in **Table 3.**

Value	Titanium	Gallium	Total
Atomic %	98%	2%	100%
Weight Gm	19.423gm	0.577gm	20gm

Table 1.
Atomic and weight percentage of metals for Ti-Ga alloy.

Value	Gallium	Silicon	Titanium	Total
Atomic %	15%	5%	80%	100%
Weight gm	4.169 gm	O.577 gm	15.271gm	20gm

Table 2.
Atomic and weight percentage of metals for Ga-Si-Tialloy.

Figure 1.
Vacuum arc furnace.

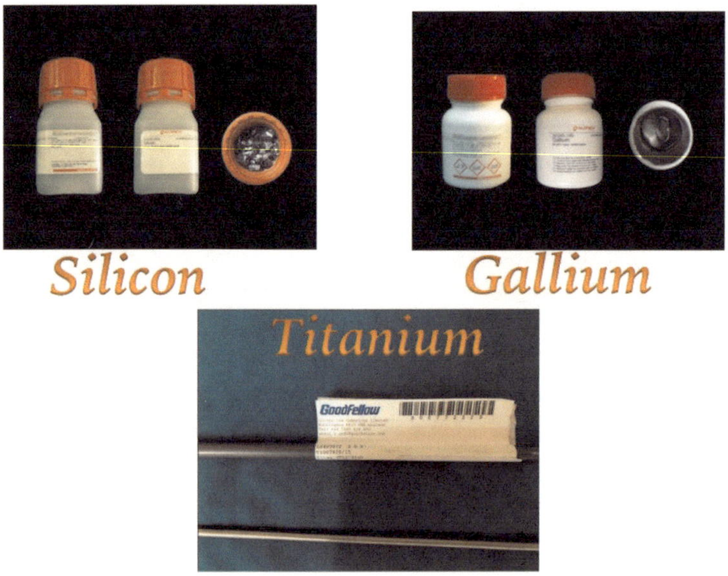

Figure 2.
Metals used for alloy preparation.

Triplets in each sample of Titanium, Titanium Gallium, Gallium Silicon Titanium alloys were made into a sample size of 1 mm length x 5 mm diameter (**Figure 3**). Samples were polished with sandpaper no: 60, 220, 240, 320, 400, 600, 1200 and 2400, cleaned with alcohol and then bathed with distilled water as described by [4]. Samples were divided into three groups, viz. Ti, Ti-Ga and Ga-Si-Ti groups for experimentation.

1. Metal pieces are weighed as per the alloy composition	2. The copper mold is cleaned using Acetone	3. Metal pieces are placed at the centre of the mold
6. Switch ON the diffusion pump at 10^{-3} mbar	5. Ensure that there is no leakage or open gap	4. Chamber is evacuated using a rotary pump
7. Let vacuum reach 10^{-5} mbar inside the chamber	8. Switch ON chiller for water flow to mold & electrode	9. Switch ON the power supply
12. Strike arc again & melt the pieces to form a button	11. Melt Ti getter to consume O_2 in chamber (if any)	10. Strike the arc using the electrode and mold
13. Flip ingot and rebuild vacuum to desired level	14. Repeat the process for 2–3 times to obtain an ingot	15. Perform suction casting if required

Table 3.
Process of a vacuum arc melting.

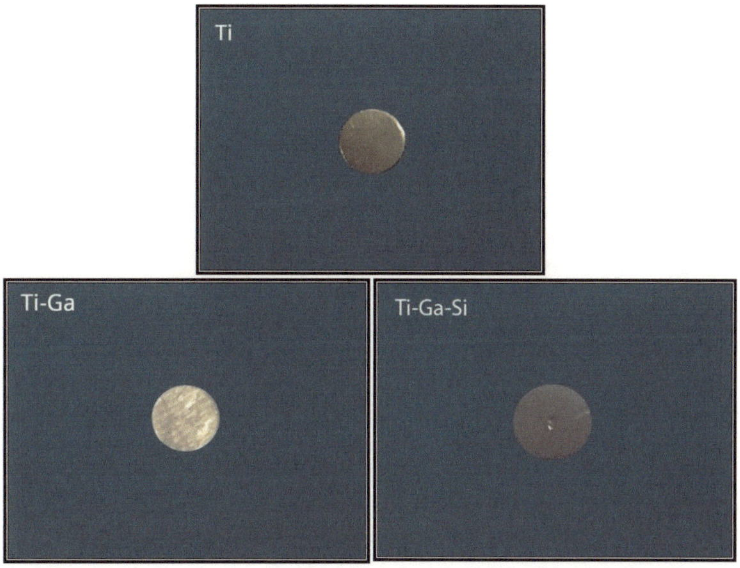

Figure 3.
Sample size used for in-vitro *study.*

2.2 Characterization of the mechanical and structural properties of Ti, Ti-Ga and Ga-Si-Ti alloys

The microstructural phases and visualization of the microstructure of the alloys and the mechanical and structural properties of the experimental alloys were studied.

2.2.1 Determination of microstructural phases

The microstructural phases of the alloys Ti, Ti-Ga and Ga-Si-Ti were assessed by X-ray diffraction (XRD). The X-ray diffraction was done in Bragg-Brentano θ-2θ Geometry method using X-Ray Diffractometer (Smart Lab, 9 kW- Rikagu, Japan instrument) as per [5].

2.2.2 Visualization of microstructure of the alloys

The samples were mirror polished with colloidal silica for 10 minutes before etching. Eventually etching was done with Kroll's reagent (Sigma Aldrich) for 4–5 seconds, reagent consist of 5% Nitric acid, 10% Hydrofluoricacid and 85% water and visualized with an inverted metallurgical microscope. De winter victory model no 4100, Germany as per the procedure of [2].

2.2.3 Scanning electron microscopic study

Triplet samples from Ti, Ti-Ga and Ga-Si-Ti alloys groups were dried at room temperature and fixed in glutaraldehyde 2%. The sample was immersed with alcohols (50%, 75%, 95% and 100%) for dehydration. The sample was placed over absorbent paper for 48 hours and analyzed with SEM (Scanning Electron Microscope) and EDX (Energy Dispersive X-ray Analysis)as per the procedure used by [6, 7].

2.2.4 Tensile strength test for the alloys

The tensile strength of the alloys was measured by using ASTM E8 as per the following specification. The samples were cut by an Electric Discharge Machine (EDM), rinsed with acetone and degreased with NaOH and distilled water and then measured with a Vernier caliper. The sample was held at the tensile fixture and pulled at the velocity of the speed of 1 mm per minute and the results were measured as described by [8].

2.2.5 Compression test for the alloys

Compression testing was done on the samples as per ASTM E9 specification and the results were recorded. The model was held in the compression fixture as sample edges were flattened and the flatness was checked with the dye gauge and the results were recorded as per [9].

2.2.6 Microhardness test for the alloys

In the VHN test, the applied force was smooth without causing any effect and held in contact for a duration of 10–15 seconds. ASTM 384 VHN was tested at 200gF/1.96133 N using cylindrical specimens of 11 mm in height of the sample. 6 readings were taken for each specimen with 'The Wilson VH1150' digital micro-hardness tester machine and the values were recorded as described by [8].

2.3 Characterization of the alloy for biological compatibility in the physiological system

1 mm long x 5 mm width in size Ti, Ti-Ga, Ga-Si-Ti alloys samples were prepared for in-vitro studies (**Figure 3**). The samples were polished with sandpaper and were cleaned with alcohol followed by distilled water. The samples were sterilized through steam autoclave 121°*C* for a duration of 30 minutes 15 pounds per square inch of pressure.

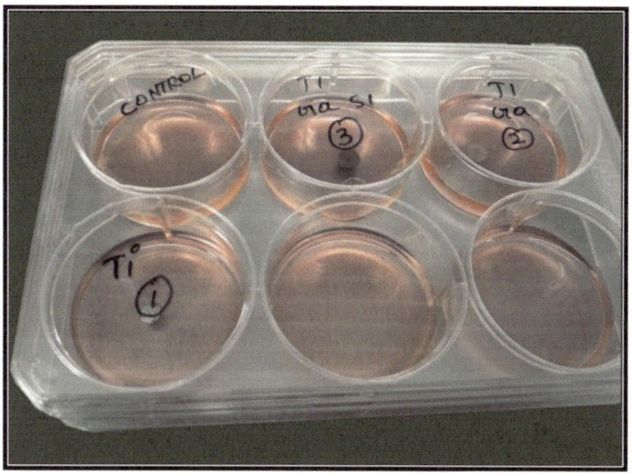

Figure 4.
Triplet sample in simulated body fluid In 6 well plate.

2.3.1 Study of the bioactivity of the Ti, Ti-Ga and Ga-Si-Ti alloys using simulated body fluid in-vitro *study*

Simulated body fluid is prepared according to [9]. The pH of 1.0 SBF/liter fluid was adjusted to pH 7.4 using 1NHCL, (i.e.) 1NHCl ——— > 8.3 ml of HCL in 91.7 ml H_2O and Six-well plates were used for the samples of all the three groups. The triplet sample from each group was soaked in the prepared simulated body fluid (**Figure 4**). Each plate of 10 ml of SBF fluid was treated with albumin and kept at room temperature and the fluid was changed every day after 24 hours for 21 days. Using SEMand EDX spectrometry, the sample surfaces were assessed qualitatively.

2.3.2 Evaluation of cytotoxicity of the Ti, Ti-Ga and Ga-Si-Ti alloys using SaoS-2 cell line in-vitro *study*

SaoS-2 cell line purchased from NCCS PUNE, INDIA was cultured with Mc Coy's 5A medium in 10% FBS (Gibco, In-vitrogen Bio-Services India, Bangalore, India) and 1% penicillin-streptomycin (HIMEDIA) and were maintained in this at 37°C with 5% $CO_{2.}$ The cell line was used for assessing cell interaction and alloys with regard to cell adhesion, viability and proliferation. Their transformation of SaoS-2 cells into osteocytes and calcium deposition at the surface of the alloy was measured through qualitative and quantitative evaluation methods.

Before seeding of cells, the samples were treated in a culture medium for 24 hours. Then the Cells were separated using a standard protocol of [10] for cell density and incubated at 37°C with 5% CO_2 in an 8x12 well plate. The cell line and alloys of three combinations in triplicate were incubated as per the procedure of [11] which is shown in **Figure 5**.

2.3.3 Alamar blue assay

The vitality and cytotoxic character of the samples were determined using the cell viability reagent alamar blue. The pH McCoy's 5A medium was adjusted to the pH of 7.0 to 7.4 and incubated at 37°C in sealed plates to prevent evaporation. Since the Alamar blue is photosensitive, so the incubation was done under darkness as described by [12].

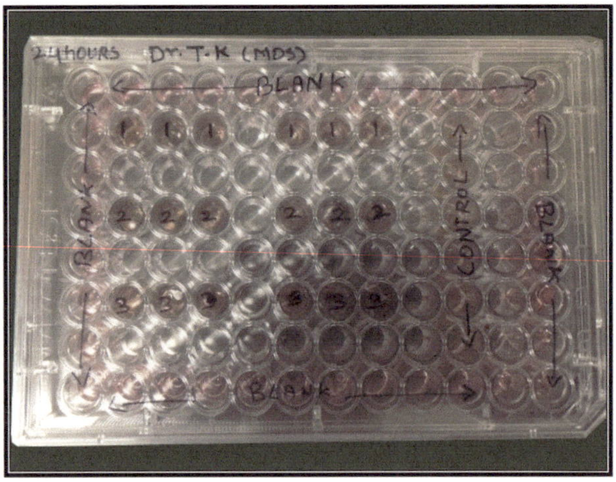

Figure 5.
Triplet sample in SaoS-2 cells in 8x12 well plates.

2.3.3.1 Cell adhesion study

After 24 hours of cell seeding, adhesion was assessed, by adding 10% Alamar blue dye to the cell medium. After 4 hours, the dyed media was aspirated and the same was measured for their adhesion to alloys at 570 nm and 600 nm as per the procedure adopted by [13].

2.3.3.2 Eosin staining of the alloys

The staining of the samples was carried out for three alloy groups. The samples cleaned with PBS were dipped in 10% formalin for duration of ten minutes. Followed by eosin staining for one minute, washed with water, absolute ethanol was used for drying the samples for three minutes and observed in bright field microscopy to observe the surface of the alloys with cell adhesion.

2.3.3.3 Viability and proliferation study

Viability and proliferation were measured at 48 hours and 96 hours after cell seeding. 10% alamar blue dye included in the medium whenever medium was changed. Six hours after adding the dye the medium containing alamar blue dye was aspirated and measured at 570 nm and 600 nm in a spectrometer.

The proliferation rate was estimated by comparing the percentage decrease in matching wells after 96 hours to 48 hours using the formula below.

Proliferation rate, $P = \frac{R96-R48}{R48} \times 100$ R96 and R48 are percentage reductions of viability tests performed at 96 and 48 hours, respectively.

2.3.3.4 Study of the samples using SaoS-2 cell line for mineralization in-vitro *study*

Mineralization of SaoS-2 Cells with the surface of the alloy for cell seeding density was done according to the technique followed by [14]. The cells have been cultured for 3 weeks and the medium was changed every three days. The mineralization procedure was carried out as per the procedure of [15]. The calcium deposits in the cell culture were measured using alizarin red dye. We utilized 1 milliliter of a 40 mM solution that had a pH of 4.2 and left it at room temperature for 10 minutes. In order to remove the dye, the wells were washed with phosphate buffer saline and distilled water five times before being examined under a stereomicroscope to obtain the images.

2.4 *In-vivo* study for evaluation of Ti, Ti-Ga and Ga-Si-Ti alloy as biomaterial implants

Study Design: An implant designed to promote bone growth may require at least two months and upto six months for bone regeneration and to study the localized tissue reaction. With the current crossover research study, the goal is to evaluate the newly synthesized Ga-Si-Ti implant for its biocompatibility in comparison with Ti and Ti-Ga implants. This removes the anatomical variation and avoids bias in evaluating osseointegration. The test was carried out to assess for reaction around the implants as per ISO-Standard No: 10993–6 and 10,993–11.

2.4.1 Implants

Non-threaded cylindrical implants were fabricated according to the surgical anatomy and the diameter of the femur bone of the rat. The implants were surgically placed on the right side femur of the rats. Three types of implants viz. (i) Ti (ii) Ti-Ga and (iii) Ga-Si-Ti in size of 1.5 m diameter and 3 mm length are shown in **Figure 6**. SEM images of Ti, Ti-Ga and Ga-Si-Ti implants are shown in **Figures 7–9**.

2.4.2 Characteristic details of the experimental groups

The study was conducted on male Wistar rats (Rattus /Norvegicus) of about 8 weeks of age with the body weight of 200–250 gm. In total, 18 rats were studied in this research. The experimental rats randomly split into three groups in which the Ti, Ti-Ga and Ga-Si-Ti metal alloys were implanted on the right femur bone and the rats were kept under observation for normal health.

The IAEC of Sathyabama Institute of Science and Technology issued the ethical clearance with the IAEC Number: SU/CLATR/IAEC/XV/152/2020. The experimental animals were housed in the Sathyabama Institute of Science and Technology animal experimental laboratory in Chennai, Tamil Nadu, India. The rats were fed a standard maintenance rat diet and water ad libitum.

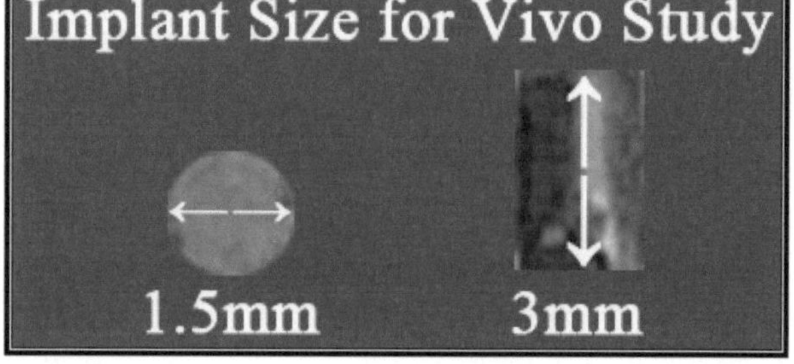

Figure 6.
Sample size used for in-vivo *study.*

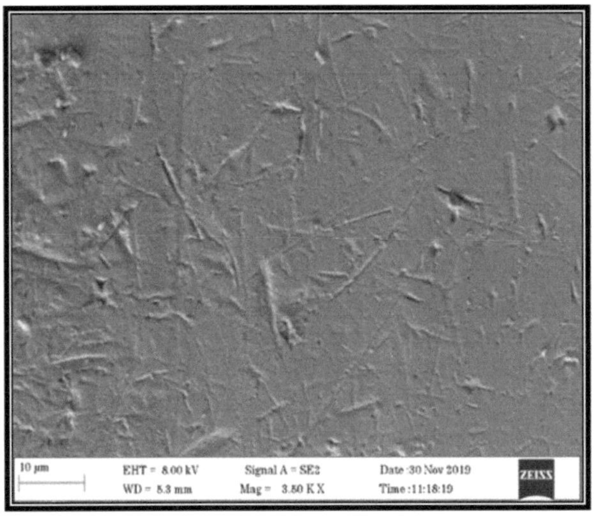

Figure 7.
SEM images of Ti.

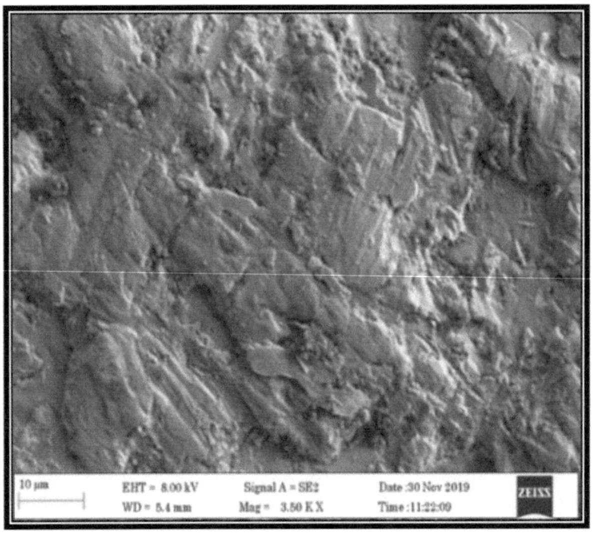

Figure 8.
SEM images of Ti-Ga.

Experiments were performed according to CPCSEA guidelines and conforming to NIH guidelines for the animal.

2.4.3 Surgical procedures

Surgical procedures (**Figure 10**) were carried out under general anesthesia, as per the procedure adopted by [16] with Ketamine (40-90 mg) and Xylazine 5-10 mg (Rompun) by the intraperitoneal path. The surgical site was tonsured and disinfected with povidone-iodine solution & an injection of Lidocaine with Adrenaline (1:80000) was infiltrated at the site of incision to prevent bleeding.

Figure 9.
SEM images of Ga-Si-Ti.

A stab incision was made along the imaginary line between the hip joint and along the lateral aspect of the thigh. The medial aspect of the femur bone was exposed on the right side. With a 1.5 mm surgical drill, a 2 mm x 3 mm defect was created in the medial aspect of the femur and under irrigation with normal saline.

The implants were placed in the defect and secured with a 3–0 silk suture around the femur bone. Then the muscles were secured followed by a skin suture (**Figure 10**). The skin wound was protected by Healex Spray contains benzocaine 0.36%w/w and cetrimide 6.5%w/w (Shreya life sciences Pvt. limited) and Nebasulf powder contains Bacitracin (250.0 IU), sulfacetamide (60.0 mg) and Neomycin (5.0 mg) (Abbott healthcare PVT LTD). Gentamycin 20 mg/kg/day for five days and Buprenorphine 0.04 mg/kg/day for 3 days) were given intra-peritoneal as post-operative antibiotic and analgesic respectively. Hydration of animals was maintained with 5 ml of dextrose saline on the day of surgery.

2.4.4 Radiographic analysis

Radiographic evaluation was carried to assess the position of implant with the surrounding bone and to conduct a study to evaluate the new bone formed along the implants.

2.4.4.1 Digital radiography

Two types of digital radiography were done in the present study.

2.4.4.1.1 Posterior anterior view of the whole body

Posterior Anterior view of the whole body [17] using Sirona orthophos Xg radiographic machine, a direct digital radiograph was taken for comparing the bone density of the femur with implant and normal femur bone of the same animal at a different

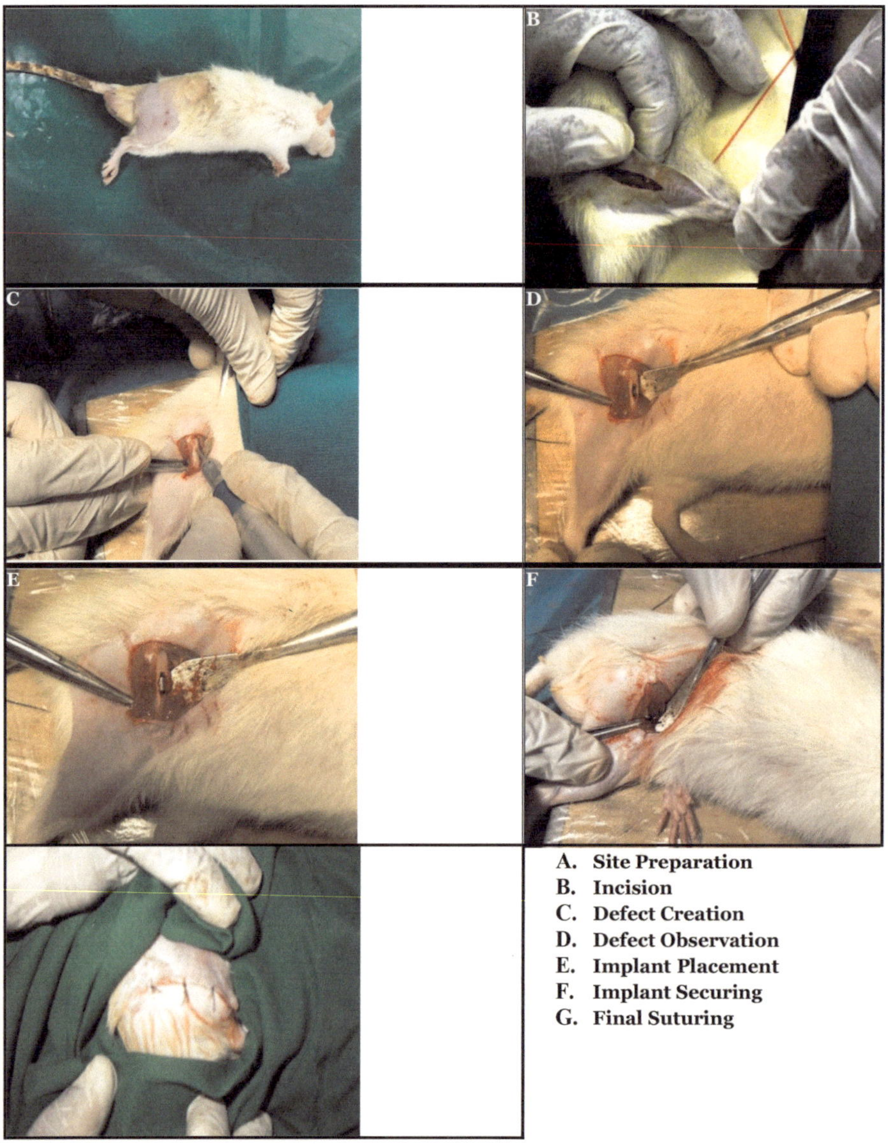

Figure 10.
Animal surgery and implant placement.

time interval, i.e. 14th, 28th, 42nd and 56th day. Once the standard images were produced and the post-implantation were aligned by selecting reference points on the implant images as per [18] following the selection of a region of interest on the images, the follow-up images were subtracted from the baseline image and negative images used as digital subtraction of radiographic images for evaluating the bone density changes. The quantitative assesment of newly formed bone and the mineralization around implants was analyzed using Sidexis Xg software for calculating the density of the bone using grayscale. Qualitative assessment was done using the images which showed decreased grayscale inferring mineral bone loss and increased grayscale inferring bone mineral gain.

2.4.4.1.2 Paralleling technique

The second method of digital radiography used paralleling technique to the film (phosphor plates) and implanted femur bone of rat by using x-mind x-ray unit setting with 60kVp, 4 Ma, with an exposure time of 0.4 Seconds to produce the highest quality of digital image with most negligible radiation. Soredex Digora Optime processing unit was used in which best contrast at lower exposure digital image can be achieved. The photon detector used in the system is a phosphor plate with the following characteristics with a dimension of 31 mm x 41 mm and the image pixel of 1034x1368 according to ISO 10993-1and ISO10993-5 non-toxic & non-irritative to biological systems, so the hygienic bag made of latex-free food-grade polyethylene was used over phosphorylates while taking a radiograph. Linear measurements of bone densitometric values were obtained using Gray values [19].

Areas of specific sites visualized through digital images were selected for densitometric analysis. In this study, linear, cortical (CBD), trabecular (TBD) bone densities and Newly Formed Bone Density (NFBD) were measured for qualitative evaluation by serial radiograph. Quantitative assessments were done by one measurement in each area per image along with bone density around implants and the point density was measured using Gray values.

2.4.4.2 Computed tomography study

Computed Tomography was done using the Somato go-now scanner in which the beam is collimated using the material ultra-fast ceramic collimation and the CT detectors of the scan utilizing a CT X-ray source of 80–120 Kv with strong filtration to produce a monochromatic beam. The width of the collimator is 0.6 mm and the true resolution of the machine is 1 mm.

In the present study, 300–450 window levels and 1300–1500 window widths were used. The data reconstruction was done using the software, to calculate the distance, pixels and intensity for image analysis. Thirty two images were recorded per rotation of 16.0 x 0.7 collimation. The spiral accusation with pitch factor is 0.8 ratios. All the specimens were scanned with the same parameters and data were collected by a computer. The image was reconstructed to evaluate the linear attenuation, co-efficient of each voxel in a slice assigned with CT number [16] or Hounsfield number to each voxel. The mean Hounsfield Units were calculated and statistically analyzed. CT scan was obtained on the 28th and 56th days to evaluate the osseointegration in terms of the newly developed bone. After the implantation, there were disparities in the density of the trabecular and cortical bones. Computed Tomographic scans have been assessed in two and three-dimension formatted images for studying the bone mineralization in response to implant material placed in critical defect created for evaluating the biocompatibility of the implant with the bone.

2.4.5 PolyFluoroChrome study (PFCs)

Experimental rats of all the three groups were injected with PFC according to the procedure described by [20]. Post-operatively the animals were administered with intra-peritoneal injected with Alizarin Red S (30 mg/kg body wt) on the 7th and 14th day, Tetracycline (12 mg/kg body wt) on 21st and 28th day and finally Calcein (4 mg/kg body wt) on 35th and 42nd day respectively.

2.4.6 Sacrifice of the experimental animals

The animals were sacrificed on the 56th day and femur bones with implants were retrieved and stored in 70% ethanol at 4°C for histopathological study and ex-vivo Micro-CT.

Euthanasia: All experimental rats were euthanized by giving a 5x dose of Ketamine \Xylazine (500/50 mg/kg) intraperitoneally. The femur bones with implants were retrieved and subjected to histopathological study. The carcass of the experimental animals was disposed of as per the Biological Waste management program of the Institution through M/s GJ Multiclave (India) Pvt. Ltd., Tambaram Sanatorium, Chennai - 600,047.

2.4.7 MICRO-CT study

Micro-CT scan of rat femur with explant was done according to [16] .The scanner used was a SCANCO Medical CT 40. The following values were used in an easily configurable measuring methodology: 12 μm spatial resolution, 70 kVp beam energy, 114 A beam intensity, 300 ms integration time and 1024 × 1024 image matrix. Explants from rats were isolated, washed with PBS and stored in ethanol in an Eppendorf tube. The density of bone mineral was calculated in $g.cm^{-3}$. BMD refers to combined density of defined volume with bone and soft tissue. The study's calibration of bone mineral density was done by 2 mm–4 mm phantom rod pairs inside a tube of water matching the animal's bone diameter in the ex-vivo study. After scanning the phantom rods, a bone sample with three different implant materials ROI was observed for BMD for calibration.

2.5 Histomorphometric analysis of bone remodeling using Polyfluorochrome (PFC) dyes in-vivo and *ex-vivo*

Specimen Preparation for Histology and Histomorphometric analysis was done as per the procedure adopted by [21].

Reagents: 10% NBF, Alcohol (70%, 80%, 96% and 100%), CH_3COCH_3 (Sigma Aldrich, Bangalore), MMA (Sigma Aldrich, Bangalore), $C_{14}H_{10}O_4$ (Sigma Aldrich, Bangalore), 5% NaOH, KMnO4 (Merck, Maharashtra) Methylene Blue (Thermo Fischer, Mumbai), Distilled water, Acid Fuschin (Sigma Aldrich, Bangalore) Saturated Picric Acid (Sigma Aldrich, Bangalore), Cyanoacrylate glue (Alteco chemicals, Japan), Stevenel's blue and Van Gieson's Picrofuchsin stain.

Specimens were fixed in 10% NBF, dehydrated in increasing ethanol concentrations (70–100%), cleared using acetone alcohol mixture and embedded in MMA. Resin blocks of specimens were prepared after polymerization in MMA under vacuum conditions for 7–10 days at room temperature. The PMMA block was split into wide sections (70-100 m) using a saw microtome (ACCUTOME 100, Struers, Denmark). It was adhered to a glass slide and polished using a grinder polisher (ECOMET 3000, Buehler, Germany). The sections were stained with Stevenel's blue and Vangieson's picro fuchsin and examined in a trinocular transmitted light microscope (Nikon Ni-E). Photomicrographs were taken with a camera (Nikon DS Ri1) attached to the trinocular microscope (Nikon Eclipse) for histomorphometry and examined in a fluorescence microscope.

The samples were observed in violet, blue and green illumination in Leica fluorescent microscope attached to a computer with Leica system software installed. Images were acquired in 10x magnification. For every sample, different illuminations were used for the same focused field to obtain the standard images. Thus, the obtained images were superimposed in the order of green illumination followed by a violet after blue, which corresponds to the excitation of dyes in the same order as administered to the rats. This was accomplished by using GIMP version 2 software. The dimensions inside individual images were measured using NIH-Image J software according to techniques described by [22].

2.5.1 Histomorphometry

Bone Implant Contact (BIC), Rate of Bone Apposition (RBA) and Osteoblast proximity were the parameters evaluated and all the measurements were made with NIH-Image J. The marking techniques were done according to [23]. The total perimeter of implant areas of implant-bone contact was marked and BIC was estimated as contacting area percentage to the whole implant perimeter as per the formula adopted by [24]. Kulak and Dempster, [25] calculated MAR by dividing the mean distance between fluorescence labels by the time interval between them.

2.6 Statistical analysis

Quantitative data obtained from in-vitro, cytotoxic test done with alamar blue assay to evaluate the adhesion, viability and proliferation with SaoS-2 cell line, in which two-tailed test, which is used in null hypothesis testing and statistical significance bone calcium mineralization and *in-vivo* radiographic analysis of the bone were statistically analyzed utilizing the SPSS software (version 21.0; IBM corporation, Armonk, NY, USA). Significance level was fixed at 5% ($\alpha = 0.05$).

For the in-vitro model, mineralization occurs at different rates in each study group when different materials are used. So, to understand mineralization, calcium nodules deposited were evaluated for the 4th, 7th, 8th, 11th, 14th and 21st day. To compare the mean values between variables recorded at a different time points, repeated measures ANOVA was applied to find the group means.

For the *in-vivo* model, linear bone density of PA view of the whole body was compared between all the three groups of Ti, Ti-Ga and Ga-Si-Ti implanted bone with adjacent side normal bone. Medial cortical, lateral cortical, trabecular bone and new bone formation were recorded on 14th, 28th, 42nd and 56th day on both the implanted side as well as adjacent normal bone. Non-parametric The implanted bone was compared to normal bone using the Mann-Whitney U test (similar to the post hoc Tukey test) and the implanted bone was compared between groups using the Kruskal-Wallis test.

Digital subtraction linear bone density of the PA view of the body was compared between the three implanted bone groups of Ti, Ti-Ga and Ga-Si-Ti. On the 14th and 28th days, the first indications of newly formed bone was documented using the digital subtraction approach. To examine the changes within each group, the Wilcoxon signed-rank test was used. For intergroup comparison, the Kruskal-Wallis test was used.

For the *in-vivo* model, to evaluate the point density in digital radiography using paralleling technique, the non-parametric statistical test of the Kruskal-Wallis test was applied.

Mean difference intergroup comparison concerning Hounsfield Units of the CT scan and mean/density in micro-CT, both were analyzed using the non-parametric Kruskal Wallis test and one-way Anova test respectively.

Histomorphometry analysis of bone was evaluated through BIC; it is described as the amount of contacting area to the total perimeter of implant in percentage, osteoblast proximity which is the distance between the first osteoblast and the implant and finally, the Rate of Bone Apposition (RBA). The BIC (Bone Implant Contact), RBA and Osteoblast proximity were analyzed using Kruskal Wallis Test.

3. Results

In the microstructure evaluation of Ti, Ti-Ga and Ga-Si-Ti exhibited 'α' phase.

In the mechanical study, the Ga-Si-Ti alloy showed a higher Ultimate tensile strength (780 MPa), Yield strength (595 MPa) and a higher percentage of elongation (18.76%) and more or less equal to Young's modulus (1.16x105 MPa) and Compressive at fracture (1696 MPa) when compared with the Ti and Ti-Ga alloy.

In the Simulated Body Fluid study, among the three samples, the Ga-Si-Ti alloy showed more precipitation on its surface and a higher Ca/P ratio (1.8%) in the precipitation.

In-vitro evaluation of cytotoxicity of the Ti, Ti-Ga and Ga-Si-Ti alloys revealed the ability of SaoS-2 cells to adhere to the surfaces of all three alloys, which indicated the recognition of the experimental alloys by the cells as surfaces to grow on. No statistical differences observed in the adhesion, viability and proliferation properties between the groups. In mineralization in-vitro study, Alizarin dye exposed the Ca and Phosphate mineralization on day 7th, 14th and 21st day of the existence of spherulites on the surfaces of all three group alloys was investigated and confirmed. Quantitative analysis of mineralization revealed that the Ga-Si-Ti alloy group showed more mean number (212 ± 2.28) of calcified nodules than cpTi (130 ± 4.36) and Ti-Ga (137 ± 2.65) group alloys.

In the in-vivo study, a PA view of whole-body radiography all groups exhibited a substantial difference in the linear bone density of newly formed bone. However, compared with Ti and Ti-Ga groups, the Ga-Si-Ti group had the highest mean bone density of newly formed bone.

The CT study revealed an increase in bone density in terms of higher Hounsfield Units for trabecular bone with Ga-Si-Ti group (802.92 ± 226) on day 56 when compared to Ti (673 ± 73.37) and Ti-Ga (708.43 ± 166.54) groups and higher Hounsfield Units of new bone density in Ga-Si-Ti, followed by Ti-Ga group and then Cp-Ti group.

Ga-Si-Ti group showed the highest bone mineral density (459.9792 mg/cm3) than Ti and Ti-Ga group in the micro CT ex-vivo study. In addition, the study exhibited a significant difference between the groups and the Ga-Si-Ti group implants in the proportion of cortical bone volume to trabecular bone volume BV/TV in percentage showed a clinical significance with the highest BV/TV bone volume of 25.07%.

In the present study, Poly Fluorochrome tracers (Alizarin red, Tetracycline and Calcein) clearly showed the remodeling pattern. When compared to other implants, Ga-Si-Ti alloy implants promoted cell proliferation and differentiation and improved the BIC. Ga-Si-Ti alloy implants exhibited ~98% BIC and much closer osteoblasts presence to the surface of the implants.

4. Discussion

Titanium is a "miracle metal" in dentistry and its alloys have become the preferred metals for preparing bone biomaterials. The most popular implant materials are Ti,

Ti-6Al-4 V, Ti-6Al-7Nb and Ti-13Nb-13Zr. Since Cp-Ti is relatively pliable, V and Al are hazardous and Nb and Zr are brittle, there is considerable interest in the development of superior Ti alloys. Considering the vast potential of Gallium and Silicon, a ternary Ti alloy containing 15% Gallium and 5% Silicon was synthesized in order to enhance this scenario and expand the clinical application of Titanium. In addition, there were few investigations conducted on the structural and mechanical properties of Ga-Si-Ti alloys and there is currently no information on the in vitro and in vivo biocompatibility of Ga-Si-Ti alloy as a biomaterial. This has been the motivation for the present study.

4.1 Microstructural properties

Microstructure analysis is useful to interpret an alloy's mechanical and corrosion resistance properties.

X-ray diffractometry was used to evaluate the structural features of Ti, Ti-Ga and Ga-Si-Ti alloys. The results revealed a single α-Ti phase with a partially stable HCP crystal structure. In Ti-Ga and Ga-Si-Ti alloys, the α - peak could be observed. Similar outcomes were reported for Cp-Ti alloy [26]; Ti-Ga alloy [2] and Ga-Si-Ti alloy [3, 26]. According to [2] Ga is a stabilizing element that, when dissolved in Ti, elevates the eutectic transformation temperature from α (hcp) to β (bcc). Ti-Ga alloys, on the other hand, have a short solidification interval and a solubility of up to 20% in the α - Ti phase. Despite the presence of β-stabilizing Silicon in the Ga-Si-Ti alloy, the Ga-Si-Ti alloy exhibited α-phase because the Si content was only 5%. Consequently, all three alloys exhibited α-phase, which was validated by XRD. This result concurs with the microstructural tests conducted by [27] on the Ti-15Nb-5Si alloy. SEM images of Ti, Ti-Ga and Ga-Si-Ti alloys indicated irregular grain boundaries in the present investigation. These uneven grain boundaries were attributed by [2] to the non-equilibrium cooling process during casting.

The EDAX analysis and mapping validated the distribution of Ga in the Ti-Ga alloy and the distribution of Ga and Si in the Ti-Si alloy. Vizureanu et al., [28] performed a similar study of Ti-Mo-Si alloys to validate the presence of Si and Mo in Ti-Mo-Si alloys.

4.2 Mechanical properties

Some of the properties like compression strength, tensile strength, modulus and elongation are important mechanical properties to be assessed in newly synthesized metals for use as biomaterials. In this study, the above properties of Ti, Ti-Ga and Ga-Si-Ti were done to assess the suitability of Ga-Si-Ti alloy as an implant, particularly as a dental implant.

The ultimate compression strength of Ti, Ti-Ga and Ga-Si-Ti alloys was found to be similar, viz. 1656 MPa, 1690 MPa and 1696 MPa respectively. The results obtained in this study are comparable to the compression strength recorded by [29] for Commercial pure-Titanium (1074 MPa) and Ti-64 alloy (1661.6 MPa).

Regarding the selection of an alloy for usage as a biomaterial, the tensile strength and modulus of the alloy are to be evaluated for assessing its fatigue strength and ductility. In the present experiment, the Ti, Ti-Ga and Ga-Si-Ti alloys showed 579 MPa, 612 MPa and 780 MPa as ultimate tensile strength and 1.10×10^5 MPa, 1.13×10^5 MPa and 1.16×10^5 MPa as Young's modulus, respectively. The percentage of elongation was observed as 17.9%, 17.6% and 18.76% for the Ti, Ti-Ga and Ga-Si-Ti alloy respectively. The findings of this study are in conformity with the findings of other investigations, which found that alloys made of Cp-Ti, Ti-Ga and Ga-Si-Ti have comparable levels of strength, modulus and elongation [2, 3, 29].

The observations made using Ti-Ga alloy are comparable to those made by [2, 30]. The present study revealed a higher ultimate tensile strength and percentage of elongation and more or less equal Young's modulus to Ga-Si-Ti alloy when compared to those mechanical properties shown by Ti and Ti-Ga alloys.

Comparing the tensile strength, yield strength, percentage of elongation and compressive at fracture of Ga-Si-Ti alloy with those of Cp-Ti and Ti-Ga alloy reveals that Ga-Si-Ti alloy has superior mechanical properties.

Similarly, the Vicker's hardness values observed in the present study for Ti, Ti-Ga and Ga-Si-Ti were 177.5 ± 4.18, $271.6 \pm 16.49$and 240.3 ± 9.89. The hardness values for Ti and Ti-Ga alloys found in this study are similar to those found by [2]. The Ti-15NbxSi alloy was given a hardness value of 313 ± 7 by [27]. This is a slightly higher than the hardness of the Ga-Si-Ti alloy from this study, where they used two β-stabilizing alloys. According to [31], the addition of Tantalum and Zirconium to Titanium improved the wear resistance of the alloy and the addition of Zirconium boosted the micro-hardness of the Titanium zirconium composite. In comparison, Choi et al., [5] correlated the increased value of hardness of Ti alloy with the melting procedure used for the alloy preparation. The mechanical properties observed in the current investigation after the addition of 2% Ga in Ti-Ga and 15% Ga in Ga-Si-Ti alloy have increased the strength and micro-hardness monotonically. This observation is related to solid solution strengthening of Ga as observed in [2] study on the evaluation of mechanical properties of Ti-Ga binary alloy. Antonova et al., [3] demonstrated that Ga can be employed effectively to improve the mechanical properties of Ti alloys, particularly Ti-Si alloys. A perusal of the literature revealed that there is a paucity of information on the effect of the addition of Ga and the combined effect of Ga and Si on Ti alloy and the present study contributed to a better understanding of the role of Ga and Si in the composition of Ti alloy. Both the addition of 2% Ga to Cp-Ti and the addition of 15% Ga and 5% Si to Cp-Ti alloy increased the mechanical properties of Ti alloy, according to the results of this study.

However, the Young's modulus reported in this research for the Ti, Ti-Ga and Ga-Si-Ti alloys is somewhat more than that of human bone.. The modulus value obtained for the Ga-Si-Ti alloy of this study is more or less similar to the best known and comm only used Ti and Ti-6Al-4 V alloy and it was considered as close to the modulus of human bone. Thus the results obtained in the study revealed that the elements Gallium and Silicon can be added to the Titanium alloy for improving its mechanical properties. Hence, the results obtained as a dental implant biomaterial, a novel Ga-Si-Ti alloy with improved structural and mechanical capabilities as compared to the widely used Ti and Ti-based alloys was proposed in this study.

4.3 Simulated body fluid

In the present study, the bioactivity of the Ti, Ti-Ga and Ga-Si-Ti alloy the samples were tested by immersing them for 21 days in Simulated Body Fluid (SBF). Immersion of the biomaterials in simulated body fluid (SBF) is an extensively used method to investigate the bioactivity of Ti and its alloys biomaterial. Kokubo, [32] opined that the bioactivity of the biomaterial can be predicted based on the formation of apatite on the surface of biomaterials when immersed in SBF and by the analysis of the apatite so formed. In the present investigation, SBF was prepared as per the procedure of [33] which is comparable to human plasma. After the immersion in SBF for 21 days, the visualization of Ti, Ti-Ga and Ga-Si-Ti alloy samples using SEM revealed increased precipitation on their surfaces. However, based on the SEM images no conclusion can be drawn on the bioactivity of the alloys since there were no differences observed between the samples in the apatite formation activity.

The element of analysis of such formed precipitate on the Ti, Ti-Ga and Ga-Si-Ti alloy samples revealed the presence of a maximum percentage of Titanium and the presence of Calcium, Phosphorus, Sodium and Oxygen along with Gallium in Ti-Ga and Gallium and Silicon in Ga-Si-Ti alloy sample immersion study. When the Ca/P ratio was calculated, it was higher in the Ga-Si-Ti alloy sample than in the Ti alloy sample. and Ti-Ga alloy samples after their immersion in SBF for 3 weeks duration. Stenport et al., [34] studied the Ca/P ratio for biomaterials immersed in SBF for 2 weeks and suggested increase in immersion duration for proper assessment of the bioactivity of the biomaterials. The bioactivity seen in terms of precipitation for the Ti-based alloys samples in this study is consistent with the findings of [35] where they demonstrated that the Titanium implants have become bioactive once they get oxidized in a calcium hydroxide solution.

Takadama et al., [36] concluded that the evaluation of the production of apatite on the surface of the biomaterials submerged in SBF, as well as the elemental characterization of apatite by EDX spectrometry, were both qualitatively and quantitatively effective in predicting the biomaterial's in vivo bone bioactivity [37]. Using the SBF model, they observed a variation in the development of calcium phosphates on Titanium implants with different types of surfaces. Qiu et al., [2], reported that the incorporation of Ga into pure Ti enhanced resistance to corrosion in both regular artificial saliva and saliva with fluoride. The bioactivity property observed with Ti, Ti-Ga and Ga-Si-Ti in the current study also agrees with the previous studies of [38, 39]. In these studies, they correlated the bioactivity characteristics of Cp-Ti and Ti-based alloys with their good corrosion resistance and biocompatibility properties.

Tadashi Kokubo and Hiroaki Takadama, [33] also opined that the SBF method of study could be of use in the development of new bioactive materials and also for screening and forecasting their biocompatibility property in animal testing. In a review by [40], it was concluded that the elements like Oxygen, Nitrogen, Calcium, Phosphorus and Sodium implanted into Titanium alloys improved the mechanical properties, bioactivity and the cyto-compatibility of Titanium and its alloys. The increased precipitation and the elemental analysis of the apatite in the present study were indicative of the Ti, Ti-Ga and Ga-Si-Ti alloys bioactivity. Moreover, among the three alloy samples, the Ga-Si-Ti alloy sample found to be superior in bioactivity in SBF, since it showed both the increased precipitation on its surface and a higher Ca/P ratio in the precipitate. Hence based on this SBF model in-vitro study, it is suggested that the Ti, Ti-Ga and Ga-Si-Ti alloys are suitable for in vivo study in experimental animals.

4.4 Cytotoxicity of Ti, Ti-Ga and Ga-Si-Ti alloys

Newly synthesized biomaterials should undergo rigorous study to determine their biocompatibility before their use in human body. *In-vitro* cytotoxicity test is used as one of the screening tests for evaluating the biocompatibility of biomaterials and it is often conducted using cell line such as SaoS-2, L929 and MG63. Prideaux et al., [15] demonstrated the usefulness of SaoS-2 cells in their *in-vitro* cytotoxicity study for determining the biocompatibility of biomaterials. Przekora, [41] reviewed the *in-vitro* cytotoxicity assay using SaoS-2 cell line in *in-vitro* model and concluded that SaoS-2 cell line was more suitable to evaluate the biocompatibility of biomaterials.

In the present study the properties such as adhesion, the viability and proliferation rate of SaoS-2 cells on the surfaces of Ti, Ti-Ga and Ga-Si-Ti alloy samples were studied. The adhesion of SaoS-2 cells on the surfaces of the alloys clearly indicated that the biocompatibility of these alloys using *in-vitro* cell line models. The SEM images of the study revealed that the cells were well attached over the surfaces of all the three sample of the alloys.

The Alamar Blue viability assay and the proliferation assay in this study suggested that the Ti, Ti-Ga and Ga-Si-Ti alloy samples can support and promote the cell growth on their surfaces. Even though there were no difference observed in the adhesion and proliferation of SaoS-2 among the three samples a slight decrease in the viability assay was observed with Ga-Si-Ti alloy sample and this could be related to the alkaline phosphatase activity of SaoS-2 cell line as reported by [42]. It is demonstrated in the present study that all the SaoS-2 cells which were viable proliferated throughout the cell culture experiment. This indicate the all the three alloys are not cytotoxic, among the three Ti showed more proliferation than Ga-Si-Ti and Ti-Ga.This observation of the present study is in agreement with [42].

This finding concurred with the observation of [43] where a decreased SaoS-2 cells viability in the presence of SiO_2. Many studies confirmed that the alloy Ti is non-toxic due to their unique ability that its surfaces can form TiO_2 has a stable passive film layer, making it superior in terms of biocompatibility. In a review by [44] mentioned that the Ti based biomaterials such as Ti and Ti 64 alloys exhibited good cell adhesion in the *in-vitro* studies and they opined that the Ti and its alloys are more promising biomaterials. Chandler et al., [45] evaluated the cytotoxicity of Gallium and reported that Ga has no significant cytotoxicity. The present *in-vitro* cytotoxicity study also revealed that the metals Gallium and Silicon were also non-toxic since the study revealed that the samples Ti-Ga and Ga-Si-Ti showed adhesion and proliferation similar to Ti alloy sample.

These observations were in agreement with the *in-vitro* cytotoxicity studies of [2] for Ti-Ga alloy and [28] for Ti-Mo-Si alloy. Cochis et al., [46] also confirmed that addition of the 2% and 20% Gallium to Titanium alloys containing 3% Silicon were nontoxic in their cytotoxicity evaluation study. Even though review of literatures revealed that the Ti, Ti based alloys biomaterials containing Gallium and Silicon are biocompatible, there is no in vitro studies to demonstrate the biocompatibility aspect of 15% Ga and 5% Silicon in Titanium alloy for comparison of Ga-Si-Ti alloy of the present study.

4.5 Mineralization

Mineralization is a process by which cells deposit inorganic calcium and phosphate on the surface of biomaterials and this mechanism of mineralization in the in-vitro research is focused on the formation of bone nodules and around the implants [47]. In the present study the Alizarin Red dye exposed the Calcium Phosphate mineralization of SaoS-2 cells at 4th, 7th, 14th and 21st days of culture with Ti, Ti-Ga and Ga-Si-Ti alloy samples in the cells. The SEM EDX examination indicated the presence of spherulites on the surfaces of each of the three alloys and the quantitative analysis with Leica stereomicroscope showed an enhanced number of calcified nodules on the Ga-Si-Ti alloy (212 ± 2.8) than Ti (130 ± 4.36) and Ti-Ga (137 ± 2.65) alloys on 21st day of observation and during the different days of study period. The statistical analysis showed a significant difference between Ti and Ti-Ga as well as between Ti and Ga-Si-Ti in calcified nodule formation. A considerable difference was also detected between Ti-Ga and Ga-Si-Ti alloys. In this study, it is clearly observed that the Ga-Si-Ti alloys has a maximum calcified nodules at the end of 21st day of experiment. Hence it is inferred from the present study that the presence of Silicon in Ga-Si-Ti alloy has enhanced the proliferation and Calcium, Phosphate nodule formation. He et al., [43] obtained similar results in the presence of Silica nanoparticles with SaoS-2 cells. Martinez-Ibanez et al., [48] also confirmed the increase in proliferation and mineralization with Si coated Ti alloy in their in-vitro study using mesenchymal stem cells.

The current findings with Ga-Si-Ti also concurred with the work of [49] where they observed a decreased viability and significantly enhanced proliferation and mineralization nodule formation in the presence of silicate ions. Hence based on the observations of this *in-vitro* study on the adhesion, viability proliferation and Ga-Si-Ti alloy with enhanced mineralization characteristics is in-vitro biocompatible and a potential alloy for the manufacture of dental biomaterials.

4.6 In-vivo radiographic study

To evaluate the position of the implants in the implant bed, regular radiographs were taken. Qualitatively by using radiographic images. The decreased grayscale is indicating bone mineral loss, while an increase in grayscale indicates bone mineral gain. Macroscopic variations in bone mineral density for a localized change can be visualized in normal radiography if it is more than 12%. Hence the digital radiograph was done to evaluate the bone density changes that can be detected with follow-up radiograph by using digital radiograph even if there is 1–5% (Matteson et al. 1996).

Khojastepour et al., [50] and Han et al., [51] assessed the implant position and new bone formation qualitatively using digital radiographs.

In the present study, the bone mineral density around the implant was evaluated using the Posterior-Anterior view of whole-body radiography and Paralleling radiography qualitatively and quantitatively. The quantitative evaluation was done by using density calculation software.

In the posterior-anterior view of whole-body radiographic study on bone density of medial cortical, lateral cortical, Trabecular bone and new bone formation along the implant were calculated for the implanted bones with adjacent side normal bone on the 14th, 28th, 42nd and 56th day after implantation in Ti, Ti-Ga and Ga-Si-Ti groups and compared.

The PA view of the whole-body radiographic study on the bone changes around the implants showed a substantial alteration in bone mineral density and the analysis revealed the highest bone mineral density in the Ga-Si-Ti group when compared to Ti-Ga and Ti group. The changes were markedly noted along with the trabecular bone and newly formed bone densities between Ti and Ti-Ga on 28th and 42nd day and Ti and Ga-Si-Ti on 28th, 42nd and 56th day, whereas during the investigation, there were no significant differences in bone density between Ti-Ga and Ga-Si-Ti, indicating that there is little change in the newly formed bone between the two groups.

When compared with implanted and adjacent side non implanted bone, no change was observed in the medial cortical and lateral cortical bone mineral density, since there was no variation in the nutrition and management of experimental rats.

Whereas with trabecular bone mineral density assessment, the Ti groups showed decreased density on the 14th, 28th and 42nd days. This observation is correlated with inflammation changes, whereas the radio-opaque appearance clinically on the 56th day proved that there is healing and osteocyte formation. This is in agreement with the observation of [52] in which the whole body of rat radiographic evaluation was done.

However, the radio-opaque region was observed with Ti-Ga and Ga-Si-Ti groups on the 28th day itself. This difference between Ti-Ga and Ga-Si-Ti group alloys is attributed to the presence of Ga elements in Ti-Ga group alloy and Ga and Si elements in Ga-Si-Ti group alloy and their role in osseointegration. This is in agreement with the observation of [53].

The assessment of newly formed bone in Ti, Ti-Ga and Ga-Si-Ti group implants revealed no mineralization changes on the 14th day and it was observed clinically as radiolucent legion. Whereas the changes in newly formed bone in the Ti group were found to be significantly increased on the 28th, 42nd and 56th day. However, the density observed on the 56th day was near approximate to the cortical bone of non-implanted bone on the adjacent side, whereas this observation was observed on the 42nd day in the Ti-Ga group and the 28th day within the Ga-Si-Ti group respectively. This is similar to the observation in the *in-vitro* study on Si plays in bone anabolic regenerative approaches by [54].

The bone mineral density study in the present experiment is also similar to the study conducted by [50] in mice and based on this study it is suggested that the PA view radiograph technique can also be used for the evaluation of the bone mineral density of new biomaterials.

4.7 CT study

The bone mineral density of cortical, trabecular and newly formed bone around the implants was measured in the Hounsfield unit in this CT investigation. On the 28th day, there was no significant difference between groups in cortical, trabecular, or newly formed bone. On the 56th day, the Ti-Ga and Ga-Si-Ti implants had a high bone mineral density in the cortical, trabecular and newly formed bone, whereas the Ti-Ga and Ga-Si-Ti implants had a high bone mineral density in the cortical, trabecular and newly formed bone. It is evident from the study that Ti-Ga and Ga-Si-Ti showed increased bone mineral density, which infers that there was good osseointegration in Ga-Si-Ti and Ti-Ga group implants than Ti implants. This is in agreement with the observation of [55].

In the present CT Study, the study provided in-situ details of mineralization of cortical, trabecular and newly formed bone and the osseointegration of the implants. Hence it is inferred that the increased mineralization in trabecular bone observed in the study is indicative of promotion of osseointegration by the presence of metal elements Gallium and Silicon in Ga-Si-Ti alloy implants. This is in agreement with the observation of [56].

4.8 Micro-CT study

However, the observation on trabecular volume (BV/TV), bone mineral density, trabecular thickness, mean trabecular number, mean trabecular separation revealed a better osseointegration in Ga-Si-Ti implant and it was also can be correlated with histo-morphometric study. The observation on increase in trabecular bone volume around the Ga-Si-Ti and other implants is in agreement with micro-CT study of [57] Further in the present study the BV/TV is 25% and it corresponds with increase in trabecular bone which is indicative of better osseointegration and it is good in all the three groups.

4.9 Histology and Histomorphometric study

Histological analysis revealed the new bone formation between the implant surface and the existing bone. Good osteoblastic activity and mononuclear cells are observed in Ti-Ga alloy implants. New bone formation around implants with Chondrogenic foci, with lacunae and chondrocyte were seen. Intra-peritoneal injections of Alizarin, Tetracycline and Calcein made it possible to evaluate the kinetics of bone growth and bone remodeling histologically around the implants. A higher mineral apposition was detected at 4 weeks was greater than that found at 8th weeks in all three groups. Bone Implant Contact (BIC), Mineral Apposition Rate (MAR) and Osteoblast proximity were the parameters evaluated. Ga-Si-Ti alloy implants have exhibited maximum BIC $\sim$ 98%. Ga has antiresorptive effects on bone and bone fragments, preventing osteoclast resorption.

5. Conclusion

It is evident from the present investigation that the addition of Gallium and Silicon elements with Cp-Ti alloy has improved the mechanical characteristics of Cp-Ti alloy. Studies conducted in vitro demonstrated that all three alloys are bioactive and non-cytotoxic. In the animal experiment, the enhanced osseointegration property of Ga-Si-Ti alloy implants is related to the synergistic anti-resorptive action of Gallium and osteoblastic property of Silicon, in addition to the benefits of Cp-Ti alloy.

This research concludes by stating that the combination of elements Gallium and Silicon to Titanium has not only enhanced the mechanical and physical properties of Cp-Ti alloy, but also the biological qualities of Cp-Ti alloy and that the newly synthesized Ga-Si-Ti alloy implant may be a promising biomaterial in the dental field.

Conflict of interest

The authors declare no conflict of interest.

Author details

Tharanikumar Sivakumar[1*], Chandrasekaran Krithika[2] and Nandikha Tharanikumar[3]

1 Sathyabama Dental College and Hospital, Chennai, India

2 MAHER University, Chennai, India

3 SRM Institute of Science and Technology Vadapalani Campus, Chennai, India

*Address all correspondence to: dr.tharanikumar1977@gmail.com

References

[1] Resnik RR, Misch CE. Classification of dental implant complications. In: Misch's Avoiding Complications in Oral Implantology. St. Louis, Missouri: Elsevier; 2018. pp. 1-12

[2] Qiu KJ, Lin WJ, Zhou FY, Nan HQ, Wang BL, Li L, et al. Ti-Ga binary alloys developed as potential dental materials. Materials Science and Engineering. 2014; **34**:474-483

[3] Antonova NV, Bankovsky OI, Firstov SA, Kulak LD, Tretyachenko LA, Velikanova TY. Structure and mechanical properties of the Ga-Si-Tialloys in the Ti-rich corner. Journal of Materials Science. 1999;**34**(14):3413-3416

[4] Parka JH, Olivares-Navarreteb R, Baierc RE, Meyerc AE, Rina Tannenbaumd E, Boyana BD, et al. Effect of cleaning and sterilization on titanium implant surface properties and cellular response. Acta Biomaterialia. 2012;**8**(5):1966-1975

[5] Choi M, Hong E, So J, Song S, Kim BS, Yamamoto A, et al. Tribological properties of biocompatible Ti-10W and Ti-7.5 TiC-7.5 W. Journal of the Mechanical Behaviour of Biomedical Materials. 2014;**30**(5):214-222

[6] Bernardi S, Bianchi S, Tomei AR, Continenza MA, Macchiarelli. Microbiological and SEM-EDS evaluation of titanium surfaces exposed to periodontal gel: in vitro study. Materials. 2019;**12**(9):1-12

[7] Cavalcanti IMG, Ricomini Filho AP, Lucena-Ferreira SC, Da Silva WJ, Leme AP, Senna PM, et al. Salivary pellicle composition and multispecies biofilm developed on titanium Nitrided by cold plasma. Archives of Oral Biology. 2014;**59**(7):695-703

[8] Kuruvilla M, Srivatsan TS, Petraroli M, Park L. An investigation of microstructure, hardness, tensile behaviour of a titanium alloy: Role of orientation. Sadhana. 2008; **33**(3):235-250

[9] McCracken M. Dental implant materials: Commercially pure titanium and titanium alloys. Journal of Prosthodontics. 1999;**8**(1):40-43

[10] Vandrovcova M, Jirka I, Novotna K, Lisa V, Frank O, Kolskacz, et al. Interaction of human osteoblast-like Saos-2 and MG-63 cells with thermally oxidized surfaces of a titanium-niobium alloy. PloSOne. 2014;**9**(6):1-15

[11] Raghavan RN, Somanathan N, Sastry TP. Evaluation of phytochemical-incorporated porous polymeric sponges for bone tissue engineering: A novel perspective. Proceedings of the Institution of Mechanical Engineers, Part H: Journal of Engineering in Medicine. 2013;**227**(8):859-865

[12] O'Brien J, Wilson I, Orton T, Pognan F. Investigation of the Alamar blue (resazurin) fluorescent dye for the assessment of mammalian cell cytotoxicity. European Journal of Biochemistry. 2000;**267**(17):5421-5426

[13] Al-Nasiry S, Geusens N, Hanssens M, Luyten C, Pijnenborg R. The use of Alamar blue assay for quantitative analysis of viability, migration and invasion of choriocarcinoma cells. Human Reproduction. 2007;**22**(5):1304-1309

[14] Choong PF, The HX, Teoh HK, Ong HK, Choo KB, Sugii S, et al. Heterogeneity of osteosarcoma cell lines led to variable responses in reprogramming. International Journal of Medical Sciences. 2014;**11**(11):1154-1160

[15] Prideaux M, Wijenayaka AR, Kumarasinghe DD, Ormsby RT, Evdokiou A, Findlay DM, et al. SaoS2 osteosarcoma cells as an in vitro model for studying the transition of human osteoblasts to osteocytes. Calcified Tissue International. 2014; **95**(2):183-193

[16] Raghavan N, Dare BJ, Somanathan N, Sastry TP. Evaluation of phytochemical-incorporated porous polymeric sponges for bone tissue engineering: A novel perspective. Trends in Biomaterials & Artificial Organs. 2015;**29**(3):188-192

[17] Ehrnthaller C, Huber-Lang M, Nilsson P, Bindl R, Redeker S, Recknagel S, et al. Complement C3 and C5 deficiency affects fracture healing. PLoS One. 2013;**8**(11):e81341

[18] Webber RL, Ruttimann UE, Groenhuis RAJ. Computer correction of projective distortions in dental radiographs. Journal of Dental Research. 1984;**63**(8):1032-1036

[19] Clavijo D, Casadiego-Torrado C, Perea-Covarrubias A, Galindo-Zamora V. Biofunctionalization of titanium to improve the response in patients with osteoporosis requiring implants. Chemical Engineering Transactions. 2018;**64**(6):643-648

[20] He T, Cao C, Zhiguo X, Li G, Cao H, Liu X, et al. A comparison of micro-CT and histomorphometry for evaluation of osseointegration of PEO-coated titanium implants in a rat model. Scientific Reports. 2017;**7**(1):1-11

[21] Maniatopoulos C, Rodriguez A, Deporter DA, Melcher AH. An improved method for preparing histological sections of metallic implants. International Journal of Oral & Maxillofacial Implants. 1986;**1**(1):31-37

[22] Egan KP, Brennan TA, Pignolo RJ. Bone histomorphometry using free and commonly available software. Histopathology. 2012;**61**(6):1168-1173

[23] Schliephake H, Scharnweber D, Dard M, Robler S, Sewing A, Huttmann C. Biological performance of biomimetic calcium phosphate coating of titanium implants in the dog mandible. Journal of Biomedical Materials Research Part A: An Official Journal of The Society for Biomaterials, The Japanese Society for Biomaterials and The Australian Society for Biomaterials and The Korean Society for Biomaterials. 2003;**64**(2):225-234

[24] Babuska V, Moztarzadeh O, Kubikova T, Moztarzadeh A, Hrusak D, Tonar Z. Evaluating the osseointegration of nanostructured titanium implants in animal models: Current experimental methods and perspectives. Biointerphases. 2016;**11**(3):030801

[25] Kulak CAM, Dempster DW. Bone histomorphometry a concise review for endocrinologists and clinicians. Arquivos Brasileiros de Endocrinologia & Metabologia. 2010;**54**(2):87-98

[26] Cordeiro JM, Beline T, Ribeiro ALR, Rangel EC, da Cruz NC, Landers R, et al. Development of binary and ternary titanium alloys for dental implants. Dental Materials. 2017;**33**(11):1244-1257

[27] Tavares AMG, Souza EAD, Silva MSCD, Matos GRL, Batista WW, Souza SASDA. Role of silicon in the microstructural development and properties of Ti-15Nb-xSi alloys for biomedical applications. Materials Research. 2021;**24**:1-10

[28] Vizureanu PS, Le Yamaguchi PTM, Baltatu MS. Biocompatibility evaluation of new TiMoSi alloys. Acta Physica Polonica A. 2020;**138**(2):283-286

[29] Alaraby HA, MMA I, Ahmed T. A study of mechanical properties of titanium alloy Ti-6Al-4v used as dental implant material. International Journal of Scientific Reports. 2017;**3**(11): 288-291

[30] Elias CN, Fernandes DJ, Moura F, de Souza E, dos Santos M, de Biasi RS. Mechanical and clinical properties of titanium and titanium-based alloys (Ti G2, Ti G4 cold worked nanostructured and Ti G5) for biomedical applications. Journal of Materials Research and Technology. 2019;**8**(1):1060-1069

[31] Faria AC, Lapria RC, Rodrigues S, Claro APRA, da Gloria M, de Mattos C, et al. Wear resistance of experimental titanium alloys for dental applications. Journal of The Mechanical Behaviour of Biomedical Materials. 2011;**4**(8): 1873-1879

[32] Kokubo T. Bioactive glass ceramics: Properties and application. Biomaterials. 1991;**12**(2):155-163

[33] Tadashi Kokubo & Hiroaki Takadama. How useful is SBF in predicting in vivo bone bioactivity. Biomaterials. 2006;**27**(15):2907-2915

[34] Stenport VF, Olander J, Kjellin P, Currie F, Sul YT, Anna A. Precipitation of calcium phosphates in the presence of collagen type I on four different bioactive titanium surfaces: An in vitro study. Journal of Oral & Maxillofacial Research. 2015;**6**(4):1-11

[35] Frojd V, Linderback P, Wennerberg A, Chávez de Paz L, Svensater Davies JR. 'Effect of nanoporous TiO2 coating and anodized Ca2+ modification of titanium surfaces on early microbial biofilm formation', BioMed Central Oral Health. 2011;**11**(1):1-9

[36] Takadama H, Kim HM, Kokubo T, Nakamura T. Mechanism of biomineralization of apatite on a sodium silicate glass: TEM–EDX study in vitro. Chemistry of Materials. 2001;**13**(3): 1108-1113

[37] Arvidsson A, Franke-Stenport V, Andersson M, Kjellin P, Sul Y-T, Wennerberg A. Formation of calcium phosphates on titanium implants with four different bioactive surface preparations. An in vitro study. Journal of Materials Science: Materials in Medicine. 2007;**18**(10):1945-1954

[38] Valencia-Alvarado R, De la Piedad-Beneitez A, López-Callejas R, Barocio SR, Mercado-Cabrera A, Pena-Eguiluz R, et al. Oxygen implantation and diffusion in pure titanium by an rf inductively coupled plasma. Vacuum. 2009;**83**:264-267

[39] Krupa D, Baszkiewicz J, Rajchel B, Barcz A, Sobczak JW, Bilinski A, et al. Effect of calcium-ion implantation on the corrosion resistance and bioactivity of the Ti6Al4V alloy. Vacuum. 2007; **81**(10):1310-1313

[40] Sobieszczyk S. Surface modifications of Ti and its alloys. Advances in Materials Science. 2010;**10**(1):29-42

[41] Przekora A. The summary of the most important cell-biomaterial interactions that need to be considered during in vitro biocompatibility testing of bone scaffolds for tissue engineering applications. Materials Science and Engineering. 2019;**97**:1036-1051

[42] Giner M, Santana L, Costa AF, Vazquez-Gamez MA, Colmenero-Camacho MA, Olmo FJ, et al. Biocompatibility and osseointegration study of new prosthetic materials. Revista De Osteoporosis Y Metabolismo Mineral. 2020;**12**(3):92-97

[43] He W, Dina A, Mosselhy YZ, Feng Q, Li X, Yang X, et al. Effects of silica–gentamicin nanohybrids on osteogenic differentiation of human osteoblast-like SaOS-2 cells. International Journal of Nanomedicine. 2018;**13**:877

[44] Geetha AK, Singh A, Gogia AK. Ti based biomaterials, the ultimate choice for orthopedic implants a review. Progress in Materials Science. 2009; **54**(3):31-37

[45] Chandler JE, Messer HH, Ellender G. Cytotoxicity of gallium and indium ions compared with mercuric ion. Journal of Dental Research. 1994;**73**(9): 1554-1559

[46] Cochis A, Azzimonti B, Chiesa R, Rimondini L, Gasik M. Metallurgical gallium additions to titanium alloys demonstrate a strong time-increasing antibacterial activity without any cellular toxicity. American Chemical Society Biomaterials Science & Engineering. 2019;**5**(6):2815-2820

[47] Postiglione L, Di Domenico G, Ramaglia L, Montagnani S, Salzano S, Di Meglio F, et al. Behaviour of SaOS-2 cells cultured on different titanium surfaces. Journal of Dental Research. 2003;**82**(9): 692-696

[48] Martinez-Ibanez M, Juan-Díaz MJ, Lara-Saez I, Coso A, Franco J, Gurruchaga M, et al. Biological characterization of a new silicon-based coating developed for dental implants. Journal of Materials Science: Materials in Medicine. 2016;**27**(4):80-89

[49] Han P, Chengtie W, Xiao Y. The effect of silicate ions on proliferation, osteogenic differentiation and cell signalling pathways (WNT and SHH) of bone marrow stromal cells. Biomaterials Science. 2013;**1**(4):379-392

[50] Khojastepour L, Mohammadzadeh S, Jazayeri M, Omidi M. In vitro evaluation of the relationship between gray scales in digital intraoral radiographs and Hounsfield units in CT scans. Journal of Biomedical Physics & Engineering. 2017; **7**(3):289-295

[51] Yi YA, Park YB, Choi H, Lee KW, Kim SJ, Kim KM, et al. The evaluation of osseointegration of dental implant surface with different size of TiO_2 nanotube in rats. Journal of Nanomaterials. 2015;**2015**:1-11

[52] Bahraminasab M, Arab S, Safari M, Talebi A, Kavakebian F, Doostmohammadi N. In vivo performance of Al 2 O 3-Ti bone implants in the rat femur. Journal of Orthopaedic Surgery and Research. 2021;**16**(1):1-14

[53] Ballo A, Xia W, Lindahl C, Lausmaa J, Engqvist H, Thomsen P. Bone responses to ions-substituted apatite/titanium dioxide coating on titanium implants. In: International Association for Dental Research (IADR). Rijeka, Croatia: IntechOpen; 2010

[54] Costa-Rodrigues J, Reis S, Castro A, Fernandes MH. Bone anabolic effects of soluble Si: in vitro studies with human mesenchymal stem cells and CD14+ osteoclast precursors. Stem Cells International. 2016;**2016**:1-12

[55] Ozturk F, Babacan H, Gumus CM. Effects of zoledronic acid on sutural bone formation: A computed tomography study. The European Journal of Orthodontics. 2012;**34**(2): 141-146

[56] Rico H, Gallego-Lago JL, Hernandez ER, Villa LF, Sanchez-Atrio A, Seco C, et al. Effect of silicon supplement on osteopenia induced by

ovariectomy in rats. Calcified Tissue International. 2000;**66**(1):53-55

[57] Butz F, Ogawa T, Chang TL, Nishimura I. Three-dimensional bone-implant integration profiling using micro-computed tomography. International Journal of Oral & Maxillofacial Implants. 2006;**21**(5): 687-695

Chapter 10

Diagnosis and Treatment Planning for Multiple Implants in Esthetic Zone

Divya Krishnamoorthi

Abstract

A new era in dentistry dawned when successfully osseointegrated implants were used to restore edentulous jaws by Branemark et al. This important discovery opened up avenues previously unimaginable in edentulous rehabilitation. The predictability and longevity of implant restoration have been irrevocably cemented in dentistry. Even with state-of-the-art technologies for implant planning and placement, the rehabilitation of the anterior maxilla has posed a particular problem because of the high esthetic demand and complex anatomy of the region. Since these two parameters vary highly in patients, a careful evaluation of the clinical scenario and comprehensive history-taking should precede meticulous treatment planning to ensure accurate diagnosis and successful treatment outcomes. Both fixed and removable options are available for the rehabilitation of the edentulous anterior maxilla. This chapter aims to compare these options and prosthetic designs with an emphasis on diagnosis and treatment planning. A comprehensive checklist for easy decision-making regarding treatment planning will be presented.

Keywords: dental esthetics, dental implants, diagnosis, edentulous jaw, maxilla

1. Introduction

> *'One of the most dangerous diseases is diagnosis'.*
>
> *- George Bernard Shaw*

Successful rehabilitation of partially edentulous arch by means of osseointegrated implants was popularized by Branemark et al. [1–3]. This launched a new era of management for the partially edentulous predicament. The rehabilitation of anterior maxillary teeth to date remains one of the most challenging situations in dentistry due to a number of esthetic and functional aspects of the restoration [2–4]. Both fixed and removable implant-supported prosthesis are used to treat this condition. High esthetic demand coupled with exacting patient expectations deems it necessary to invest considerable time in accurate diagnosis and treatment planning. This would ensure predictable outcomes and patient satisfaction. This chapter aims to compare available treatment options and prosthesis design for edentulous maxilla with emphasis on diagnosis and treatment planning. This would be accomplished through an evidence-based review of factors influencing the clinical decision-making process. Criteria for decision-making parameters with regard to removable v/s fixed prosthesis will also be put in place.

2. Special condition in maxilla

Historically the use and research of implant prosthesis were conducted predominantly on the mandibular arch, due to reduced denture-bearing surface and function of the tongue contributing to the instability of mandibular dentures. A multitude of designs was evaluated from implant-supported complete dentures to subperiosteal implants to provide better patient stability and comfort [5].

The accommodation to maxillary denture was easier than the mandibular counterpart, in part due to better retention, stability, support, and esthetics. When implant rehabilitation was attempted, the principles established for mandibular implants were adopted. This meant screw-retained cantilever pontics were fabricated in spaces with excessive resorption and long standard abutments were installed with prosthesis on top. Though acceptable in the mandible, in the maxilla it left open interproximal spaces compromising esthetics [5].

This is complicated by factors like (1) the resorption pattern of the maxilla, where following extraction the horizontal bone resorption is twice as pronounced as vertical (2) anatomical limitation for implant placement due to the vertical distance between the alveolar crest and the nasal sinuses in the anterior maxilla and (3) pneumatization of maxillary sinuses which limits implant placement in posteriors [6]. The long-term prognosis for implants in the maxilla is less secure than that of the edentulous mandible. The poor bone quantity, quality, and high esthetic demands complicate diagnosis, treatment planning, and treatment in the anterior maxillary region [7].

3. Diagnosis

The most important decision to be made while rehabilitating an edentulous maxilla is whether to restore it with a fixed or removable prosthesis. All diagnostic criteria must be evaluated before formulating a treatment plan. Parameters such as bone quality and quantity, lip support, lip line and esthetic demands, etc., to be evaluated are described in detail by Zitzmann and Marinello [8] implant placement should be deferred until a definitive diagnosis and treatment plan have been framed.

Checklist for implant rehabilitation:

1. Patient factors

   a. Patient preference

   b. Phonation

   c. Ability to effectively perform oral hygiene

   d. Economics

2. Extra oral examination [9]

    a. Facial and lip support

    b. Esthetic plane

    c. Maxillomandibular relationship

    d. Smile line

    e. Lip length

    f. Vestibular space

    g. Horizontal tooth display

3. Intraoral examination

    a. Quality of mucosa

    b. Thickness of mucosa

    c. Quantity and quality of available bone

    d. Incisal papilla position

    e. Interarch space

    f. Speech disruption

    g. Tooth size to arch size discrepancies

3.1 Patient factors

If the patient presents with an existing denture, it could act as one of the best diagnostic tools. The denture must be examined thoroughly objectively for retention, support, stability, vertical dimension, the tooth selection and arrangement, phonetics and esthetics. Patients option on the same should also be carefully noted down.

Tendency to prefer fixed restorations over removable have been demonstrated previously, but the responsibility of determining the best treatment option lies with the restorative dentist after careful evaluation of all the parameters [8].

3.2 Extra-oral examination

3.2.1 Facial and lip support

The resorption pattern of maxilla generally proceeds in a cranial and medial fashion, rendering a retruded position of maxilla in anterior region. Assessment must be done with and without the existing denture from both front and profile view.

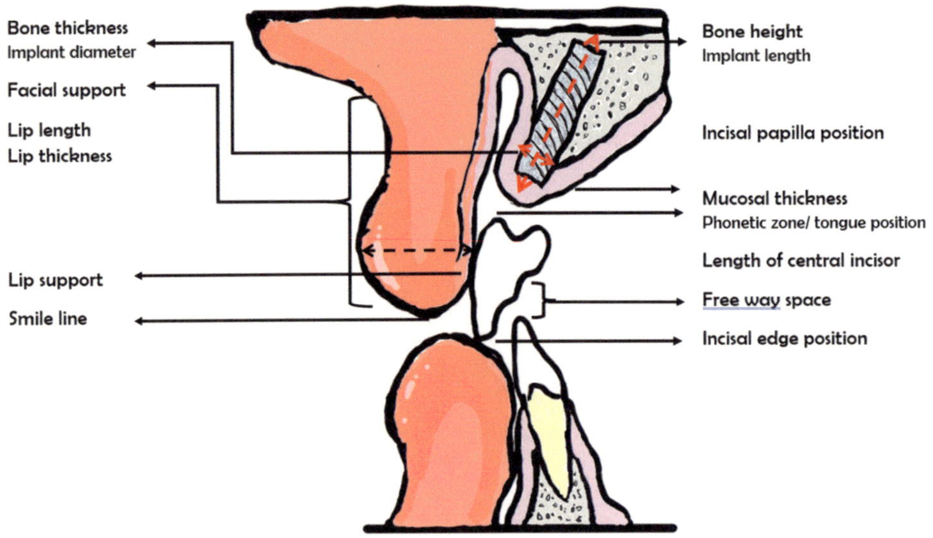

Figure 1.
Extraoral and intraoral factors.

Inadequate facial support may be corrected by buccal flange of removable restoration. Lip support is provided by shape of alveolar ridge and cervical crown contour of the anterior teeth (**Figure 1**).

A diagnostic set up must be prepared to better evaluate anterior teeth, position and relation to lip. Often depending on the severity of the resorption, there may be discrepancy between the anticipated position of teeth and the ridge. This must be considered while planning the implant position as so to satisfy patient expectation of function, esthetics and phonation. Which trying to overcome larger discrepancies, extensive grafting and removable prosthesis with flange may be of use, the associated limitations and risks must be clearly explained to the patient [9].

3.2.2 Esthetic plane and maxillomandibular relationship

"E" plane represents a line drawn from the nose tip to the chin tip. For a pleasing appearance the upper lip and lower lip must be 4 m and 2 mm behind the line respectively. There are mild changes amongst ethnic groups. It may be generalized that closer the lips are to the "E" plane the more dominate the teeth and lips will appear, the farther behind the "E" plane the lips are, the more dominate the nose and chin appear.

Based on this for a patient with Class 2 maxillomandibular relationship, where maxilla is prognathic, having convex profile care must be taken not to anterior teeth large or too white as to make them further dominant. And vice-verse should be considered for a concave profile patient, having Class 3 maxillomandibular relationship, where the mandible is prognathic. Lighter shades may benefit these patients by balancing the stronger nose and chin [9].

3.2.3 Smile line and lip length

The orofacial muscular tonicity and lip mobility during speech and smile determines the smile line. An average smile as described by Tjan et al. displays 75–100% of maxillary incisors and interproximal gingiva. Low smile line displays less than 75% of the incisors, and high smile displays additional gingiva [9, 10]. Vig and Brundo classified patients into 5 groups based on upper lip length and presented the corresponding tooth display [11]. They concluded that average lip length of 21–25 mm displayed 2.18 mm and 1 mm of maxillary and mandibular teeth. Short lip length of 10–15 mm exposed 4 mm of maxillary teeth and 0.7 mm of mandibular teeth. Whereas long lip length of 31–35 mm showed only 0.25 mm maxillary incisor and 2.25 mm of mandibular incisors.

Ridge display during smile presents a challenge in rehabilitation as gingiva- restoration junction would be in visible zone. Lip and support must be evaluated as they influence the tooth exposure during smile and speech. A long lip in most instances covers the anterior reducing the exposure during smile, favorable for rehabilitation of anterior. Whereas a short lip exposes anterior and meticulous planning with be required to satisfy esthetic parameters.

3.2.4 Vestibular space and horizontal tooth display

When a patient presents with an existing denture, the smile must be evaluated with and without the denture. The movement of lips and display zone of smile must be noted. The lips function as curtains than frame the teeth during smile. Smile esthetics is affected by the amount of tooth and soft tissue displayed both in vertical and horizontal dimension. Assessment must be made include posed/voluntary smile which is static and an involuntary or dynamic smile. The lip is often much more animated in an involuntary smile. Excessive display or gummy smile is less attractive and often challenging to rehabilitate as the junction between the prosthesis and the soft tissue may be in the smile zone and discernable. Use of removable prosthesis in such cases with pink acrylic seamlessly forming a scaffolding of interdental papilla and attached gingival would be much preferred [12].

Frush and Fischer described buccal corridor space as the space between the buccal surfaces of posterior teeth and corners of the mouth in smile. This negative space has influence on smile esthetics as well. The presence of this space creates a much more natural smile as opposed to denture like smile created by elimination of the space all together. It is suggested that a minimal buccal corridor space is more attractive [13, 14].

3.3 Intra-oral examination

3.3.1 Quality and quantity of mucosa

Loss of tooth is accompanied with the resorption of interseptal bone, and bone remodeling around the socket. Often the interdental papilla is flattened or is missing due to lack of interproximal contact. The loss of interdental papilla in most scenario is an irreversible damage [15]. Achieving pre-extraction papillary architecture is unlikely and the patient must be made acutely aware of this.

The quality of mucosa should be evaluated by palpation, radiographs and sounding [16]. Gingival biotype was classified by Seibert and Lindhe as thick, thickness of more than or equal to 2 mm and thin, having thickness less than 1.5 mm [17]. Thin gingival biotype presents consider challenge in molding and thoughts must be directed towards soft tissue grafting to change the biotype to thick. Thick biotype is easier to mold and helps hide the abutment margins and facilitates proper emergence of the prosthetic crown.

The loss of interdental papilla may be compensated by manipulating the soft tissues by the use of ovate pontic. Or by the use of gingival colored porcelain or acrylic resin.

3.3.2 Bone quality and quantity

The resorption in maxilla follows a centripetal pattern. In severe resorption cases this results in maxilla which is superior and palatal to the original position of the dentate alveolar ridge. This creates a biomechanical disadvantage while placing anterior implant where prosthesis must most definitely be placed in pre-extraction tooth position with a labial cantilever. This also poses a problem in posterior region where the ridge may be in cross bite condition. This would result in prosthesis in cross bite or with excessive facial cantilever. Hence a sound assessment of the bone quality and quantity should precede any treatment planning [18].

A classification based on post extraction ridge changes of mandible was proposed by Atwood [19], though quantitative, lacking enough details to help in detailed treatment planning, lead to classification by Lekholm and Zarb to be popularized [5]. Both maxilla and mandible was classified into 5 shapes (A–E) based on the degree of resorption and 4 patterns (1–4) based on quantity and type of cortical bone present. Shape A represents minimal resorption and E severe resorption, type 1 having thick cortical area and type 4 thinnest. The Quality of bone often dictates the treatment planning. Edentulous maxilla generally comprises of type 3 or 4 bone quality, may prescribe placement of additional implants, in anticipation of failure of one or more. Another A Therapeutically Oriented Classification- HVC Ridge Deficiency Classification delivers therapeutic recommendations for both hard and soft tissue deficiencies [20].

As visual inspection cannot reveal the density, and volume of the bone underneath the soft tissue, radiographical examination is mandatory. Cone beam computer tomography (CBCT) reveal three-dimensional architecture for precise planning of surgical and prosthetic phases. The CBCT may be performed with gutta percha or metal markers for maximum advantage from the scan. These markers are placed perpendicular to the occlusal plane in acrylic resin duplicates of diagnostic denture. The exact trajectory of the bone can be evaluated using CBCT making treatment planning easier and unfailing [21].

Long standing edentulism of posterior maxilla may result in pneumatization of maxillary sinuses, reducing the quantity if bone available for implant placement in this region. Corrective procedure like maxillary sinus lift and ridge augmentation can be performed [22]. The donor site for the augmentation mostly likely depends on the volume of bone required. Commonly utilized site is iliac crest when augmenting both sinuses, requiring additional surgical procedure and less acceptance amongst patients. Alternatively, information from the CBCT may be used to direct the implant away to alternative sites with voluminous bone making the procedure minimally invasive [23]. Zygomatic, maxillary tuberosity and pterygomaxillary implants can be used as an alternative treatment option. Zygomatic implants engage the zygomatic bone inferolateral to the orbital rim, along with anterior implants they can provide anchorage to maxillary prosthesis.

3.3.3 Incisal papilla position

Anterior maxillary implant placement is further complicated due to the anatomical variations in dimension of incisive canal and foramen. The canal has two openings—the incisive foramen and nasopalatine foramen. The nasopalatine nerves and vessels transverse the canal. Accidental damage to this anatomical structure may cause nervous tissue injury, sensory dysfunction and even non osseointegration of implants. The size, position, shape and number of foramina are varied in population. The presence of wider foramina along with thin alveolar bone may require surgical intervention (**Figure 1**). Careful clinical and radiographic assessment assisted by CBCT must be made [8].

3.3.4 Inter arch space

Available interarch space commands the prosthetic design. Different designs, different types of prosthesis require different dimensional tolerance. The resorption pattern of alveolar ridge has been a considerable problem. Accurately mounted casts on semi-adjustable articular with facebow transfer can be used to study the interarch distance and make decisions regarding the prosthetic design type best situated for the situation (**Figure 1**).

A conventional overdenture requires 12–16 mm, of which 2–3 mm is reserved for the heat cure acrylic resin to provide sufficient bulk, and rest for the prosthetic teeth. When it is planned to connect, he implants additional 2–3 mm space is required to accommodate he superstructures. A screw retained prosthesis requires only 10–12 mm space limitations may require reestablishing patient's vertical dimension or change to occlusal plane [24].

3.3.5 Incisal edge position

Guidelines used in conventional complete denture construction can be used to determine the incisal edge position of the future prosthesis. Visibility of 2 mm anteriorly determines the anterior occlusal plane. Additionally, 'F', 'V' and 'S' sounds can be used to determine the correct vertical and horizontal placement of the incisors [25]. The average length of central incisor is 10.5 mm, clinical crown height may increase in elderly due to gingival recession. The in clinical of the anterior teeth should be determined based on the lip support required. After determining the crown height and angulation the available space from the crestal bone height should be calculated (**Figure 1**).

In case of minimal resorption, the cervical edge of the anterior teeth would coincide with the crestal soft tissue level, in which case a fixed restoration would be ideal. A try- in should be done without the flange of the aid in this assessment. In case of large discrepancy between the crest and cervical aspect of try teeth, but lip support, pink ceramic can be used to overcome this discrepancy. To optimize the existence of both vertical and horizontal discrepancies removable prosthesis with flange may be indicated [25].

4. Radiographic diagnosis

As mentioned earlier the accurate assessment of bone quality and quantity is paramount for successful implant treatment. It affects implant selection, position and angulation. A CBCT with a radiographic stent or diagnostic set up with radiographic markers incorporated depicting the ideal implant position can be used. The incisal edge of the radiographic marker represents the ideal position of the screw access channel of a screw retained fixed implant prosthesis located at the cingulum or the central fossa. All possible implant positions are marked with titanium pins (7 mm in length) after determining the ideal angulation using a surveyor. The markers should be incorporated so as to be perpendicular to the occlusal plane and end apically at the height of the planned clinical crown margin [24].

5. Treatment planning

A foolproof treatment plan should be devised after careful consideration of patient needs and expectations, and all the diagnostic findings addressed above. It is paramount that the treatment plan addresses the chief complaint of the patient. The expected final outcome should be communicated with the patient with utmost clarity. To avoid disappointment after undergoing an expensive and time-consuming treatment patient education, informed consent after understanding and acknowledging the advantages, disadvantages and limitations of the plan is vital. Also, any expected modifications to the plan should be explained and signed off by the patient prior to the treatment [12].

Jivraj, Chee and Corrado have listed the factors to be considered in treatment planning for edentulous maxilla:

1. Esthetics and patient desires

2. Type of prosthesis

3. Number of implants

4. Implant distribution

5. Economics

5.1 Esthetic and patient desire

Understanding the patient's motivation for seeking treatment is integral to the satisfying these needs. Promises must not be made before thorough diagnosis. in case a demand to replace a removable prosthesis with fixed is made by the patient, consequence of missing denture flange must be explained to the patient. The flange may compensate the lip to teeth horizontal discrepancy, or vertical discrepancy of brought about because of severe resorption. In such cases a complete denture provides better esthetics by reproducing the lost soft tissue especially interdental papilla in pink acrylic. This may be lost entirely when replacing with a fixed prosthesis, and patient may end up with unesthetic black triangles, due to missing interdental papilla in the esthetic zone. An effort can be made to create an illusion of interdental papilla by altering the contact point between the teeth and staining, or utilizing pink ceramics, the results may or may not be to the satisfaction of the patient. This must be communicated with the patient explicitly [12].

5.2 Type of prosthesis

Implant prosthesis may derive support either from the underlying implant alone or from both the implant and foundation tissues. Implant prosthesis can be categorized as fixed or removable. Fixed prosthesis is metal ceramic prosthesis that are supported by the underlying implants and based on the discrepancy between the planned tooth position and the crest of the soft tissue may or may not require pink ceramic to simulate gingival contour. Removable prostheses are categorized as implant supported overdentures and implant retained overdentures [24]. An implant retained prosthesis is supported by foundation tissues and retained by the implant through ball or magnets attachments, bar and clips or precision milled components [12].

5.3 Number of implants

Consideration must be giving to a multitude of factors when planning the number of implants to be placed, some of them are:

1. Quality of bone

2. Anticipated force to be placed on the restoration

3. Relationship between the shape of the residual ridge and the dental arch form.

5.3.1 Quality of bone

Success of implant rehabilitation in maxilla is less than mandible. Recommendations for rehabilitation edentulous maxilla with fixed prosthesis is to place 6–8 implants at a distance of 10–15 mm, and with removable restoration 4–6 implants for both implant and implant and tissue supported restoration [24]. Maxilla is frequently composed of type 3 or 4 quality bone. When the quality of the bone is too poor for conventional drill, over engineering of the maxilla may be called for.

Long term edentulism in posterior maxilla leading to pneumatization of sinuses and refusal to ridge augmentation procedures, may drive the restorative dentist to consider either cantilever restorations or short dental arch concept [26, 27].

5.3.2 Anticipated forces

Facial and oral muscle tonicity must be taken int. o consideration when planning implant prosthesis. Any hypertrophy of oral musculature must be noted, especially of the masseter. Pronounced antegonial notch indicative of bruxer and likely high forces on the restoration must be anticipated. An over engineering would be advised in such cases.

The opposing arch must be carefully examined for anticipated occlusal contacts, presence of restoration or prosthesis. A rehabilitation opposing natural dentition is much likely to exert larger force than one opposing a complete denture. Larger the anticipated force more the number of implants should be allowed for [12].

5.3.3 Relationship between shape of residual ridge and dental arch form

Dental arches and the residual ridge are classified as ovoid, square and tapering. An esthetic rehabilitation may require the restored dental arch form to deviate from the residual ridge shape. The arch form is determined by the last tooth on the pre maxilla and not the residual ridge form, which may in cases have to be placed facially to attain better esthetics.

An ovoid dental arch requires placement of one implant each at the canine position and at least one additional implant, preferably in the central incisor position. The premaxilla to accommodate 3 implants may have to undergo augmentation procedures. The addition of the anterior implant in the central incisor position provides better resistance to forces and provides biomechanical stability to the prostheses design. This reduces the forces on the abutment screw, reducing the chances of screw loosening. In a square arch the canine, lateral and central incisor are placed more or less on the same plane. This reduces the forces between the cantilevered lateral and central incisors, and placement of one implant each at canine region may suffice when splinted with posterior implants to restore the square arch [12].

Rehabilitating a tapered arch is the most challenging situation, as the anterior implants witness the greatest forces, because the anterior teeth are cantilevered far facially. This dictates placement of four implants in the premaxilla to replace the six missing anterior teeth.

5.4 Implant distribution

Implant distribution and placement will be the determining factor in restoration's emergence profile. Patients planned to receive removable or fixed restoration with considerable gingival ceramic can be forgiving to incorrect or placement of implants in interproximal position. The flange or pink porcelain will disguise the improper placement. For rehabilitation with fixed prosthesis, it is imperative that implant placement be precisely planned and executed using surgical guides. Incorrect placement or placement in interproximal position causes esthetic and hygiene maintenance problems [4].

The distribution of implant along the arch is crucial for better load sharing. Antero posterior spread of the implant should be optimized and cantilever minimized. Splinting advantageous as it improves biomechanics of the prosthesis design. In case of implant retained prosthesis the placement becomes crucial as straight-line placement would be required to successfully execute a connection for a bar, that does not impinge the palatal tissues. The distribution should be as to accommodate a clip to be placed in the denture [24].

5.5 Economics

Implant rehabilitation is a costly affair, and both fabrication and maintenance phase should be taken into account. Fixed restoration requires implant components and meticulous laboratory support and are hence costlier than their removable counterpart. Though the maintenance phase requires less investment. While overdentures fabrication is cheaper, the maintenance is consistent, and hence it is questionable whether the economical indication of overdenture be justified. The cost factor of both phases must be made aware to the patient [12].

Table 1 summarizes the checklist to be followed for treatment planning in edentulous maxilla for implant prosthesis.

Checklist for implant rehabilitation:		
	Fixed implant prosthesis	**Removable implant prosthesis**
1. Patient factors		
a. Patient preference	Most preferred	With minimal palatal coverage is preferred
b. Phonation	Most problems	Fewer problems
c. Ability to effectively perform oral hygiene	Most demanding	Easier
d. Economics	More expensive initially	Less expensive initially
2. Extra oral examination		
a. Facial and lip support	Present	Needed from the prosthesis
b. Esthetic plane	Convex profile	Concave profile
c. Maxillomandibular relationship	Angle class I/ II	Angle class III
d. Smile line	Low	High
e. Lip length	Long	Short
f. Vestibular space	Less visible	Increased visibility
g. Horizontal tooth display	Minimal	Maximum
3. Intraoral examination		
a. Quality of mucosa	Keratinized	Non keratinized
b. Thickness of mucosa	Thick	Thin
c. Quantity and quality of available bone	Shape A/B Type 2/3	Shape C/D/E Type 3/4
d. Incisal papilla position	Palatal	Crestal/buccal
e. Interarch space	10–12 mm	>12 mm
f. Tooth size to arch size discrepancies	Absent	Present

Table 1.
Checklist to be followed for treatment planning in edentulous maxilla for implant prosthesis.

6. Conclusion

Diagnosis and treatment planning is a crucial step in rehabilitation of edentulous maxilla with implant supported or retained prosthesis. A myriad of factors affects the type of prosthesis design, implant number and distribution. Factors to consider before making a final treatment plan include parameters such as patient related factors like expectation and desire, extraoral factors like facial and lip support, smile line, and intraoral factors such as bone and soft tissue quality and quantity. And all these must be carefully considered, clinical and radiographic evaluation is must be done with at most sincerity and a treatment plan conceived must a line with patient expectation.

Author details

Divya Krishnamoorthi
Department of Prosthodontics and Crown and Bridge, Vinayaka Mission's Sankarachariyar Dental College, A Constituent College of Vinayaka Mission's Research Foundation (Deemed to be University), Ariyanur, Tamil Nadu, India

*Address all correspondence to: drdivyakrishna@vmsdc.edu.in

References

[1] Lekholm U, Gunne J, Henry P, Higuchi K, Lindén MSU, Bergström DDSC, et al. Survival of the Brånemark implant in partially edentulous jaws: A 10-year prospective Multicenter study. The International Journal of Oral & Maxillofacial Implants. 1999;**14**:639-645

[2] Adell R, Lekholm U, Rockler BBP. A 15 year study of osseointegrated implants in the treatment of the edentulous jaw. International Journal of Oral Surgery. 1981;**10**(6):387-416

[3] Engquist B, Nilson HAPS. Single tooth replacement by osseointegrated Branemark implants: A retrospective study of 82 implants. Clinical Oral Implants Research. 1995;**6**(4):238-245

[4] Misch CE. Contemporary Implant Dentistry. 3rd ed. St. Louis: Mosby. pp. 397-428

[5] Lekholm U, Zarb G. Patient selection and preparation. In: Branemark P-I, Zarb GA, Albrektsson T, editors. Tissue-integrated Prostheses: Osseointegration in Clinical Dentistry. Chicago: Quintessesnce; 1985. pp. 199-209

[6] Tallgren A. The reduction in face height of edentulous and partially edentulous subjects during long-term denture wear a longitudinal roentgenographic cephalometric STUDY. Acta Odontologica Scandinavica. 1966;**24**(2):195-239

[7] Bosse LP, Taylor TD. Problems associated with implant rehabilitation of the edentulous maxilla. Dental Clinics of North America. 1998;**42**(1):117-127

[8] Zitzmann NU, Marinello CP. Treatment plan for restoring the edentulous maxilla with implant-supported restorations: Removable overdenture versus fixed partial denture design. The Journal of Prosthetic Dentistry. 1999;**82**(2):188-196

[9] Tjan AHL, Miller GD, The JGP. Some esthetic factors in a smile. The Journal of Prosthetic Dentistry. 1984;**51**(1):24-28

[10] Manjula WS, Sukumar MR, KishorekumarS, GnanashanmugamKMK. Smile: A review. Journal of Pharmacy & Bioallied Sciences. 2015;**7**(Suppl. 1):S271

[11] Vig RG, Brundo GC. The kinetics of anterior tooth display. The Journal of Prosthetic Dentistry. 1978;**39**(Table I):502-504

[12] Jivraj S, Chee W. Treatment planning of implants in the aesthetic zone. British Dental Journal. 2006;**201**(2):77-89

[13] Tikku T, Khanna R, Maurya RP, Ahmad N. Role of buccal corridor in smile esthetics and its correlation with underlying skeletal and dental structures. Indian Journal of Dental Research. 2012;**23**(2):187-194

[14] Khava NEA, Arab S, Shamshiri A, Osle IMM. The effect of the buccal corridor and tooth display on smile attractiveness. Australian Orthodontic Journal. 2015;**31**(2):195-200

[15] Tarnow D, Elian N, Fletcher P, Froum S, Magner A, Cho SC, et al. Vertical distance from the crest of bone to the height of the interproximal papilla between adjacent implants. Journal of Periodontology. 2003;**74**(12):1785-1788

[16] Kan JY, Rungcharassaeng K, Umezu K, Kois JC. Dimensions of

peri-implant mucosa: An evaluation of maxillary anterior single implants in humans. Journal of Periodontology. 2003;**74**(4):557-562. DOI: 10.1902/jop.2003.74.4.557

[17] Seibert JL, Lindhe J. In: Lindhe J, editor. Esthetics and Periodontal Therapy. 2nd ed. Copenhangen, Denmark: Munksgaard; 1989. pp. 477-514

[18] Wilson DJ. Ridge mapping for determination of alveolar ridge width. The International Journal of Oral & Maxillofacial Implants. 1989;**4**(1):41-43

[19] Atwood DA. Postextraction changes in the adult mandible as illustrated by microradiographs of midsagittal sections and serial cephalometric roentgenograms. The Journal of Prosthetic Dentistry. 1963;**13**(5):810-824

[20] Wang HL, Al-Shammari K. HVC ridge deficiency classification: A therapeutically oriented classification. International Journal of Periodontics and Restorative Dentistry. 2002;**22**(4):335-343. Available from: http://www.ncbi.nlm.nih.gov/pubmed/12212680

[21] Schwarz MS, Rothman SL, Rhodes ML, Chafetz N. Computed tomography: Part II. Preoperative assessment of the maxilla for endosseous implant surgery. The International Journal of Oral & Maxillofacial Implants. 1987;**2**(3):143-148

[22] Seibert S. Alveolar ridge preservation and reconstruction. Periodontology 2000. 2000;**11**(1996):69-84

[23] Henry PJ. A review of guidelines for implant rehabilitation of the edentulous maxilla. The Journal of Prosthetic Dentistry. 2002;**87**(3):281-288

[24] Misch CE, editor. Contemporary Implant Dentistry. 3rd ed. India: Mosby, Elsevier; 2007

[25] Robert LE. Complete-denture esthetics. Dental Clinics of North America. 1996;**40**(1):71-84

[26] Armellini D, Von Fraunhofer JA. The shortened dental arch: A review of the literature. The Journal of Prosthetic Dentistry. 2004;**92**(6):531-535

[27] Witter DJ, Van Palenstein Helderman WH, Creugers NHJ, Käyser AF. The shortened dental arch concept and its implications for oral health care. Community Dentistry and Oral Epidemiology. 1999;**27**(4):249-258

Chapter 11

Simple Surgical Methods for Soft Tissue Management around Fresh Extraction Sockets during Socket Preservation

Fares Kablan

Abstract

Soft tissue volume and quality are considered important factors for functional and esthetic long-term outcomes around natural teeth and dental implants. However, achieving them is challenging for oral surgeons. Healing of an extraction site is combined with normal physiological ridge resorption and loss of interdental papillae scaffold. Therefore, the rehabilitation of these ridges with dental implants or pontic site of fixed dental prosthesis usually necessitates soft tissue management to achieve natural-looking tooth replacement. The aim of this chapter is to introduce two surgical topics that are used to preserve the soft tissue quality, volume, and architecture during teeth extraction procedure. The first topic is the "transient coronectomy" that is used to save the interdental papilla during teeth extraction, and the second topic is "The back-cut technique" that is used to enhance the socket seal of post-extraction sites.

Keywords: socket preservation, local flaps, interdental papilla, coronectomy, back cut

1. Introduction

The physiological healing process following tooth extraction results in hard and soft tissue losses and ridge contour deformation and has a detrimental effect on the subsequent treatment with dental implants or conventional prostheses [1–4]. Preservation of post-extraction socket has some clinical relevance in preventing physiological bone resorption of the extracted site. Augmentation of post-extraction sockets requires a primary closure of the wound to promote proper ridge regeneration. To achieve that goal, different materials and methods are used to enhance the socket seal such as autogenic free soft tissue grafts [5–10], extraction-site granulation tissue [11], acellular dermal matrix [12, 13], collagen matrix and sponge [14], resorbable and non-resorbable barrier membranes [15, 16], periosteal scraping, and local flaps [17, 18].

Surgical resection of skin lesions results in skin defects that can be treated with local randomized flaps [19]. The back-cut incisions have been used in skin surgeries to enhance the rotation and the advancement of the designed flap for easier closure of an open wound [20] cresent. Flap advancement and rotation are also used in the oral surgery for reconstruction of soft tissue deficiency around teeth, dental implants, and bone grafts [17, 18].

The modern dentistry is alert to both esthetic and function of the treatment outcomes. Loss of the interdental papilla has a great impact on the final restorative treatment outcomes, especially in the esthetic zones. Therefore, treatment plans should consider maintaining and restoring the interdental papilla anatomy and architecture around dental implants and fixed partial denture pontic sites. Teeth extractions usually lead to sockets collapse, ridge resorption, and loss of interdental papilla volume and anatomy. Reconstruction of interdental papilla in the edentulous sites is considered one of the greatest treatment challenges facing the multidisciplinary dental team [21–25].

Several surgical and nonsurgical treatment modalities have been described for papilla preservation and reconstruction around natural teeth and around implants with their comorbidities and predictability [26–30]. Nonsurgical techniques include orthodontics that is usually used for teeth and root alignment [31–33] and restorative treatment aimed to correct the position of the contact point between two adjacent teeth that has a critical effect on the papilla [34]. Among the surgical techniques, free connective tissue graft and pedicle flaps are used for interdental papilla management. Azzi et al. in 1998 reported the use of subepithelial graft for papilla reconstruction [35] and the use of connective tissue graft that was harvested from the maxillary tuberosity and the palate in 2001 [36]. Socket grafting procedures have also been considered to preserve the ridge contour and to support the soft tissue in the extraction site [37]. As an alternative to socket preservation, the root submergence technique was described and used in order to preserve the periodontal tissue at the pontic site of fixed dental prosthesis [38, 39]. In the past, coronectomy was performed leaving submerged roots inside the ridge, in order to preserve the dimensions of the alveolar ridge and therefore to improve the retention and the stability of conventional removable prostheses [40]. In addition, coronectomy is also performed in surgical extractions of impacted wisdom teeth at the lower jaw as an alternative treatment to minimize the risk of nerve injury and to reduce complications [41–43].

Additional use of coronectomy is described by the author as a novel method, which is delineated as "transient coronectomy." This method includes the removal of the clinical crown of the subsequent extracted tooth by leaving the root submerged inside the ridge in order to prevent the collapse of the interdental papilla during the preparation of the temporary fixed prosthesis. The submerged root will be removed at the end of the treatment appointment.

The present chapter introduces two surgical tips for soft tissue management during tooth extraction. The first method is the "back-cut technique" that enhances mobilization of the flap at a coronal direction to obtain the primary closure of the bone-augmented extraction sites. This technique can be used for single or multiple extraction sockets without affecting the vestibular depth. The second tip is the "transient coronectomy" that carries out intraoperatively in order to preserve the interdental papillae during the treatment appointment and to enhance natural tooth locking around dental implants and natural teeth.

2. Back-cut technique

2.1 Technique presentation

A full thickness flap with intrasulcular incision is carried out around the extracted tooth with two releasing incisions obliquely, mesial and distal, that are extended 3–4 mm

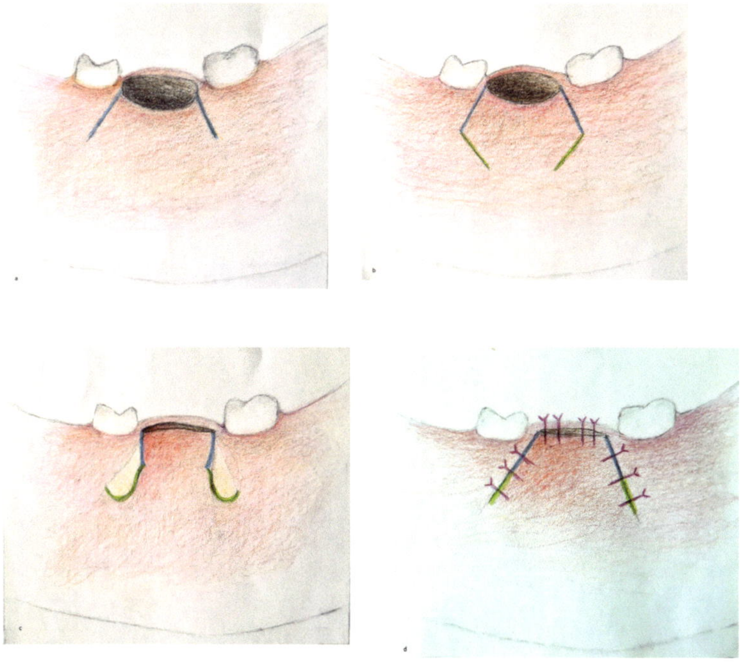

Figure 1.
Schematic view of the "back-cut technique". (a) Trapezoid flap; intrasulcular incision, and two releasing incisions. (b) Two additional incisions anterior and posterior performed as back cuts at the base of the trapezoid flap form the back-cut technique design. (c) Rotation and advancement of the flap at coronal direction covering the sockets as a result of the addition of the two back cuts. (d) The socket seal was obtained with a wide base flap following the final suturing.

apically to the mucogingival junction (trapezoidal flap) (**Figure 1a**). Two additional incisions are performed at the base of the flap; the first incision is angulated distally 110–120 degrees at the anterior releasing incision, and the second incision with the same angulation is performed continuously with the distal releasing incision and angulated mesially. Those two additional incisions are delineated by the author as the "back cuts" (**Figure 1b**), and the flap design is delineated as the "back-cut technique". This technique is performed in order to enhance the rotation and the advancement of the flap coronally covering the socket without periosteal-releasing incisions. As a result of the flap advancement and rotation, all the incisions (two releasing incisions and two back cuts) are dislocated coronally and create enough soft tissue for the socket seal (**Figure 1c**). The tooth might be extracted before or after the flap elevation. The next step is socket grafting with bone particles and the suturing of the flap. The sutures are first performed at the socket area (occlusally), followed by sutures at the releasing incisions of both the mesial and the distal, including the two back cuts in their new location. A wide base flap configuration is achieved at the end of the suturing job and looks like trapezoid flap (**Figure 1d**).

2.2 Clinical cases

2.2.1 Case 1

A 35-year-old patient was referred for the extraction of two maxillary teeth and socket preservation. Clinical and radiographic examinations showed extensive crown

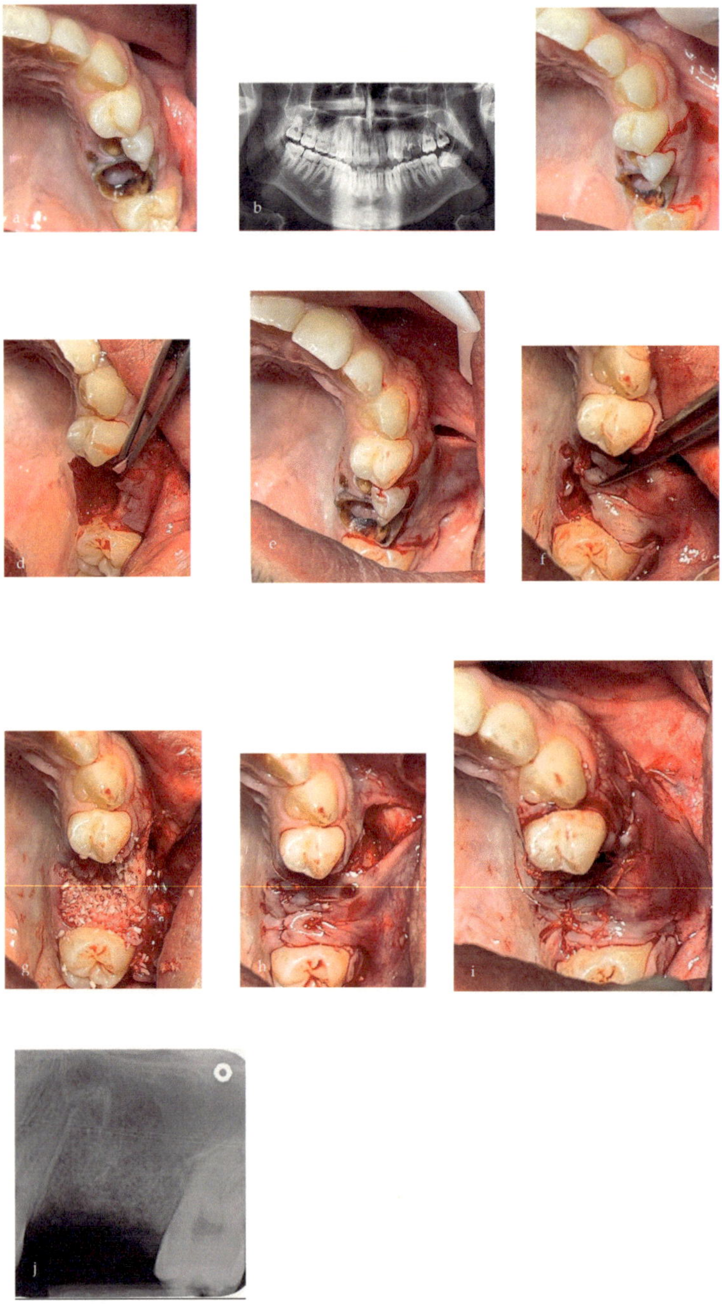

Figure 2.
Back-cut technique, case 1. (a and b) Clinical and radiographic views, crown destruction with periapical lesions involved the left maxillary second premolar, and the first molar. (c)A trapezoid flap was designed and obtained by intrasulcular and two releasing incision. (d) Intraoperative view demonstrates the disability of the flap to cover. (e) Intraoperative view shows that the flap rotation and advancement coronally obtained by the back cuts will enhance the sockets seal. (f) Particulate bone substitute is used to fill the socket. (g) The sockets' preservation with particulate bone graft. (h) The sutures at the occlusal part of the sockets obtained by flap rotation and advancement demonstrate the primary closure. (i) Primary closure at the bone-augmented socket without affecting the vestibular depth; a wide base flap was achieved. (j) Periapical view of the treated site at four months after the surgery demonstrates the bone graft.

destruction of the second premolar and the first molar at the left maxilla accompanied with periapical lesions (**Figure 2a** and **b**). Trapezoid flap was performed by intrasulcular incision followed by mesial- and distal-releasing incisions (**Figure 2c**). Thereafter, the teeth were extracted, and followed by meticulous debridement of the apical lesions. However, the soft tissue covering of the fresh extraction sockets was not obtained by the trapezoid flap (**Figure 2d**), therefore, two back cuts were performed (**Figure 2e**) to enhance the flap rotation and for the advancement in enhancing the soft tissue seal of the sockets (**Figure 2f**). The sockets' preservation was then carried out with an allogenous particulate bone (**Figure 2g**). The back-cut technique was utilized for the primary closure of the augmented socket. The back-cut technique flap design allowed the socket seal, preventing the exfoliation of the bone particles from the socket (**Figure 2h**). Primary closure was obtained at the end of the suturing job (**Figure 2i**). During 4-months follow-up postoperatively, the healing process went very well without any dehiscence of the surgical site or loss of the bone graft (**Figure 2j**). Thereafter, the patient was referred to his oral surgeon for implant placement.

2.2.2 Case 2

A 42-year-old patient was referred for the extraction of her first right mandibular molar and socket preservation (**Figure 3a**). Following trapezoid flap, two back-cut incisions were performed, the first anterior and the second posterior, at the extraction site. After tooth extraction and meticulous debridement of the socket, human-derived bone particles were used to fill the socket (**Figure 3b** and **c**). The extraction open

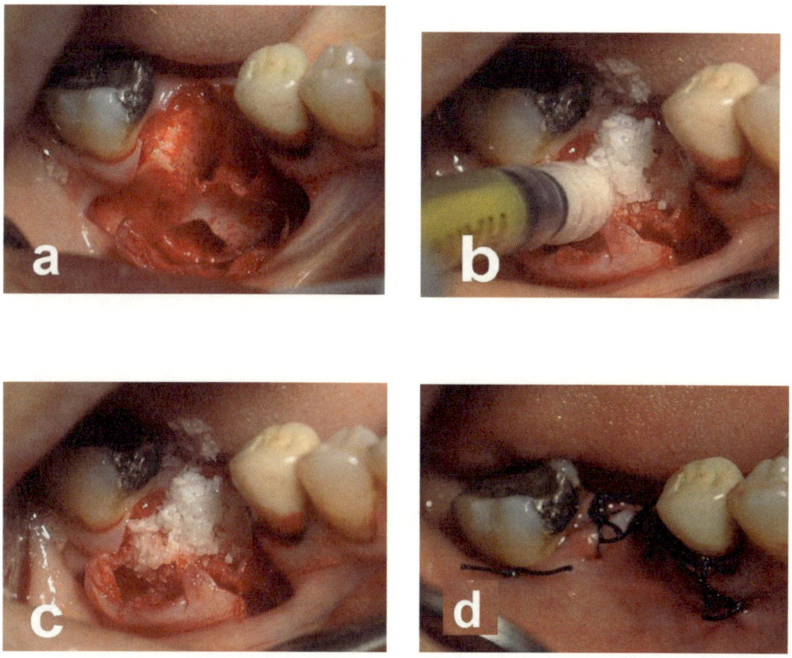

Figure 3
(a) Extraction socket of the first right mandibular molar with two back cuts: anterior and posterior. (b and c) Socket preservation with human-derived bone substitute. (d) Primary closure of the bone-augmented socket, via coronally rotation and advancement of the flap without affecting the vestibular depth.

wound was closed primarily by rotation and advancement of the flap (**Figure 3d**). The earlier designed back cuts enhanced the rotation and advancement of the flap without periosteal scrub and without distortion of the vestibular depth.

3. Transient coronectomy

Transient coronectomy, which is described herein, is indicated when maxillary or mandibular teeth at the esthetic zones should be extracted and reconstructed by temporary fixed prosthesis. The subsequent final rehabilitation might be dental implant-supported or conventional fixed prosthesis.

3.1 Technique

Illustration case: A 50-year-old patient was referred for extractions of his six anterior maxillary teeth that had been used as abutments for an old porcelain-fused-to-metal (PFM) fixed prosthesis and subsequent placement of dental implant. The treatment plan included the extraction of the teeth, socket preservation, and temporary restoration during the healing period. After 4–5 months, dental implants were placed for dental implant-supported fixed prosthesis rehabilitation. The transient coronectomy technique was used during the teeth extractions and the socket preservation procedure and is illustrated in the following 6 steps.

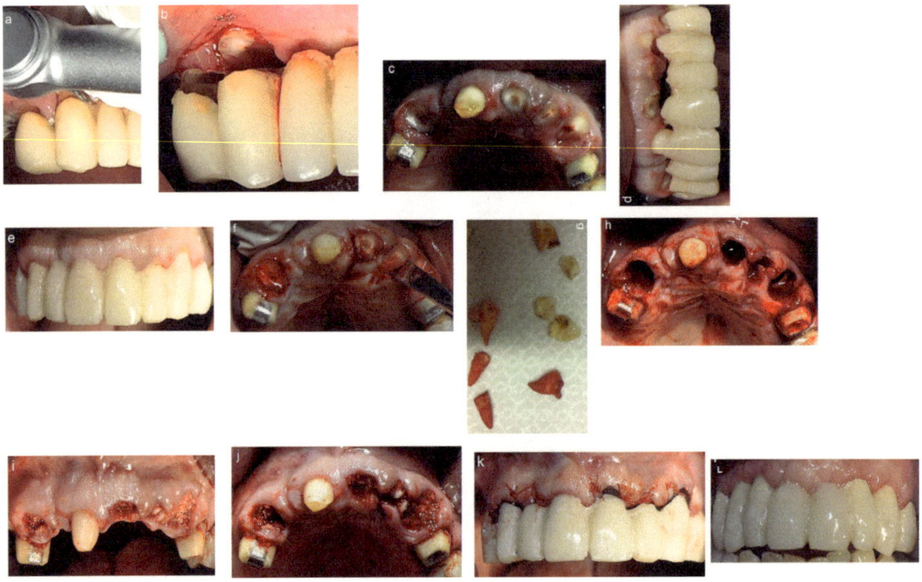

Figure 4
(a) High speed is used to cut the clinical tooth crown (CTC). (b) The clinical tooth crown (CTC) was cut over the level of the gingiva. (c) Grinding of the roots in a concave shape, and the interproximal root shelf was designed as the papilla outline. Abutment teeth were prepared. (d) Try in of the temporary bridge. (e) Passive setting of the temporary bridge inside the submerged roots, keeping the original interdental papilla outline. (f) Atraumatic extraction of the submerged roots. (g) 4 crowns were transected at the beginning of the treatment. (h) Sockets after the extraction of the submerged roots with minimal injury to the soft tissue. (i and j) Socket preservation; excellent preservation of the interdental papilla and the socket soft tissue. (k) Cementation of the temporary bridge, immediately after the suturing. (l) 3 months' follow-up: good ridge contour. The interdental papilla was preserved in its original shape and volume.

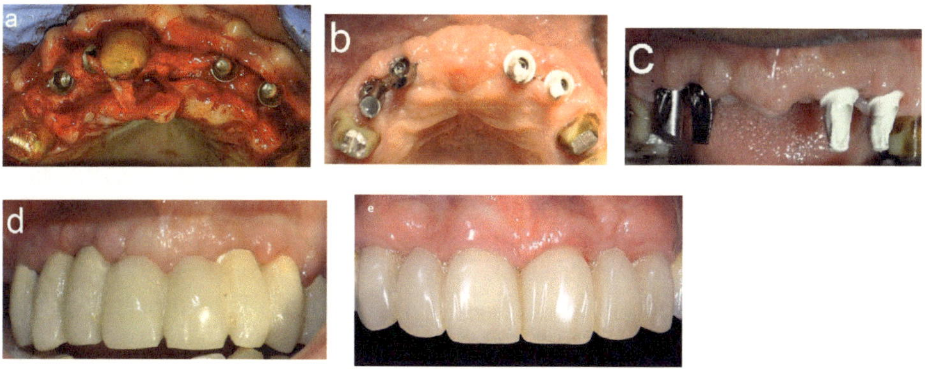

Figure 5
(a) Implant placement (at 5 months). (b and c) Implants follow-up (after 4 months). (d) Temporary bridge over the implants. (e) Follow-up of 24 months: fixed prosthesis over the implants with good esthetic outcome.

Step 1: Decortication of the clinical tooth crown (CTC) was carried out at 1–2 mm above the gingival margins in order to prevent injury to the soft tissue and resulted in submerged roots (**Figure 4a** and **b**). Step 2: The submerged roots underwent a careful concave shaving and trimming, leaving the interproximal thin walls to support the interdental papilla, without injury to the soft tissue (**Figure 4c**). Step 3: Relining of the prefabricated temporary fixed prosthesis was carried out to fit the designed submerged roots outline; this maneuver supports the papillae and preserves the gingival architecture (**Figure 4d** and **e**). Step 4: Atraumatic extraction of the submerged roots was carried out (**Figure 4f** and **g**). Step 5: Socket preservation and soft tissue suturing were carried out (**Figure 4h–j**). Step 6: Cementation of the already relined temporary prosthesis was carried out (**Figure 4k**). The patient had been followed up frequently during the healing time during 4–5 months after the extractions (**Figure 4l**).

Dental implants were placed 5 months after the first surgery (**Figure 5a**), and fixed prosthesis supported dental implants was performed after an additional 4 months. The outcomes have been satisfying during the subsequent follow-up (**Figure 5b–e**).

In cases that the rehabilitation of the extracted tooth will be performed by conventional fixed prosthesis supported abutment teeth, the final rehabilitation was performed 4 months after the socket preservation surgery.

4. Discussion

4.1 Back-cut technique

Extraction of one molar or more than two adjacent teeth creates an open wound at the extraction site. Spontaneous physiological healing of the socket will usually lead to socket wall resorption and soft tissue contracture. This can be prevented by bone grafting of the extraction socket. Bone grafts in the extraction sockets requires a primary closure to promote proper ridge regeneration. Primary closure of the socket is necessary to prevent postoperative flap dehiscence and exposure and migration of the bone particles. Several methods have been described to achieve this goal [6–16]. Landsberg and Bichacho were the first to describe the socket seal surgery, utilizing soft tissue graft or biomaterials to restore the soft tissue volume and architecture [7]. Tal reported the use of an autogenous masticatory mucosal graft to seal the extraction sockets [8]. Others have used connective tissue grafts [9, 10]. Mardinger et al. described the use of intra-socket soft tissue to cover the augmented sockets [11]. Rotated palatal flaps with full or partial thickness have been reported for the primary coverage of extraction sites [17, 18]. Moreover, alternative non-autogenous grafts are widely used in the socket seal surgeries. Kim et al. in 2011 described the use of collagen sponge and xenogenic bone grafts for socket preservation [14]. Additional materials are also successfully used such as acellular dermal matrix [12, 13] and different types of collagen membranes [14–16]. Rotation flaps with or without advancement are widely used in the cosmetic medicine to close open skin wounds. Rotation flaps with back cuts are also used to enhance the flap length and its movement over the primary defect. Nasseri and Murray in 2015 reported the crescentic back cut that contributes to the better mobilization of the flap, reduces the scar length, and shifts tensions away from the direction of the primary defect closure, while eliminating tissue redundancy at the flap's base [20]. The back-cut technique presented in this chapter is used to achieve the primary closure of the bone-augmented fresh sockets. It provides the oral surgeon similar advantages as those reported in the cosmetic surgery. They increase the mobilization and the sliding of the flap over the open socket wound, lengthen the flap, and enhance the tension free primary closure over the augmented bone. The designed incisions of the flap and their suturing method convert again the narrow base flap resulting from the back cuts into a wide base flap. This improves the blood supply to the flap base and eliminates the possible redundancy of the flap. The addition of mesial and distal back cuts, as described herein, to the original incisions of the trapezoid flap for socket seal surgeries eliminates the need for periosteal-releasing incisions in the inner side. Periosteal-releasing incisions enhance only the flap advancement but miss the rotation ability. In addition, they jeopardize the blood supply and affect the healing of the operated site negatively. The ability to close the extraction open wound without periosteal scrub has several additional advantages that include: preservation of the periosteal integrity and blood supply, the flap thickness and circulation, reduction of infections during the healing period. Moreover, the vestibular depth will not be affected, which will improve the oral hygiene throughout the subsequent site rehabilitation. It is well established in the oral surgery literature that primary soft tissue closure over the surgical site increases the predictability of bone regeneration at the treated sites [44]. Additional indications for this technique include soft tissue primary closure during immediate dental implant placement in fresh sockets, open sinus augmentation combined with teeth extractions, and guided bone regeneration procedures.

In those cases where soft tissue augmentation is not mandatory at the extraction site and is used only to achieve primary closure, the back-cut technique can be used predictably, which eliminates the need for additional surgical procedures for harvesting soft tissue grafts and their accompanying suffering such as pain, discomfort, hemorrhage, or infection. Moreover, the back-cut technique also eliminates the need for non-autogenous covering materials such as barrier membranes, mucografts, and acellular dermal matrix, and as a result, it reduces the potential side effects as foreign materials and their costs. The grafted bone volume at the sockets that are sealed by the back-cut technique was preserved during the healing period and subsequently during the dental implant surgery in the second stage. The inserted implants had an adequate new bone volume and quality at the recipient sites.

4.2 Transient corenectomy technique

The presence of interdental papilla has a crucial role in gingival and teeth esthetics and the patient's smile. The etiology of interdental papilla loss is multifactorial, including teeth extractions. Preservation or reconstruction of the lost interdental papilla as a result of teeth loss or extractions has been challenging for the dental treatment team [26]. In this chapter, the "transient coronectomy" is described as a novel approach to prevent papilla collapse during teeth extractions. The collapse of the interdental papilla and loss of the gingival contour occur immediately after the extraction of the teeth, so immediate provisionalization is recommended to optimally preserve the tissue during the surgical procedure and thereafter. In 2006, Margeas reported the use of natural teeth as provisional following implant placement to achieve peri-implant gingival esthetics [45]. In 2008, Taleghani et al. reported the use of a temporary bridge with an ovate pontic at the site of extraction to support the proximal papilla and the facial soft tissue and to enhance the healing gingival tissue [46]. For the rational to preserve the soft and hard tissues and to maintain the periodontal attachment complex, the root submergence technique has been reported and used in esthetic implant therapy with favorable esthetic outcomes [38, 39]. However, several complications were encountered with the applications of this technique such as periapical lesions, external root resorption, root caries, and late eruption of the submerged root that can result in the extraction of submerged root. As a result, late failure of the treatment that required more complicated reconstruction procedures may occur [47]. The transient coronectomy technique, as a novel concept, for the use of the submerged root intraoperatively is defined by the author as the "coronectomy" of the hopeless tooth clinical crown, and its extraction at the end of the treatment appointment is defined as "transient." According to this approach, the clinical tooth crown (CTC) is transected at 1–2 mm above the gingival level in order to prevent injury to the soft tissue; thereafter, the final coronectomy procedure is accomplished by shaving and trimming of the residual tooth root carefully in a concave form to the level of the surrounding soft tissue, leaving the interproximal thin walls of the root to support the interdental papilla. According to this work platform, the outline of the trimmed submerged root will support the interproximal papilla during the relining of the prefabricated temporary fixed prosthesis (crown, bridge) with 100% fitting to the original location and shape of the interdental papilla. Moreover, the provisionalization will inhibit immediate papilla collapse that will otherwise occur rapidly if it is performed after entire tooth extraction. The removal of the roots later during the same appointment and socket preservation eliminates the risk of late failure that might be accounted in the scenario of leaving the root submerged. The transient coronectomy technique used by the author predictably maintains the interdental papilla topography and the gingival emergence profile during the treatment appointment, the healing period, and the subsequent treatments with final favorable cosmetic results.

An additional benefit of extraction of the submerged root after the relining of the temporary fixed prosthesis for the dental treatment team is the working in a clean environment. If the entire tooth is extracted at the beginning of the treatment, it will cause oozing during the treatment on the one hand, and the relining of the temporary prosthesis can contaminate the fresh sockets on the other hand.

5. Conclusion

Primary closure of the bone-augmented sockets was obtained and maintained using the back-cut technique with minimal costs and minimal postoperative morbidities. This technique is advocated in cases of socket seal surgery that involve extraction of adjacent teeth or even only one molar. The transient coronectomy technique during tooth extraction is a simple tool to accelerate interdental papilla preservation with long-term outcomes. Both surgical methods provide the patients and the dental treatment team with simple facilities to enhance the esthetic outcomes of rehabilitation of the extractions sites with minimal morbidity and costs.

Acknowledgements

Thanks to Dr. Abir Abu Subeh for technical support.

Thanks to Ms. Eiman Zriek for her illustration drawing of beautiful pictures.

Author details

Fares Kablan[1,2]

1 Department of Oral and Maxillofacial Surgery, Galilee Collage of Dental Sciences, Galilee Medical Center, Naharyia, Israel

2 The Azzriele Faculty of Medicine, Bar-Ilan University, Safed, Israel

*Address all correspondence to: kablanp1@gmail.com

References

[1] Arau'jo MG, Lindhe J. Dimensional ridge alternations following tooth extraction: An experimental study in the dog. Journal of Clinical Periodontology. 2005;**32**:212

[2] Van der Weijden GA, Dell' Acqua F, Silt DE. Alveolar bone dimensional changes of post extraction socket in human: A systematic review. Journal of Clinical Periodontics. 2009;**36**(12):1048-1058

[3] Mecall RA, Rosenfeld AL. Influence of residual ridge resorption patterns on fixture placement and tooth position. 2. Presurgical determination of prosthesis type and design. The International Journal of Periodontics & Restorative Dentistry. 1992;**12**(1):32-51

[4] Mecall RA, Rosenfeld AL. Influence of residual ridge resorption patterns on fixture placement and tooth position. Part III: Presurgical assessment of ridge augmentation requirements. The International Journal of Periodontics & Restorative Dentistry. 1996;**16**(4):322-337

[5] Lekov V, Kenney EB, Weinlaender M, et al. A bone regenerative approach to alveolar ridge maintenance following tooth extraction. Report of 10 cases. Journal of Periodontology. 1997;**68**(6):563-570

[6] Misch CE, Dietsh-Misch F, Misch CM. A modified socket seal surgery with composite graft approach. The Journal of Oral Implantology. 1999;**25**:244-250

[7] Lansberg C, Bichacho N. A modified surgical/prosthetic approach for optimal single implant supported crown. Part 1—The socket seal surgery. Practical Periodontics and Aesthetic Dentistry. 1994;**6**(2):11-17

[8] Tal H. Autogeneous masticatory mucosal grafts in extraction socket seal procedures; a comparison between sockets grafted with demineralized freeze-dried bone and deproteinized bovine bone materials. Clinical Oral Implants Research. 1999;**10**(4):289-269

[9] Edel A. The use of a connective tissue graft for closure with over an immediate implant covered with occlusive membrane. Clinical Oral Implants Research. 1995;**6**(1):60-65

[10] Kablan F, Laster Z. The use of free fat tissue transfer from the buccal fat pad to obtain and maintain primary closure and to improve soft tissue thickness at bone-augmented sites: Technique presenation and report of case series. The International Journal of Oral & Maxillofacial Implants. 2014;**29**(2):e220-e231

[11] Mardinger O, Chaushu G, Ghelfan O, Nissan J. Intrasocket reactive soft tissue for primary closure after augmentation of extraction sites with severe bone loss before implant placement. Journal of Oral Maxillofactory Surgery. 2009;**67**(6):1294

[12] Novaes AB Jr, Souza SL. Acellular dermal matrix graft as a membrane for guided bone regeneration: A case report. Implant Dentistry. 2001;**10**(3):192-196

[13] Batista EL, Batista FC, Novaes AB. Management of soft tissue ridge deformities with acellular dermal matrix, clinical approach and outcome after 6 months of treatment. Journal of Periodontology. 2001;**72**(2):265-273

[14] Kim YK, Yun PY, Lee HJ, Kim SG. Ridge augmentation using collagen sponge and xenogenic bone grafts. Implant Dentistry. 2011;**20**:267-272

[15] Nunes FAS, Pignaton TB, Novaes AB Jr, Taba M Jr, Messora MR, Palioto DB, et al. Evaluation of a bone substitute covered with a collagen membrane for ridge preservation after tooth extraction. Clinical and tomographic randomized controlled study in humans. Clinical Oral Implants Research. 2018;**29**:424-433

[16] Pedroso Bergamo ET, Bordin D, Arbex L, Konrad D, Gil LF, et al. Ridge architecture preservation following minimally traumatic exodontia techniques and guided tissue regeneration. Implant Dental. 2019;**28**:319-328

[17] Nemcovsky CE, Artzi Z, Moses O. Rotated split palatal flap for soft tissue primary coverage over extraction sites with immediate implant placement. Description of the surgical procedure and clinical results. Journal of Periodontology. 1999;**70**(8):926-934

[18] Tal H, Bichacho N, Kornowski Y, Nemcovsky CE. Rotated palatal flaps: A functional and aesthetic solution in endentulous sites. Practical Procedures & Aesthetic Dentistry. 2004;**16**(8):599-606

[19] Baker SR. Local Flaps in Facial Reconstruction. 2nd ed. Philadelphia, PA: Elsevier; 2007

[20] Nasseri E, Murray C. Crescentic back cut: A novel technique for the management of flap/defect discrepancies in rotation flaps. Journal of Cut Medical and Surrounding. 2015;**201519**(6):588-591

[21] Gruder U. The inlay graft technique to create papilla between implants. Journal of Esthetic Dentistry. 1997;**9**(4):165-168

[22] Tarnow DP, Eskow RN. Considerations for single-unit esthetic implant restorations. Compendium of Continuing Education in Dentistry. 1995;**16**(8):1778-1784

[23] Chen MC, Liao YF, Chan CP. Factors influencing the presence of interproximal dental papilla between maxillary anterior teeth. Journal of Periodontology. 2010;**81**(2):318-324

[24] Kim TH, Cho YJ, Lee JY, Kim SJ, Choi JI. An analysis of the factors responsible for relative position of interproximal papilla in healthy subjects. Journal of Periodontal & Implant Sinica. 2013;**43**(4):160-167

[25] Singh UP, Uppoor AS, Nayak DG, Shah D. Black triangle dilemma and it's management in the esthetic dentistry. Dental Research Journal. 2013;**10**(3):296-301

[26] Prato GP, Rotundo R, Cortellini P, Tinti C, Azzi R. Interdental papilla management: A review and classification of the therapeutic approaches. The International Journal of Periodontics and Restorative Dental. 2004;**24**(3):246-255

[27] Takei HH. The interdental space. Dental Clinics of North America. 1980;**24**(2):169-176

[28] Zetu L, Wang HL. Management of inter-dental/inter-implant papilla. Journal of Clinical Periodontology. 2005;**32**(7):831-839

[29] Cortellini P, Prato GP, Tonetti MS. The modified papilla preservation technique. A new surgical approach for interproximal regenerative procedures. Journal of Periodontology. 1995;**66**(4):261-266

[30] Cortellini P, Prato GP, Tonetti MS. The simplified papilla preservation flap. A novel surgical approach for the management of soft tissues in regenerative procedures. The

International Journal of Periodontics and Restorative Dentistry. 1999;**19**(6):589-599

[31] Blatz MB, Hurzeler MB, Stub JR. Reconstruction of the lost interproximal papilla: Presentation of surgical and nonsurgical approaches. International Jpurnal of Periodontics and Restorative Dentistry. 1999;**19**(4):395-406

[32] Kokich VG. Esthetics: The orthodontic-periodontic restorative connection. Seminars in Orthodontics. 1996;**2**(1):21-30

[33] Inhgher JS. Forced eruption: Alternation of soft tissue cosmetic deformity. The International Journal of Periodontics & Restorative Dentistry. 1989;**9**:417-425

[34] Wu YJ, Tu YK, Huang SM, Chan CP. The influence of the distance from the contact point to the crest of bone on the presence of interproximal dental papilla. Chang Gung Medical Journal. 2003;**26**(11):822-828

[35] Azzi R, Etienne D, Carranza F. Surgical reconstruction of the interdental papilla. International Journal of Periodontics and Restorative Dental. 1998;**18**(5):467-474

[36] Azzi R, Takei HH, Etienne D, Carranza FA. Root coverage and papilla reconstruction using autogenous osseous and connective tissue grafts. The International Journal of Periodontics & Restorative Dentistry. 2001;**21**(2):141-147

[37] Nemcovsky CE, Serfaty V. Alveolar ridge preservation following extraction of maxillary anterior teeth. Report on 23 consecutive cases. Journal of Periodontology. 1996;**67**(4):390-395

[38] Salama M, Ishikawa T, Salma H, Funato A, Garber D. Advantages of the root submergence technique for pontic site development in esthetic implant therapy. The International Journal of Periodontics & Restorative Dentistry. 2007;**27**(6):521-527

[39] Mohadeb JV, Somar M, He H. Effectiveness of decoronation technique in the treatment of ankylosis: A systematic review. Dental Traumatology. 2016;**32**(4):255-563

[40] Casey DM, Lauciello FR. A review of the submerged-root concept. Journal of Prosthetic Dentistry. 1980;**43**(2):128-132

[41] O'Riordan BC. Coronectomy (intentional partial odontectomy of lower third molars). Oral Surgery, Oral Medicine, Oral Pathology, Oral Radiology, and Endodontology. 2004;**98**(3):274-280

[42] Kim YB, Joo WH, Min KS. Coronectomy of lower third molar in combination with vital pulp therapy. European Journal of Dentistry. 2014;**8**(3):416-418

[43] Porgel MA. Partial odontectomy. Oral & Maxillofacial Surgery Clinics of North America. 2007;**19**(1):85-91

[44] Moses O, Pitaru S, Artzi Z, et al. Healing of dehiscence-type defects in implants placed togother with different barrier membranes: A comparative clinical study. Clinical Oral Implants Research. 2005;**16**:210

[45] Margeas RC. Predictable periimplant gingival esthetics: Use of natural tooth as provisional following implant placement. Journal of Esthetic and Restorative Dentistry. 2006;**18**(1):5-12

[46] Taleghani M, Rashan S, Baker FL, Denson PS, Herrman AB. Nonsurgical management of interdental papillae loss

following extraction of anterior teeth. General Dentistry. 2008;**56**(4):326-331

[47] von Wowernn N, Winther S. Submergence of roots for alveolar ridge preservation. A failure (4-year follow-up study). International Journal of Oral Surgery. 1981;**10**(4):247-250

Chapter 12

Dental Implants: Immediate Placement and Loading in the Esthetic Zone

Nachum Samet

Abstract

Extraction of an anterior tooth is always a traumatic event for patients, regardless of their age. In the past, patients were required to wait long months for bone healing and for implant integration, using an uncomfortable removable denture. In the last decade or so, dentists came to the understanding that when done correctly and in the right cases, immediate placement and loading techniques can result in aesthetically pleasing temporary and long-term results, while maintaining supporting bone and soft tissues. In fact, it was found, that placing an immediate implant rather than waiting can in fact prevent some of the bone resorption which happens after extractions.

Keywords: dental implants, immediate placement, immediate loading, aesthetic zone, temporary/provisional crown, ovate pontic

1. Introduction

Anterior teeth function in eating and speaking, of course, but also have a role in shaping facial aesthetics. In fact, many people see them as part of their personal identity. For these reasons, loss of an anterior tooth is always a traumatic event for patients, regardless of their age. To relieve their physical and emotional pain, patients seek a fixed restorative solution as fast as possible.

Patients who suffered severe trauma and lost multiple teeth are usually aware of the fact that it is almost impossible to restore their teeth to their original look. With a good explanation from a dentist, their expectation level can be adjusted so that it is realistic and so that they can accept a non-ideal restoration, which is both aesthetic and functional but not identical to the way their original teeth and gums looked before the trauma.

Patient expectation level is completely different when only one anterior tooth is lost. In these cases, it is clear to patients, regardless of their condition, that it is possible to restore both the "white" and the "pink" to their original state so that both the restoration and the gum-line look exactly like the homologous existing tooth.

The problem is that when a tooth is lost, two known phenomena happen: flattening and loss of proximal papillae as well as horizontal and vertical bone loss.

Although there are techniques to restore lost soft and hard tissues, these techniques are complex, require extremely high skills and experience, require long healing periods and are expensive. And on top of this, even in the best hands, these techniques are not 100% predictable [1–3].

Two advancements opened the way to preserve both bone and soft tissue architecture in the anterior upper segment. One is the understanding that placing an implant immediately after extraction can reduce bone loss at the site, maintaining the buccal aspect above the restored tooth as close as possible to the way it looked prior to the trauma [4–6]. Another way to preserve buccal bone and gingival architecture is the "Socket-Sheild" technique. In this technique, the root is cut mesiodistally leaving its buccal aspect attached to the buccal bone, and the implant is placed behind it. Although found effective [7–9], this technique is extremely difficult to perform and requires conditions that do not always exist, therefore, it will not be discussed in this chapter.

Prior to treatment, and as in any medical and dental procedure, it is critical to start with a correct diagnosis. The entire site and its surroundings with a focus on the remaining root, the integrity of the buccal bone, the architecture of the gums around the site and any evidence of infection at the gum level or around the root must be evaluated. Taking photographs of the anterior region before treatment is highly recommended, both for further analysis and evaluation of the healing phase and medico-legal reasons.

The second most important phase is to carefully evaluate all possible treatment options and discuss them with the patient, who should be notified about the pros and cons of each option, including restoring the missing dentition with a fixed bridge, a removable device, or an implant. If an implant-based solution is chosen, the different treatment sequences: immediate placement, immediate placement and loading or late placement with or without immediate loading must also be discussed with the patient, presenting again the pros and cons of each option. And lastly, the patient expectations must be discussed, ensuring that he/she understands that no one can achieve 100% success rates in any medical or dental procedure.

Short and long-term success requires mastering both the "white stuff" (the crown) and the "pink stuff" (the gingiva surrounding the site). The control over the "white stuff" is almost always in the hands of a master dental technician. And indeed, dental technicians who are also artists can achieve amazingly wonderful aesthetic results when it comes to mimicking the natural shade, translucency, transparency and anatomical characteristics of the adjacent teeth. It is not always easy, but undoubtedly possible. In contrast, the control over the 3D position of the implant and the recreation of correct emergence profile and gingival architecture is completely in the hands of the operating dentist or the dental team. The correct positioning of the implant at the site, its angulation, depth and especially the position of its opening is critical to enable achievement of the desired aesthetic and functional result. Next, a temporary crown with a unique gingival design, which ensures proper support for the gums and papillae, must be placed [10–15]. Following a period of healing and gingival maturation, it is mandatory to transfer the exact achieved gingival architecture to the dental laboratory so that the final crown is made with the desired gingival shape to ensure long-term stability of the gumline.

This chapter aims to present the different stages of immediate placement and loading in the anterior aesthetic zone, focusing on critical influencing parameters that affect short- and long-term success.

2. Stages of immediate implant placement and loading in the aesthetic zone

The long-term aesthetic success of a restoration in the anterior zone and its gingival architecture requires a thorough analysis of the condition of the injured area. Patients need to be informed and must understand their condition so that they develop realistic expectations related to the possible final outcome.

2.1 Evaluation of the site

Bone condition and height around the socket, especially in the buccal aspect and near adjacent teeth, are critical. Loss of bone in any direction, especially in the buccal area, severely affects the ability to simply restore site's aesthetics. Loss of bone due to infection, periodontal disease, root fracture or crack or iatrogenic accidental damage to the delicate buccal bone during extraction may contra-indicate immediate placement and loading of implants in such sites. In these cases, a staged approach, including bone and/or gingival augmentation, may be required prior to implant placement. The type of gingiva is also important since it is easier to control thick biotype gingiva than thin biotype gingiva, although, with careful management, it is possible to achieve good results in any gingival type.

2.2 Implant design and drilling protocol alterations

Immediate implant placement and loading is an accepted and predictable dental procedure, with high success rates. As in other implant placement procedures, high initial stability is one of the most important factors related to their short- and long-success rates.

There is a critical difference between the way immediate implants and standard implants gain their primary stability. While standard implants gain stability throughout their length, immediate implants gain stability through the threads of their apical section. This is a result of the fact that the diameter of the extraction socket is almost always wider than the diameter of the implant (**Figure 1**). Because of this, it is highly recommended that the dentist chooses an implant with deep and sharp apical threads, (**Figure 2**), that can cut through bone to ensure high initial stability. In addition, the dentist MUST alter the standard drilling protocol, taking into account the diameter of the apical core of the implant used, so that the osteotomy is not wider than this core. In other words, high initial stability is gained through these apical threads, and mild compression of bone at the apical section of the placed implant, utilizing the accepted concept of "Under Drilling" techniques [16, 17]. Since each implant system has different diameters at the apical section and different drill diameters, the dentist must study these parameters of the system used, to ensure that the osteotomy is small enough to ensure high implant stability while being large enough to allow the core of the implant to penetrate in.

Correct planning considering these parameters will ensure high initial stability (above 35NCM) to support immediate placement of a temporary restoration that enables aesthetics without being exposed to functional occlusal loads.

2.3 Drilling depth and implant position

The specific 3D position of the implant's head is critical for an ideal restoration [18–20]. A depth of 3–4 mm is usually sufficient to allow the creation of a correct

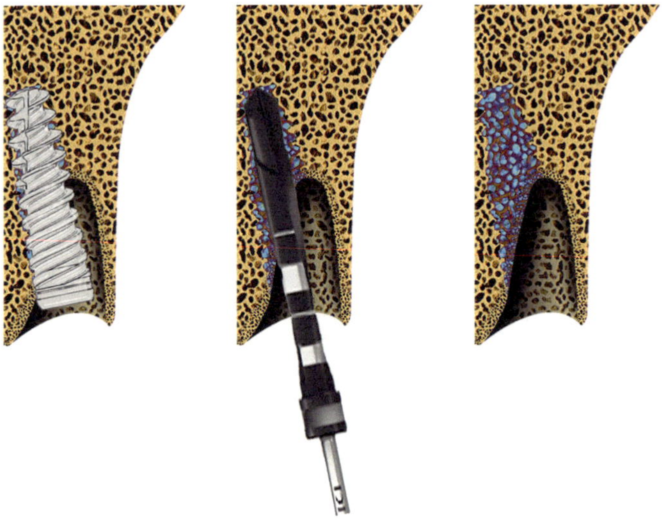

Figure 1.
Areas of support for an anterior implant.

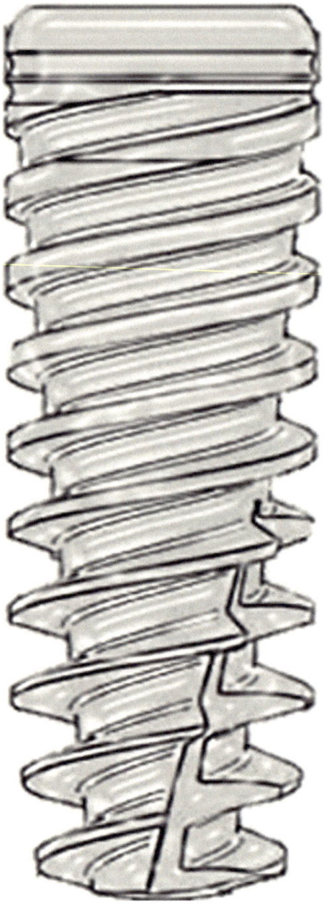

Figure 2.
An implant with wide apical screw design.

emergence profile, without creating a deep gingival pocket that is too hard to maintain. It should be clear that screw-retained restorations are by far better than cement-retained restorations as temporary restorations on immediate implants. One reason is that there is no cement involved in screw-retained restorations, so no cement may accidentally be pushed into the fresh socket. The second reason is the ease of manipulation of screw-retained restorations and the fact that no vertical force needs to be placed on the newly placed implant while preparing and while trying-in the temporary crown. This means, while placing the implant, effort must be put to place the opening of the implant at the cingular area of the restored tooth and at an angle that would enable easy access to the retaining screw from the palatal side of the crown (**Figure 3**).

Regarding implant's depth, many dentists are not aware of the difference between platform-stitching implants and standard internal hex ones when it comes to the position of the restorative platform. While in standard internal hex implants, the crown may emerge from the implant's platform or very close to it, in platform-stitching implants, there is a 1–2 mm gap between the implant's platform and the base of the restoration (**Figure 4**). This means that when using platform-switching implants, the head of the implant must be placed deeper than when using standard internal hex ones.

2.4 Creating a temporary crown with an emergence profile that supports the gingiva

The alveolar bone is critical to support the gingiva. Lack of bone at the buccal aspect of the socket, at the proximal aspect/s or even at the palatal aspect may lead to gum recession and the creation of "black holes" and flattening of the gingival architecture [20, 21]. Therefore, it is critical to ensure that the socket is intact after extraction, and that bone surrounds it all around. When this is the situation, it is beneficial to support the gum tissue immediately after extraction with a correctly

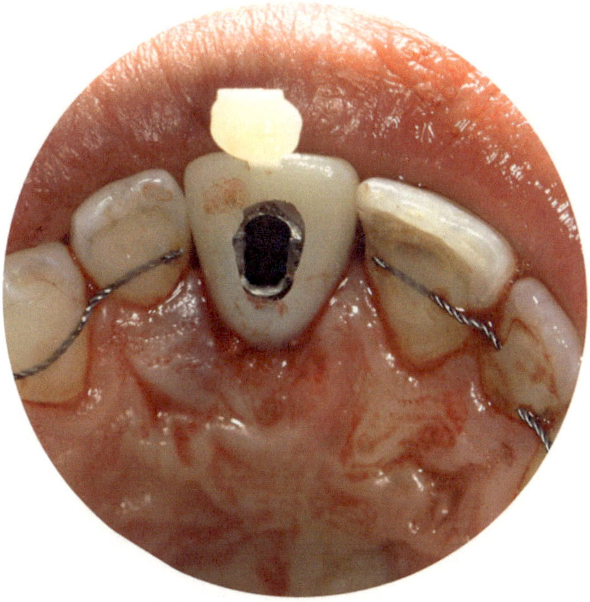

Figure 3.
Screw hole at the cingulum area.

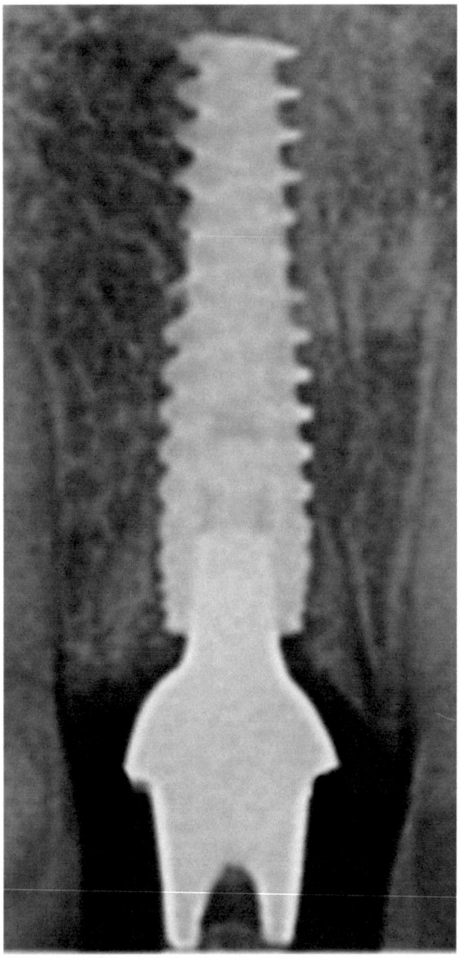

Figure 4.
Platform switching implant type.

made temporary crown, which will maintain the shape of the gums and papillae. The best and easiest way to do it is by using a screw-retained temporary crown, made with an "ovate pontic" emergence profile design at its base facing the implant (**Figure 5**) [22–25]. A standard and round-shaped healing abutment is not ideal since it creates a round opening in the gingiva that does not have the required gingival architecture (**Figure 6**). This means that when standard healing abutments are used, an additional surgical procedure may be required to achieve the desired gingival contour.

The design of a correct emergence profile requires an analysis of the 3D position of the head of the implant, its depth, and the level of support needed by the gingival tissues. 3–4 mm of depth is sufficient to create correct ovate-pontic shape that will support the gums (**Figure 7**). The gingival aspect of the temporary crown must be polished to prevent gum irritation and plaque accumulation, and the patient must be taught how to clean it. Since gingival tissues tend to shrink and change during the healing phase, some mild adjustments of the emergence profile may be required once osseointegrate is completed, to ensure aesthetically pleasing outcome.

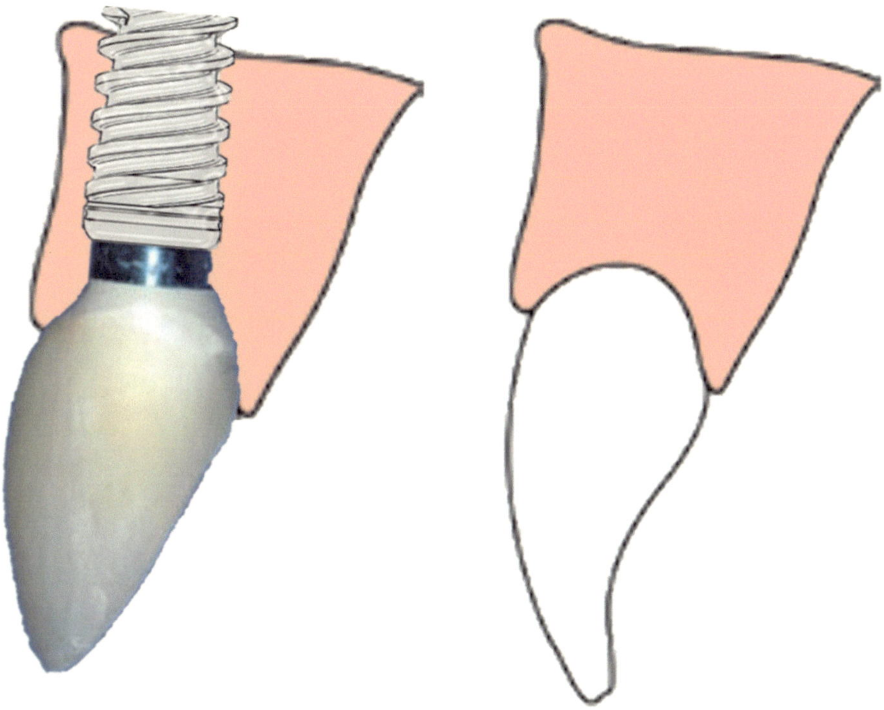

Figure 5.
Ovate pontic design.

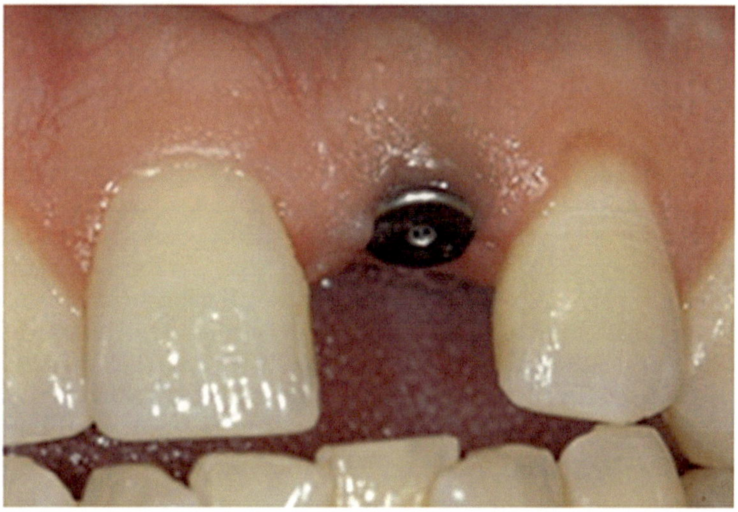

Figure 6.
Tissue around a standard healing cap.

Once the temporary crown is completed, it should be firmly attached to the implant, and the access hole must be sealed. Occlusal adjustment must ensure that minimal (if any) occlusal loads are placed on the crown during the healing phase. It is also recommended that proximal contacts are minimal and sufficient to prevent

Figure 7.
Ovate pontic emergance profile.

food impaction between the temporary crown and the adjacent teeth. It is mandatory to explain to patients that the temporary crown is made for supporting the gums and for aesthetic reasons only and that biting, chewing or any other application of forces (such as during smoking, parafunction, nail-biting etc.) may cause implant failure.

The temporary crown must not be removed during the healing and osteointegration period and sufficient time must be given to allow gingival maturation. When done correctly, the gingival architecture is preserved, as seen when removing the temporary crowns (**Figure 8a** and **b**).

2.5 Transferring gingival data to the dental laboratory

To ensure that the lab technician fabricates a crown that is identical to the temporary one, the internal shape of the healed gingival architecture must be transferred to the laboratory. There are several published methods to do that. Both silicone-based impressions and intra-oral scanning are acceptable [26, 27]. In addition, it is recommended to transfer a physical impression of the emergence profile and the shape of the temporary crown (**Figure 9**), as well as photographs from a few angles. This allows the technician to better visualize the required shape and its actual environment.

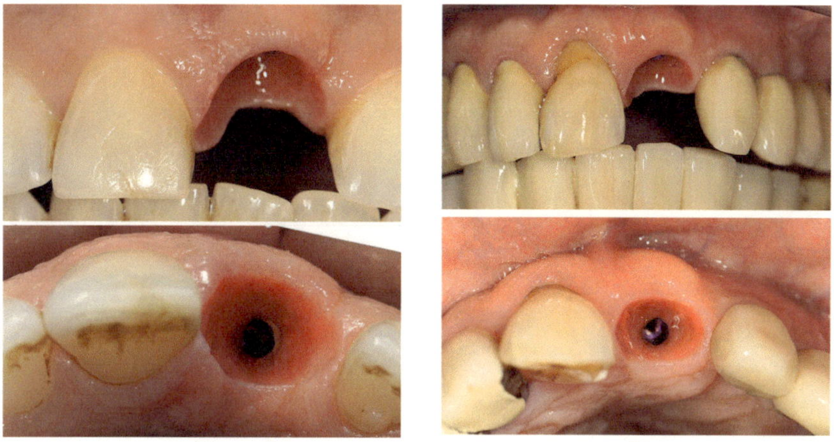

(a) (b)

Figure 8.
a. Tissue of 28 year old patient after crown removal. b. Tissue of 80 year old patient after crown removal.

Figure 9.
A technique to transfer crown shape to the dental lab.

2.6 Long-term maintenance

Like any tooth, the tissues around an implant are dynamic and constantly change and are impacted by the accumulation of plaque, calculus and food particles and time, of course. Daily cleaning must include soft brushing, the use of floss, or a super-floss underneath the ovate-pontic shaped crown. In addition, patients need to understand the need to have periodic check-up visits and treatment by a hygienist to ensure short- and long-term success of any tooth and restoration.

3. Conclusion

The restoration of a single anterior tooth is one of the most demanding procedures in dentistry. Each aspect of the treatment influences the end result: from a correct

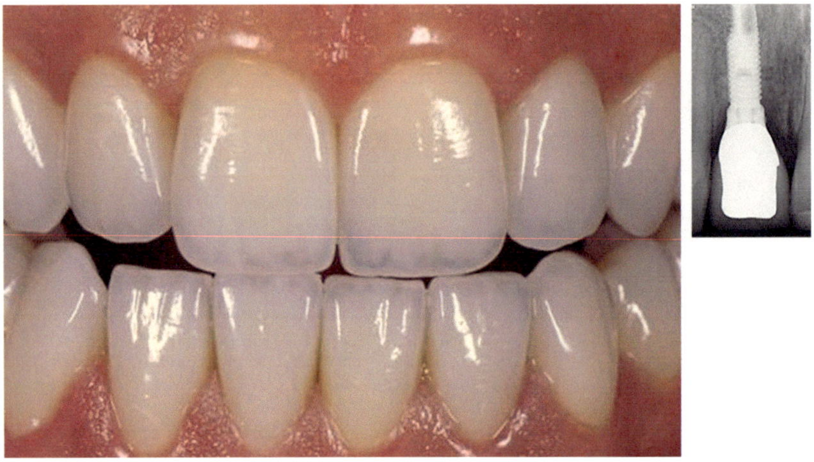

Figure 10.
Ten years follow up of an immediate loaded implant with final crown.

diagnosis to case selection, from the choice of a specific implant design to alteration of the drilling protocol, and from the clinical decision of where to three-dimensionally place the head of the implant to the shape of the gingival aspect of the temporary and final crown. Thorough understanding and correct implementation of these principles, coupled with the help of an artist dental technician can lead to long-term solutions for the benefit of our patients (**Figure 10**).

Acknowledgements

Crown 21, shown in Fig 10 was fabricated by Mr. Tamir Heker from Tamir Heker Dent. Tel-Aviv, Israel.

Author details

Nachum Samet[1,2,3,4,5]

1 Kyung Hee University, South Korea

2 Titu Maiorescu University, Romania

3 Taishan Medical University, China

4 Universidad Católica San Antonio de Murcia, Spain

5 Harvard School of Dental Medicine, Boston, USA

*Address all correspondence to: samet@drsamet.com

References

[1] Tsuda H, Rungcharassaeng K, Kan JY, Roe P, Lozada JL, Zimmerman G. Peri-implant tissue response following connective tissue and bone grafting in conjunction with immediate single-tooth replacement in the esthetic zone: A case series. International Journal of Oral and Maxillofacial Implants. 2011;**26**(2):427

[2] Hof M, Pommer B, Strbac GD, Sütö D, Watzek G, Zechner W. Esthetic evaluation of single-tooth implants in the anterior maxilla following autologous bone augmentation. Clinical Oral Implants Research. 2013;**24**:88-93

[3] Arunjaroensuk S, Panmekiate S, Pimkhaokham A. The stability of augmented bone between two different membranes used for guided bone regeneration simultaneous with dental implant placement in the esthetic zone. International Journal of Oral & Maxillofacial Implants. 2018;**33**(1):206-216

[4] Slagter KW, den Hartog L, Bakker NA, Vissink A, Meijer HJ, Raghoebar GM. Immediate placement of dental implants in the esthetic zone: A systematic review and pooled analysis. Journal of Periodontology. 2014;**85**(7):e241-e250

[5] Kuchler U, Chappuis V, Gruber R, Lang NP, Salvi GE. Immediate implant placement with simultaneous guided bone regeneration in the esthetic zone: 10-year clinical and radiographic outcomes. Clinical Oral Implants Research. 2016;**27**(2):253-257

[6] Yang X, Zhou T, Zhou N, Man Y. The thickness of labial bone affects the esthetics of immediate implant placement and provisionalization in the esthetic zone: A prospective cohort study. Clinical Implant Dentistry and Related Research. 2019;**21**(3):482-491

[7] Siormpas KD, Mitsias ME, Kontsiotou-Siormpa E, Garber D, Kotsakis GA. Immediate implant placement in the esthetic zone utilizing the "root-membrane" technique: Clinical results up to 5 years Postloading. International Journal of Oral & Maxillofacial Implants. 2014;**29**(6):1397-1405

[8] Aslan S. Improved volume and contour stability with thin socket-shield preparation in immediate implant placement and provisionalization in the esthetic zone. Int J Esthet Dent. 2018;**13**(2):172-183

[9] Bramanti E, Norcia A, Cicciù M, Matacena G, Cervino G, Troiano G, et al. Postextraction dental implant in the aesthetic zone, socket shield technique versus conventional protocol. Journal of Craniofacial Surgery. 2018;**29**(4):1037-1041

[10] Becker W, Goldstein M. Immediate implant placement: Treatment planning and surgical steps for successful outcome. Periodontology 2000. 2008;**47**(1):79-89

[11] Buser D, Chappuis V, Belser UC, Chen S. Implant placement post extraction in esthetic single tooth sites: When immediate, when early, when late? Periodontology 2000. 2017;**73**(1):84-102

[12] Den Hartog L, Huddleston Slater JJ, Vissink A, Meijer HJ, Raghoebar GM. Treatment outcome of immediate, early and conventional single-tooth implants in the aesthetic zone: A systematic review to survival, bone level, soft-tissue, aesthetics and patient satisfaction.

Journal of Clinical Periodontology. 2008;**35**(12):1073-1086

[13] Khzam N, Arora H, Kim P, Fisher A, Mattheos N, Ivanovski S. Systematic review of soft tissue alterations and esthetic outcomes following immediate implant placement and restoration of single implants in the anterior maxilla. Journal of Periodontology. 2015;**86**(12):1321-1330

[14] Evans CD, Chen ST. Esthetic outcomes of immediate implant placements. Clinical Oral Implants Research. 2008;**19**(1):73-80

[15] De Rouck T, Collys K, Wyn I, Cosyn J. Instant provisionalization of immediate single-tooth implants is essential to optimize esthetic treatment outcome. Clinical Oral Implants Research. 2009;**20**(6):566-570

[16] Cochran DL. The evidence for immediate loading of implants. Journal of evidence based dental practice. 2006;**6**(2):155-163

[17] Mijiritsky E, Mardinger O, Mazor Z, Chaushu G. Immediate provisionalization of single-tooth implants in fresh-extraction sites at the maxillary esthetic zone: Up to 6 years of follow-up. Implant Dentistry. 2009;**18**(4):326-333

[18] Tarnow D, Elian N, Fletcher P, Froum S, Magner A, Cho SC, et al. Vertical distance from the crest of bone to the height of the interproximal papilla between adjacent implants. Journal of Periodontology. 2003;**74**(12):1785-1788

[19] Cheng Q, Su YY, Wang X, Chen S. Clinical outcomes following immediate loading of single-tooth implants in the esthetic zone: A systematic review and meta-analysis. International Journal of Oral & Maxillofacial Implants. 2020;**35**(1):167-177

[20] Wheeler SL. Implant complications in the esthetic zone. Journal of Oral and Maxillofacial Surgery. 2007;**65**(7):93-102

[21] Yan Q, Xiao LQ, Su MY, Mei Y, Shi B. Soft and hard tissue changes following immediate placement or immediate restoration of single-tooth implants in the esthetic zone: A systematic review and meta-analysis. International Journal of Oral & Maxillofacial Implants. 2016;**31**(6):1327-1340

[22] Spear FM. The use of implants and ovate pontics in the esthetic zone. Compendium of Continuing Education in Dentistry (Jamesburg, NJ: 1995). 2008;**29**(2):72-74

[23] Zitzmann NU, Marinello CP, Berglundh T. The ovate pontic design: A histologic observation in humans. The Journal of Prosthetic Dentistry. 2002;**88**(4):375-380

[24] Kan JY, Kois JC. Predictable peri-implant gingival aesthetics: Surgical and prosthodontic rationales. Practical Procedures & Aesthetic Dentistry. 2001;**13**(9):691-698

[25] Zetu L, Wang HL. Management of inter-dental/inter-implant papilla. Journal of Clinical Periodontology. 2005;**32**(7):831-839

[26] Elian N, Tabourian G, Jalbout ZN, Classi A, Cho S-C, Froum S, et al. Accurate transfer of Peri-implant soft tissue emergence profile from the provisional crown to the final prosthesis using an emergence profile cast. Journal of Esthetic and Restorative Dentistry. 2007;**19**:306-314

[27] Wei D, Di P, Tian J, Zhao Y, Lin Y. Evaluation of intraoral digital impressions for obtaining gingival contour in the esthetic zone: Accuracy outcomes. Clinical Oral Investigations. 2020;**24**(4):1401-1410

Chapter 13

Implantogenomic: Conceptualizing Osseointegration toward Personalized Dental Implant Therapy

Ali K. Refai

Abstract

The spectrum of patients' needs for dental implant treatment ranges between healthy individuals to those with complex diseases and compromised jaw bones. The aim of this review chapter is to introduce the application of personalized dental medicine to dental implant field as a therapeutic strategy that is best suited to individualized patient's genetic makeup "Implantogenomics" to enhance their longevity and clinical outcomes. The ultimate goal of personalized medicine and dentistry is tailoring targeted treatment to the patient's individual genetic makeup and having predictive outcomes. This approach will transfer the traditionally known "one size fits all" to an actionable model, tailoring therapy to individuals in a homogenous stratified group. In this review chapter, in analogy to pharmacogenomics, personalized dental implant and its implantogenomics concept have been proposed as a novel application of personalized dentistry. It is conceivable that the actionable model that integrates genomics and materiomics will accelerate the production of personalized implantable biomaterials and biomedical devices. Moreover, the convergence of multi-disciplines including biological sciences, material sciences, and computational tools may underpin the application of personalized dental implant therapy in the future. This approach will unleash the potential of advancing technologies to tailor dental implants targeting different subpopulations. Despite this optimistic goal, challengeable remains ahead of us where the conduction of well-directed scientific and clinical research is needed.

Keywords: osseointegration, personalized dental implant therapy, omics, materiomics, implantogenomics

1. Introduction

For more than 50 years, dental implants were employed for the treatment of patients who are both partly and wholly edentulous. This treatment modality has been shown to improve the quality of life of large populations and demonstrated to perform very well, clinically, functionally, and esthetically [1–3]. Osseointegration is an essential factor in the success rates of dental implant-supported prostheses. The first known use of the term "osseointegration" was in the 1970s when Branemark defined it as a "Direct structural and functional link between organized, live bone tissue and the surface of a load-carrying artificial implant at the resolution level of the light microscope" [4]. At the bone-implant interface, osteoinduction and osteoconduction events are necessary for osseointegration to occur. The biological

Term	Definition
Artificial intelligence (AI)	Specific machine learning algorithms, has the ability to make decisive interpretation of "big"-sized complex data and, hence, appears as the most effective tool for the analysis and understanding of multi-omics data for patient-specific observations and complex biological process
Biomaterialomics	Integration of multi-omics data and high-dimensional analysis with artificial intelligence (AI) tools throughout the entire pipeline of biomaterials development
Materiomics	Introducing an omics approach to biomaterials research where the convergence of materials science, biological science, and technological advancement (encompassing computational methods and experimental assays) hold the promise in the production of personalized implantable biomaterials and biomedical devices
OMICs	A field of study in biology ending in *-omics*, such as genomics, proteomics, transcriptomic, or metabolomics. The related suffix *-ome* is used to address the objects of study of such fields, such as the genome, proteome, transcriptome, or metabolome, respectively
Personalizing dental implant	The integration of multi-omics data and advanced biotechnologies to tailor biofunctionalized implant system; treating both healthy and compromised jaw bone with the same success and survival rates
Personalized Medicine	Personalized medicine is an evolving field that uses diagnostic tools to identify specific biological markers, often genetic, to help determine which medical treatments and procedures will be best for each patient
Systems biology	A holistic approach to deciphering the complexity of biologic systems. It is collaborative, integrating many scientific disciplines, including biology, computer science, engineering, bioinformatics, physics, and others, to predict how these systems change over time and under varying conditions, and to develop solutions to the world's most pressing health and environmental issues

Table 1.
Shows the main terminologies used throughout the review.

mechanisms defining the three concepts are known to be connected yet distinct from one another [5].

Although osseointegration was defined as a time-dependent healing process, whereby clinically asymptomatic rigid fixation of alloplastic materials is achieved and maintained in bone during functional loading, osseointegration, a novel definition has emerged recently [6]. Albrektsson et al. defined "osseointegration as a foreign body reaction where interfacial bone is formed as a defense reaction to shield off the implant from tissues" [7]. Understanding both the old definition and the new one necessitate conceptualizing osseointegration beyond its definitions.

To achieve such goal, this review chapter proposed an actionable model integrating various fields. The convergence of scientific disciplines including advancements in molecular biology in terms of "multi-omics data," bioengineering in the term "materiomics and biomaterialomics," and proper computational tools that will hasten the production of personalized implantable medical devices. Thus, the formulation of this strategic approach is critical toward personalized dental implant as a target future therapeutics. Moreover, the introduction of osseointegration's new concept "implantogenomics" demonstrates patient-specific implants whereby reliable dental implant therapy is tailored based on the genetically unique individual. **Table 1** shows the main terms used throughout the review.

2. Synopsis of wound healing around dental implants

The process of osseointegration involves a cascade of complex physiological mechanisms similar to intramembranous ossification and direct fracture bone healing [8]. After an implant is inserted into an osteotomy site, immediately protein serum is adsorbed onto the implant surface and forms conditioning layer. It is well known that the characteristics of implant surface determine the biological and molecular profile of the adsorbed protein layer. At the micro-environment between bone and implant surface, cells interact with this layer rather than with the implant surface itself [9, 10].

The protein adsorption stage is followed by coordinating and sequentially overlapping biological events. These biological events include mesenchymal and non-mesenchymal cellular interactions (i.e., recruitment, attachment, proliferation, and differentiation), soluble proteins and growth factors secretions, osteogenic cells proliferation and differentiation, osteoid production and matrix calcification, and bone formation and remodeling [11–14].

The principal factors that regulate the biological events of endosseous wound healing are the cells dominating the osteotomy site. Cells populate each stage of the healing process and are found to coordinate and communicate with each other by producing a myriad of signaling molecules including; cytokines, chemokines, growth factors, extracellular matrix proteins, hormones, and ions, which orchestrated the osseous healing process at biointerface. The secretion of the signaling molecules is well controlled by the sequential activation of corresponding typical genes [14, 15]. In fact, the decisive underlying cellular and molecular mechanisms that regulated endosseous wound healing and osseointegration are yet to be fully understood [15, 16]. **Figure 1** summarizes the biological events and the cellular and molecular mechanism of osseointegration.

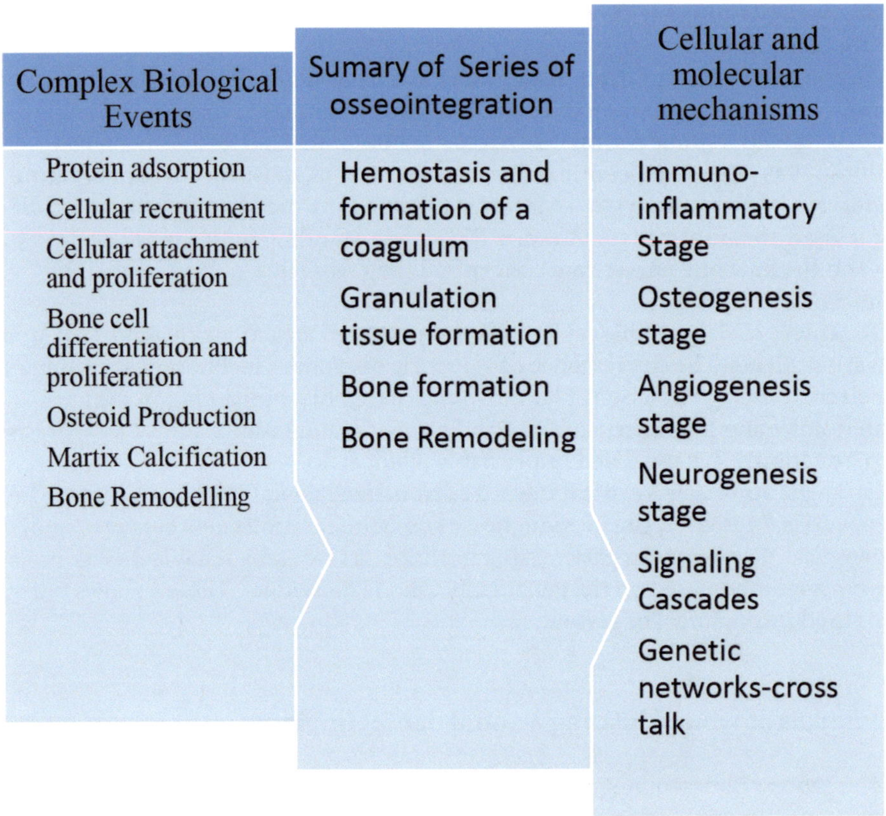

Figure 1.
Summary of the biological events and the cellular and molecular mechanisms of osseointegration.

3. Gene markers and genetic networks of osseointegration

The biological processes of osseointegration (inflammo-immuno-angio-neuro-osteogenesis processes) are directed by several genes expressed in an orderly sequence. Studies showed that gene expression profile after implant installation shifted from genes upregulated during immuno-inflammatory processes and cell proliferation to genes expression upregulated in favor of angiogenesis, osteogenesis, and neurogenesis processes [17, 18]. This alternation of biological processes may be attributed to the concept of "foreign body equilibrium." It has been proposed that osseointegration is the result of an immune system's reaction to a foreign body, which, at the right level of intensity, will balance itself out and allow osteogenesis to start on surface of the implant [19, 20].

Several in vitro and in vivo studies have demonstrated the influence of surface topography on gene expression of osseointegration. In brief, over the last two decades, the studies investigated the genetic basis of osseointegration have shown that the most observed genes were immunoinflammatory-related gene markers, extracellular matrix and osteoblast differentiation-related genes, bone formation, and remodeling-related genes, transcriptional-whole genome-proteome profiles of osseointegration related to immune-inflammo-angio-neuro-skeletogenesis biological processes, micro RNAs and signaling pathways. These biomarkers recapitulated the reported significant genetic markers and genetic networks associated with the biological processes of osseointegration [14, 16, 21–26].

Osseointegration encompasses a network of reactions that are dependent on a building system rather than a single pathway [20]. In this context, it is noteworthy that even though thorough studies have made significant progress in demonstrating the fundamental mechanisms underlying the molecular basis of osseointegration, as previously mentioned, the revolution of ongoing scientific research has highlighted the importance of other genetic networks related to osseointegration [25]. Whole genome microarray analyses have shown that the genetic networks that control osseointegration include other genes that function in coherent and/or independent pathways to complete the process of osseointegration around dental implants, in addition to the genes regulating osteogenesis and involving the expression of osteoblast marker genes. Other gene networks involved in the process of osseointegration include those that control the cross-linking of the bone collagen matrix, extracellular matrix-related genes associated with cartilage, and clock genes related to the peripheral circadian rhythm system. The modulation of those novel genetic networks was found to be topographical dependent and devoted to accelerate osseointegration [25, 27, 28].

4. The effect of implant surface properties on osseointegration

Implant surface properties including its topographical, chemical, mechanical, and physical are considered to be one essential factor to promote and maintain osseointegration. Since the foundation of osseointegration phenomenon, many bioengineering techniques have been developed to modify dental implant surface characteristics. These modification techniques have been classified into subtractive and additive treatments. Subtractive treatments include machining, Sandblasting, acid etching, and laser ablation. Additive treatments include anodization, fluoride surface treatment, nanoscale surface, plasma spraying, hydroxyapatite coating, electrophoretic deposition, biomimetic precipitation "peptides growth factors and ECM proteins", and drugs incorporated "bisphosphonates and statins" [29–33].

Recent reviews have shown that dental implant surface modifications were enormously studied aiming to promote osseointegration in the last decades [16, 26]. Mechanistically, implant surface properties are able to modulate osseous healing by influencing the cells-implant interactions and then producing several genes and proteins that serve as key regulators to inflammo-immuno-angio-neuro-osteogenesis processes [16, 26, 34].

Quit recently, new interesting approach is surface loading with cells potentially gaining more attention in the future [35]. However, this type of functionalization has not been extensively studied, as its clinical translation is more challenging due to the cell preparatory requirements and regulations. In tandem with the progression of cell-based therapy and personalized medicine, this approach may accelerate tissue repair in many clinical aspects [35].

5. Personalized medicine and dentistry

5.1 Personalized medicine

Personalized medicine is a contemporary evolving field in medicine and healthcare industries [36, 37]. Its aspirational goal is about tailoring medical treatment to the patient's individual genomic, anatomical, and physiological characteristics in order to provide a better quality of life and prolong life expectancy. Though the concept of current personalized medicine as new medical innovation was laid in tandem with the successful completion of the Human Genome Project (HGP) in the USA, in 2003 [37, 38], the term personalized medicine was coined in the published works in 1999 [39]. **Figure 2** shows the main elements of personalized medicine and dentistry concept.

The Human Genome Project (HGP) along with further projects (International HapMap Project (2002–2010), and the 1000 Genomes Project (2008–2015) authorizing personalized medicine to grow steadily in the medical practice and pharmacotherapy [37]. Among the main project achievements were discovering 1800 disease-related genes, developing numerous molecule biomarker-based tests, identifying genetic variations which lead to an individual different response to environmental factors, as well as drugs and vaccines effectiveness [37, 38, 40–42]. Personalized medicine has been extensively reviewed in the pharmacogenetics "study of genetic causes of individual variations in drug response" and pharmacogenomics (impact of multiple mutations in the entire genome that may determine the patient's response to drug therapy [43, 44].

Both pharmacogenetics and pharmacogenomics testing is involved as novel approaches in drug development and future registered therapeutic for a specific disease. For example, the implication of using pharmacogenetics and pharmacogenomics in the fields of cardiovascular, pulmonary, oncological, and bone diseases and also feature the potential economic value of their development [44, 45].

For centuries, health care providers, namely physicians treated medical conditions based on identifying the signs and symptoms and prescribed drug therapy without a diagnostic pretest. This traditional approach is so-called "One Size-Fits-Alls," which is doomed to failure. The failure has been attributed to several factors including lack of patient stratification, existing of genetic variation and treatment drug reaction, identifying risk assessment, understanding factors involved in disease development

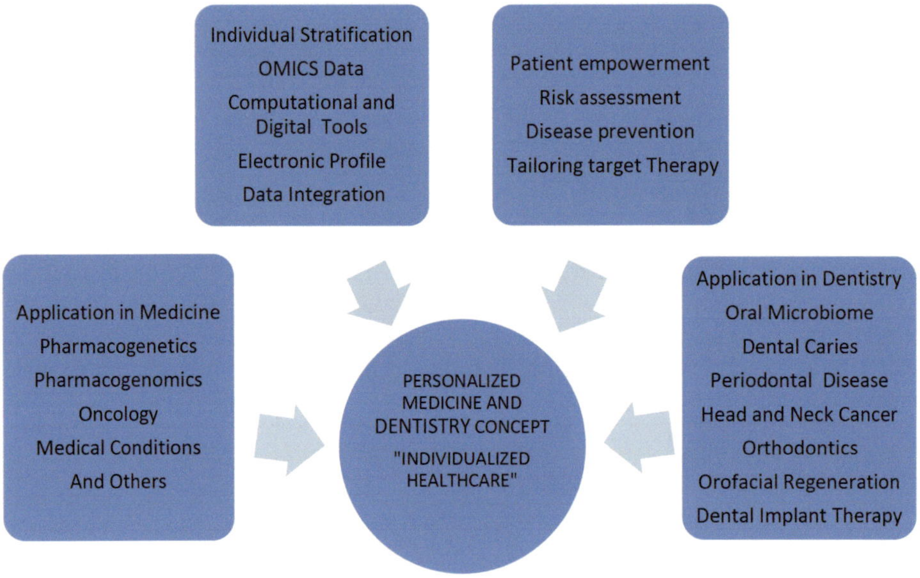

Figure 2.
Personalized medicine and dentistry concept: Main components and its applications in medical and dental fields.

"epigenetics" and taking in account patient's profile "age, sex and weight" [36, 37]. Collectively, such factors are the reason for extensive differences in response to medication.

The employing of pharmacogenetics and pharmacogenomics testing in personalized medicine will show which medication will have a positive effect and which one will have a negative effect. This approach will shift the paradigm from currently using "one size fits all" to individualizing drug target "a test and treat" approach. The consequence of such transformation is expanding the personalized medicine concept in the medical practice and healthcare industries [37, 46, 47].

Although there is no one uniform definition of personalized medicine because of lack of consensus on definitions and nomenclature, several agencies, institutes, and associations have provided a diverse range of definitions depending on the scope and the main elements of personalized medicine [37, 43]. Personalized medicine is a multifarious approach to patient care that not only improves our ability to diagnose and treat disease, but offers the potential to detect disease at an earlier stage, when it is easier to treat effectively. The full implementation of personalized medicine encompasses six domains (**Table 2**) [48].

The concept of personalized medicine [49] has been designated by many names in the literature [37, 50]; including precision medicine [51, 52], individualized medicine [53], network medicine [54], predictive medicine [55], stratified medicine [56] as well as P4 medicine [57]. These terms have a similar meaning with some slight differences between them and used interchangeably as alternative and synonyms terms [37, 50].

5.2 Personalized dentistry

Like medicine, there is a growing attention toward personalized/precision dentistry. The personalized dentistry branches from personalized medicine as an emerging concept, a multifaceted, individualized data-driven approach to oro-cranio-facial care environment. The personalized dentistry will act as an actionable model in dentistry that represents a paradigm shift in dental practice from the one size fits all approach to tailoring therapy to individuals in a homogenous stratified group. Each group of individuals sharing similar risk factor data includes but is not limited to genomic, environmental, e-health record, and lifestyle [50, 58–62].

Domain	Goal
1. Risk Assessment	Genetic testing to reveal predisposition to disease
2. Prevention	Behavior/lifestyle/treatment intervention to prevent disease
3. Detection	Early detection of disease at the molecular level
4. Diagnosis	Accurate disease diagnosis enabling individualized treatment strategy
5. Treatment	Improved outcomes through targeted treatments and reduced side effects
6. Management	Active monitoring of treatment response and disease progression

Table 2.
Domains of personalized medicine adapted from Reference [48].

Nowadays, the advances in personalized oral health focus on the creation of new digital technologies and medical devices, conceptualizing complex process align with understanding system biology, improving patient's stratifications, identifying biomarkers as diagnostic and prognostic factors related to disease development and progression. In to aforementioned, personalized oral health concentrates on developing proper treatment planning and early disease detection as preventive strategies. In fact, the main goal of personalized dental medicine is tailoring treatment taking into account its application, dosing, and response to treatment according to individual characteristics [50, 63, 64].

Bartold explained the importance of implementing personalized dentistry as a custom dental management model for the future of dental practice in two folds. First, this approach will bring the field of dentistry in tandem with current medical practice. Second, personalized dentistry stressing on patient-centered care and improving patient health outcomes [65].

In the last decade, several researchers have been investigating the application of personalized dentistry concept to various branches of dentistry. Some examples of these studies include, head and neck cancer [66, 67], periodontics [68–70], restorative [71–73], pedodontic [74–76], orthodontic [77, 78], TMJD [79], Public health and preventive care [80–84], prosthodontic [85], dental implant [16, 86–88], and oral rehabilitation and regenerative medicine [89].

In the context of advanced healthcare research in the medical field, personalized medicine is escalating, whereas, in dentistry, personalized dentistry is still in its infancy. Several challenges have been attributed to delay implementations of personalized dentistry in the practice environment. According to Polverini, these challenges include gaps in education of the current knowledge related to the basis of personalized dental medicine, addressing issues such as reimbursement for preventive dental procedures, insufficient an integrated electronic health record, difficulty in protecting patient genetic information, and differences in access to genomic medicine by certain populations. However, the dental profession should act in tandem with medical profession to pace toward innovation and harness advanced technologies to ensure the development and progression of personalized healthcare environment [62, 90].

6. Personalized dental implant therapy: Introducing Implantogenomics Concept

The emergence of advanced technologies underpins the application of personalized dental implant therapy as a novel dental approach to providing precise patient's care. In fact, the presence of a broad spectrum of patients with various conditions include systemic diseases, special needs, and syndromic and elderly with compromised jaws necessitate patient's stratifications who share similar risk factors. Having such homogeneous groups and subgroups will help in developing proper protocol, treatment decision and prediction and tailoring precise patient-centered treatment strategy.

In 2020, we have proposed the personalized dental implant along with its new concept implantogenomics [16]. The current opinion review wishes to introduce the new concept to more audience in both medical and dental fields. In analogy to pharmacogenomics, implantogenomics should accurately achieve three major components prior to be implemented as a future personalized therapeutic strategy in the realm of implantology. First, stratifying individuals into subgroups based on a person's unique genome, second, identifying specified prognostic and predictive biomarkers which characterized endosseous wound healing of each subpopulations, and third, applying genomic information to design dental implant suit each subpopulation.

It is conceivable that the actionable model which integrates multiple advancement technologies will leverage the development of new medical devices including dental implants in favor of regenerative medicine and dentistry. These include but is not limited to multi-omics data, advanced material science, biotechnologies, and contemporary bioinformatics tools along with artificial intelligence. It is suggested that the joining of bioinformatics and artificial intelligence will handle high throughput genetic data more precisely in terms of defining, categorizing and analyzing [91, 92]. The ultimate goal of personalized dental implant therapy is to have the four Rights. (1) Right treatment in terms of implant design and legitimate protocols, (2) Right patient in terms of patient stratification and minimal biological complication outcomes, (3) Right time in terms of bio-physiological events, and (4) Right Positioning in terms of Jaw status. **Figure 3** elucidates an actionable model of personalized dental implant therapy.

More recently, Albrektsson et al. proposed that personalized dental implant therapy and its implantogenomics concept may accelerate the development of genetic diagnostic tests as a target future therapeutic for residual ridge resorption (RRR) in elderly [16, 93]. In this regard, it should be mentioned that the application of personalized dental medicine to the realm of dental implant requires innovative leadership to direct wide-spectrum of scientific and clinical research as well as multidisciplinary and multicenter collaborations. Thus, the application of personalized dental medicine in the field of dental implant remains challengeable.

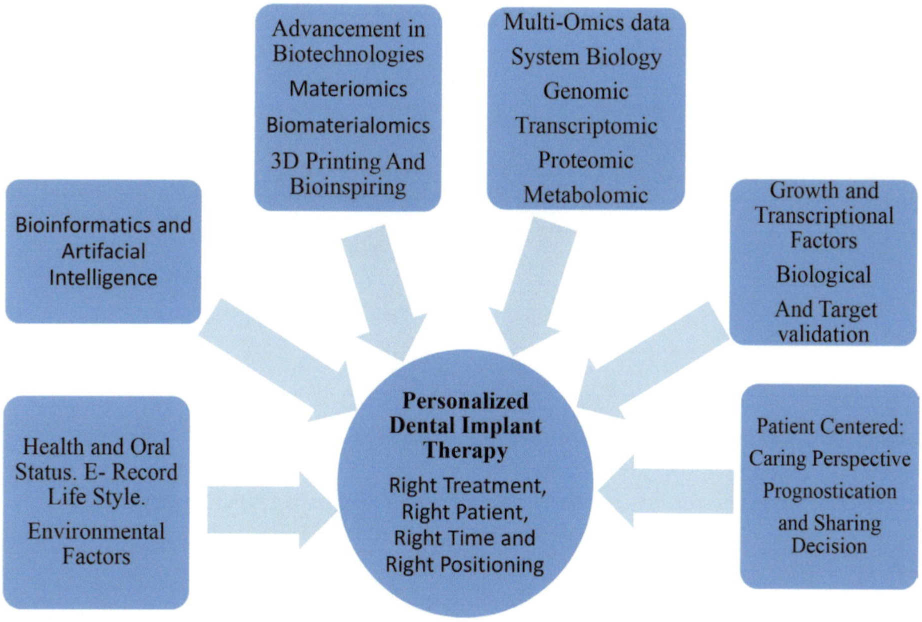

Figure 3.
An actionable model. The personalized dental implant therapy- the convergence of multiple disciplines.

7. Conclusions

The Integration of multi-disciplines including biological sciences, material sciences, and computational tools in actionable model may underpin the application of personalized dental implant therapy in the future. This approach will unleash the potential of advancing technologies to tailor dental implants targeting different subpopulations with more predictive outcomes. Despite this optimistic goal, challengeable remain ahead of us where conduction of well-directed scientific and clinical research is needed.

Author details

Ali K. Refai
College of Dentistry, King Saud University, Riyadh, Kingdom of Saudi Arabia

*Address all correspondence to: alirefai@hotmail.com; aalrefai@ksu.edu.sa

References

[1] Albrektsson T, Buser D, Sennerby L. Crestal bone loss and oral implants. Clinical Implant Dentistry and Related Research. 2012;**14**:783-791

[2] Albrektsson T, Canullo L, Cochran D, de Bruyn H. "Peri-implantitis": A complication of a foreign body or a man-made "disease". Facts and fiction. Clinical Implant Dentistry and Related Research. 2016;**18**:840-849

[3] Albrektsson T, Wennerberg A. On osseointegration in relation to implant surfaces. Clinical Implant Dentistry and Related Research. 2019 Mar;**21** (Suppl. 1):4-7

[4] Branemark PI, Hansson BO, Adell R, Breine U, Lindstrom J, Hallén O, et al. Osseointegrated implants in the treatment of the edentulous jaw. Experience from a 10-year period. Scandinavian Journal of Plastic and Reconstructive Surgery. 1977;**16**(Suppl):1-132

[5] Albrektsson T, Johansson C. Osteoinduction, osteoconduction and osseointegration. European Spine Journal. 2001 Oct;**10**(Suppl. 2): S96-S101

[6] Zarb GA, Koka S. Osseointegration: Promise and platitudes. The International Journal of Prosthodontics. 2012Jan-Feb;**25**(1):11-12

[7] Albrektsson T, Chrcanovic B, Jacobsson M, Wennerberg A. Osseointegration of implants—a biological and clinical overview. JSM Dental Surgery. 2017;**2**(3):1-6

[8] Mavrogenis AF, Dimitriou R, Parvizi J, Babis GC. Biology of implant osseointegration. Journal of Musculoskeletal & Neuronal Interactions. 2009;**9**:61-71

[9] Schwarz F, Herten M, Sager M, Wieland M, Dard M, Becker J. Histological and immunohistochemical analysis of initial and early osseous integration at chemically modified and conventional SLA titanium implants: Preliminary results of a pilot study in dogs. Clinical Oral Implants Research. 2007;**18**:481-488

[10] Dohan Ehrenfest DM, Coelho PG, Kang BS, Sul YT, Albrektsson T. Classification of osseointegrated implant surfaces: Materials, chemistry and topography. Trends in Biotechnology. 2010;**28**:198-206

[11] Puleo DA, Nanci A. Understanding and controlling the bone-implant interface. Biomaterials. 1999;**20**:2311-2321

[12] Berglundh T, Abrahamsson I, Lang NP, Lindhe J. De novo alveolar bone formation adjacent to endosseous implants. Clinical Oral Implants Research. 2003;**14**(3):251-262

[13] Davies JE. Understanding peri-implant endosseous healing. Journal of Dental Education. 2003;**67**:932-949

[14] Salvi GE, Bosshardt DD, Lang NP, et al. Temporal sequence of hard and soft tissue healing around titanium dental implants. Periodontology 2000. 2015;**68**(1):135-152

[15] Terheyden H, Lang NP, Bierbaum S, Stadlinger B. Osseointegration communication of cells. Clinical Oral Implants Research. 2012;**23**:1127-1135

[16] Refai AK, Cochran DL. Harnessing omics sciences and biotechnologies in understanding osseointegration – personalized dental implant therapy. The International Journal of Oral & Maxillofacial Implants. 2020 May/Jun; **35**(3):e27-e39

[17] Donos N, Hamlet S, Lang NP, Salvi GE, Huynh-Ba G, Bosshardt DD, et al. Gene expression profile of osseointegration of a hydrophilic compared with a hydrophobic microrough implant surface. Clinical Oral Implants Research. 2011;**22**:365-372

[18] Ivanovski S, Hamlet S, Salvi GE, Huynh-Ba G, Bosshardt DD, Lang NP, et al. Transcriptional profiling of osseointegration in humans. Clinical Oral Implants Research. 2011;**22**:373-381

[19] Albrektsson T, Dahlin C, Jemt T, Sennerby L, Turri A, Wennerberg A. Is marginal bone loss around oral implants the result of a provoked foreign body reaction? Clinical Implant Dentistry and Related Research. 2014;**16**(2):155-165

[20] Trindade R, Albrektsson T, Wennerberg A. Current concepts for the biological basis of dental implants: Foreign body equilibrium and osseointegration dynamics. Oral and Maxillofacial Surgery Clinics of North America. 2015;**27**(2):175-183

[21] Thalji G, Cooper LF. Molecular assessment of osseointegration in vitro: A review of current literature. The International Journal of Oral & Maxillofacial Implants. 2014 Mar-Apr;**29**(2):e171-e199

[22] Thalji G, Cooper LF. Molecular assessment of osseointegration in vivo: A review of the current literature. The International Journal of Oral & Maxillofacial Implants. 2013 Nov-Dec;**28**(6):e521-e534

[23] Sartori EM, Magro-Filho O, Silveira Mendonça DB, Li X, Fu J, Mendonça G. Modulation of micro RNA expression and osteoblast differentiation by nanotopography. The International Journal of Oral & Maxillofacial Implants. 2018 Mar/Apr;**33**(2):269-280

[24] Sartori EM, das Neves AM, Magro-Filho O, Mendonça DBS, Krebsbach PH, Cooper LF, et al. The role of microRNAs in the osseointegration process. The International Journal of Oral & Maxillofacial Implants. 2019 March/April;**34**(2):397-410

[25] Nishimura I. Genetic networks in osseointegration. Journal of Dental Research. 2013;**92**:109S-118S

[26] Refai AK. Assessment of the quality and the risk of bias of in vivo study models investigating the molecular basis of osseointegration: A systematic review. (Unpublished manuscript materials)

[27] Morinaga K, Sasaki H, Park S, Hokugo A, Okawa H, Tahara Y, et al. Neuronal PAS domain 2 (Npas2) facilitated osseointegration of titanium implant with rough surface through a neuroskeletal mechanism. Biomaterials. 2019 Feb;**192**:62-74

[28] Okawa H, Egusa H, Nishimura I. Implications of the circadian clock in implant dentistry. Dental Materials Journal. 2020 Mar 31;**39**(2):173-180

[29] Smeets R, Stadlinger B, Schwarz F, Beck-Broichsitter B, Jung O, Precht C, et al. Impact of dental implant surface modifications on osseointegration. BioMed Research International. 2016;**2016**:6285620

[30] Mendonça G, Mendonça DB, Aragão FJ, Cooper LF. Advancing dental implant surface technology--from micron- to nanotopography. Biomaterials. 2008 Oct;**29**(28):3822-3835

[31] Novaes AB Jr, de Souza SL, de Barros RR, Pereira KK, Iezzi G, Piattelli A. Influence of implant surfaces on osseointegration. Brazilian Dental Journal. 2010;**21**(6):471-481

[32] Gaviria L, Salcido JP, Guda T, Ong JL. Current trends in dental implants. Journal of the Korean Association of Oral and Maxillofacial Surgeons. 2014 Apr;**40**(2):50-60

[33] Jemat A, Ghazali MJ, Razali M, Otsuka Y. Surface modifications and their effects on titanium dental implants. BioMed Research International. 2015 Sep;**7**:1-11

[34] Stich T, Alagboso F, Křenek T, Kovářík T, Alt V, Docheva D. Implant-bone-interface: Reviewing the impact of titanium surface modifications on osteogenic processes in vitro and in vivo. Bioengineering & Translational Medicine. 2021 Jul 12;**7**(1):e10239

[35] Shanbhag S, Suliman S, Pandis N, Stavropoulos A, Sanz M, Mustafa K. Cell therapy for orofacial bone regeneration: A systematic review and meta-analysis. Journal of Clinical Periodontology. 2019 Jun;**46**(Suppl 21):162-182

[36] Mathur S, Sutton J. Personalized medicine could transform healthcare. Biomedical Reports. 2017 Jul;**7**(1):3-5

[37] Kichko K. Personalized Medicine as Innovation: What Can Germany Learn from the USA. 1st ed. Germany: Springer Gabler; Sep 2019

[38] The National Institutes of Health (NIH), The Department of Health and Human Services (2003):The Human Genome Project. Available from: http://report.nih.gov/NIHfactsheets/ViewFactSheet.aspx?csid=45&key=H#H

[39] Jain KK. Personalized medicine. Current Opinion in Molecular Therapeutics. 2002 Dec;**4**(6):548-558

[40] The National Institutes of Health (NIH), The National Human Genome Research Institute (2012): International HapMap Project. Available from: https://www.genome.gov/hapmap

[41] The International Genome Sample Resource (IGSR) (2015): The 1000 Genomes Project. Available from: http://www.internationalgenome.org/about

[42] Haga SB. Overview of Policy, Ethical and Social Considerations in Genomic and Personalized Medicine. Genomic and Personalized Medicine, 1, 2. Amsterdam, Netherlands: Elsevier; 2012. pp. 392-404

[43] Jain KK. Textbook of Personalized Medicine. 2nd ed. New York: Springer; 2015

[44] Dere WH, Suto TS. The role of pharmacogenetics and pharmacogenomics in improving translational medicine. Clinical Cases in Mineral and Bone Metabolism. 2009 Jan;**6**(1):13-16

[45] Weng L, Zhang L, Peng Y, Huang RS. Pharmacogenetics and pharmacogenomics: A bridge to individualized cancer therapy. Pharmacogenomics. 2013 Feb;**14**(3):315-324

[46] Collins FS. The Language of Life, DNA and the Revolution in Personalized Medicine. New York, NY: Harper Collins; 2012. p. 58

[47] Brittain HK, Scott R, Thomas E. The rise of the genome and personalised medicine. Clinical Medicine (London, England). 2017 Dec;**17**(6):545-551

[48] https://www.personalizedmedicinecoalition.org/Userfiles/PMC-Corporate/file/pmc_age_of_pmc_factsheet

[49] Van't Veer LJ, Bernards R. Enabling personalized cancer medicine through analysis of gene expression patterns. Nature. 2008;**452**(7187):564

[50] Beck JD, Philips K, Moss K, Divaris K, Morelli T, Offenbacher S. Advances in precision oral health. Periodontology 2000. 2020 Feb;**82**(1):268-285

[51] Council NR. Toward Precision Medicine: Building a Knowledge Network for Biomedical Research and a New Taxonomy of Disease. Washington, DC: National Academies Press; 2011

[52] Ashley EA. Towards precision medicine. Nature Reviews. Genetics. 2016 Aug 16;**17**(9):507-522

[53] Shastry B. Pharmacogenetics and the concept of individualized medicine. The Pharmacogenomics Journal. 2006;**6**(1):16

[54] Silverman EK, Loscalzo J. Developing new drug treatments in the era of network medicine. Clinical Pharmacology and Therapeutics. 2013;**93**(1):26-28

[55] Horesh Bergquist S, Lobelo F. The limits and potential future applications of personalized medicine to prevent complex chronic disease. Public Health Reports. 2018;**133**(5):519-522

[56] Hingorani AD, Windt DA, Riley RD, Abrams K, Moons KG, Steyerberg EW, et al. Prognosis research strategy (PROGRESS) 4: Stratified medicine research. BMJ. 2013 Feb;**5**(346):e5793

[57] Tian Q, Price ND, Hood L. Systems cancer medicine: Towards realization of predictive, preventive, personalized and participatory (P4) medicine. Journal of Internal Medicine. 2012;**271**(2):111-121

[58] Kornman KS, Duff GW. Personalized medicine: Will dentistry ride the wave or watch from the beach? Journal of Dental Research. 2012;**91**(suppl 7):s8-s11

[59] Garcia I, Kuska R, Somerman MJ. Expanding the foundation for personalized medicine: Implications and challenges for dentistry. Journal of Dental Research. 2013;**92**(suppl 7):s3-s10

[60] Slavkin HC. From phenotype to genotype: Enter genomics and transformation of primary health care around the world. Journal of Dental Research. 2014;**93**(suppl 7):s3-s6

[61] Slavkin HC. The future of research in craniofacial biology and what this will mean for oral health professional education and clinical practice. Australian Dental Journal. 2014;**59**(suppl. 1):186-190

[62] Polverini PJ. Personalized medicine and the future of dental practice. Personalized Medicine. 2018 Nov;**15**(6):449-451

[63] Ryan ME, Garcia R. Advances in precision oral health research. Advances in Dental Research. 2019 Nov;**30**(2):28-30

[64] Ryan ME, Fox CH. Advances in precision oral health research: Opportunities for the future! Journal of Dental Research. 2019 Nov;**98**(12):1285-1286

[65] Bartold PM. Personalized/precision dentistry – The future of dentistry? Australian Dental Journal. 2017 Sep;**62**(3):257. DOI: 10.1111/adj.12549 PMID: 28891244

[66] D'Silva NJ, Gutkind JS. Oral cancer: Integration of studies for diagnostic and therapeutic precision. Advances in Dental Research. 2019;**30**(2):45-49

[67] Polverini P, D'Silva N, Lei Y. Precision therapy of head and neck squamous cell carcinoma. Journal of Dental Research. 2018;**97**(6):614-621

[68] Divaris K. Searching deep and wide: Advances in the molecular understanding

of dental caries and periodontal disease. Advances in Dental Research. 2019;**30**(2):40-44

[69] Kornman KS. Contemporary approaches for identifying individual risk for periodontitis. Periodontology 2000. 2018. Oct;**78**(1):12-29

[70] Bartold PM, Ivanovski S. P4 medicine as a model for precision periodontal care. Clinical Oral Investigations. 2022 Sep;**26**(9):5517-5533

[71] Weber M, Søvik JB, Mulic A, et al. Redefining the phenotype of dental caries. Caries Research. 2018;**52**(4):263-271

[72] Baker JL, He X, Shi W. Precision reengineering of the oral microbiome for caries management. Advances in Dental Research. 2019;**30**(2):34-39

[73] Schwendicke F. Tailored dentistry: From "one size fits all" to precision dental medicine? Operative Dentistry 2018Sep/Oct; 43(5): 451-459

[74] Divaris K. Predicting dental caries outcomes in children: A "risky" concept. Journal of Dental Research. 2016;**95**(3):248-254

[75] Divaris K, Joshi A. The building blocks of precision oral health in early childhood: The ZOE 2.0 study. Journal of Public Health Dentistry. 2020 Mar;**80**(Suppl. 1):S31-S36

[76] Divaris K. Precision dentistry in early childhood: The central role of genomics. Dental Clinics of North America. 2017 Jul;**61**(3):619-625

[77] Martínez-Rodríguez C, Patricia JP, Ricardo OA, Alejandro IL. Personalized dental medicine: Impact of intraoral and extraoral clinical variables on the precision and efficiency of intraoral scanning. Journal of Personalized Medicine. 2020 Aug 17;**10**(3):92

[78] Hartsfield JK Jr. Chapter 12: Personalized orthodontics, limitations and possibilities in practice. In: Krishnan V, Davidovitch Z, editors. Biological Mechanisms of Tooth Movement. 2nd ed. Oxford, UK: Wiley Blackwell Publishing; 2015. pp. 164-172

[79] Bair E, Gaynor S, Slade GD, et al. Identification of clusters of individuals relevant to temporomandibular disorders and other chronic pain conditions: The OPPERA study. Pain. 2016;**157**(6):1266

[80] Meyer SL. Toward precision public health. Journal of Public Health Dentistry. 2020 Mar;**80**(Suppl 1):S7-S13

[81] Khoury MJ, Bowen MS, Clyne C, Dotson WD, Gwinn ML, Green RF, et al. From public health genomics to precision public health: A 20-year journey. Genetics in Medicine. 2018;**20**(6):574-582

[82] Khoury MJ, Iademarco MF, Riley WT. Precision public health for the era of precision medicine. American Journal of Preventive Medicine. 2015;**50**(3):398-401

[83] Gansky SA, Shafik S. At the crossroads of oral health inequities and precision public health. Journal of Public Health Dentistry. 2020 Mar;**80**(Suppl. 1): S14-S22

[84] Schmalz G, Ziebolz D. Changing the focus to the whole patient instead of one oral disease: The concept of individualized prevention. Advances in Preventive Medicine. 2020 May;**20**(2020):6752342

[85] Joda T, Zitzmann NU. Personalized workflows in reconstructive dentistry-current possibilities and future opportunities. Clinical Oral Investigations. 2022 Jun;**26**(6):4283-4290

[86] Dorogoy A, Rittel D, Shemtov-Yona K, Korabi R. Modeling dental implant insertion. Journal of the Mechanical Behavior of Biomedical Materials. 2017 Apr;**68**:42-50

[87] Calciolari E, Donos N. The use of omics profiling to improve outcomes of bone regeneration and osseointegration. How far are we from personalized medicine in dentistry? Journal of Proteomics. 2018 Sep;**30**(188):85-96

[88] Misch CM. Editorial: Personalised medicine: Applications for dental implant therapy. International Journal of Oral Implantology (Berlin). 2021 May 12;**14**(2):119-120

[89] Zhang QZ, Chen C, Chang MB, Shanti RM, Cannady SB, O'Malley BW, et al. Oral rehabilitation of patients sustaining orofacial injuries: The UPenn initiative. Advances in Dental Research. 2019;**30**(2):50-56

[90] Schwendicke F, Krois J. Precision dentistry-what it is, where it fails (yet), and how to get there. Clinical Oral Investigations. 2022 Apr;**26**(4):3395-3403

[91] Groen N, Guvendiren M, Rabitz H, Welsh WJ, Kohn J, de Boer J. Stepping into the omics era: Opportunities and challenges for biomaterials science and engineering. Acta Biomaterialia. 2016 Apr;**1**(34):133-142

[92] Basu B, Gowtham NH, Xiao Y, Kalidindi SR, Leong KW. Biomaterialomics: Data science-driven pathways to develop fourth-generation biomaterials. Acta Biomaterialia. 2022 Apr;**15**(143):1-25

[93] Albrektsson T, Tengvall P, Amengual-Peñafiel L, Coli P, Kotsakis G, Cochran DL. Implications of considering peri-implant bone loss a disease, a narrative review. Clinical Implant Dentistry and Related Research. 2022 Aug;**24**(4):532-543

Chapter 14

The Relationship between Dental Occlusion and "Prosthetic Occlusion" of Prosthetic Restorations Supported by Natural Teeth and Osseointegrated Dental Implants

Robert Ćelić, Hrvoje Pezo, Stanislava Senzel and Gracia Ćelić

Abstract

The concept of human dental occlusion represents much more than the mere physical contact of the biting surfaces of opposing teeth. It is not a static, unchanging, structural relationship, but rather a dynamic, real, physiological relationship between different tissue systems. It is best defined as the functional relationship between the components of the masticatory system, which includes the teeth, the periodontium, the neuromuscular system, the temporomandibular joints and the craniofacial skeleton. Biologically, occlusion represents a coordinated functional interaction between different cell populations of the masticatory tissue systems that differentiate, model, remodel, destroy and regenerate. When the functional balance of the masticatory system is disturbed or when occlusion is restored by various types of prosthetic restorations, specific goals of occlusal treatment become important, especially today with the rapid insertion of dental implants. The aim of this chapter is to highlight the characteristics of dental occlusion in relation to the characteristics and requirements of 'prosthetic occlusion' for different types of prosthetic restorations supported by natural teeth, gingiva, alveolar ridges and dental implants. A particular focus in writing the chapter is the analysis of the scientific literature on the interrelationship between the so-called occlusion concepts and the biomechanical aspects of different types of implant prosthetic restorations.

Keywords: dental occlusion, occlusal concepts/schemes, prosthetic restorations, implant prosthetic restorations, dental implants, overload, biomechanics

1. Introduction

There is no doubt that the human dentition must only be considered in the context of the human masticatory system and the whole organism. Knowledge of the characteristics of occlusion (tooth contacts at maximum intercuspation and movements of the mandible) and the interaction within the masticatory system are the basis for the development of so-called occlusal concepts, which have their diagnostic, therapeutic and clinical application in the daily practise of general dentists and specialists in dental prosthetics. The therapeutic capacity of dental occlusion is reflected in the fabrication of fillings made of different restorative materials, different types of prosthetic restorations supported by natural teeth, gingiva, alveolar bones of the maxilla and mandible, osseointegrated dental implants and other therapeutic devices such as different types of occlusal splints.

The term dental occlusion is often given the attribute "controversial" in professional and scientific literature. The contradiction is caused by different objective and subjective interpretations found in theories and empirical observations about occlusion, by the insufficient number and inadequate design of scientific studies (lower level of scientific evidence) about occlusion, up to the argument of certain authors that the characteristics of occlusion, occlusal schemes and concepts have a very low clinical significance. They justify this with the fact that in most cases the patient's masticatory system can adapt to minor changes at the level of occlusion [1–6].

However, common sense and daily clinical practise undoubtedly indicate that knowledge of the morphology and function of natural tooth occlusion cannot and should not be ignored, regardless of the current strength of scientific evidence. This knowledge, with certain modifications and specificities, certainly finds its application in the fabrication of prosthetic and implant prosthetic restorations.

2. Characteristics of dental occlusion and occlusal concepts for different types of prosthetic/implant prosthetic restorations

Mandibular positions and movements in static and dynamic occlusal relationships of the maxilla and mandible have their diagnostic and therapeutic applications, especially in prosthetic patients with partial and complete tooth loss and in patients with the clinical picture of temporomandibular disorder [7]. The joint and tooth position (retruded contact position, RCP) of centric relation (CR), maximum intercuspation tooth position (MIP) and physiological rest are diagnostic and therapeutic reference positions of the mandible used in dental prosthodontics for the fabrication of prosthetic restorations supported by natural teeth, gingiva, and alveolar bone ridges [8–10].

The impossibility of reaching a consensus on the definition of centric relation among experts (prosthodontists, orthodontists, periodontists, oral and maxillofacial surgeons) in the field of occlusion is particularly controversial. The most common conclusion of the studies [11–15] is that agreement on the definition that best or most accurately describes the position of centric relation has not been fully reached and that further and repeated research is needed. Zonnenberg et al. [16] suggest abandoning the term "centric relation" in scientific and academic communication. The authors noted that everyone has a unique relationship that cannot be described in a single term. In healthy patients, this relationship is determined by the maximum intercuspation of the teeth (MIP) and should therefore be considered biologically acceptable. However, they acknowledged that there are individual patients who present clinically with mandibular instability because they have an unstable MIP due to dental and/or skeletal injury or due to malocclusion requiring treatment. In these patients, establishing a new jaw relation is an important part of appropriate occlusal treatment procedures. There are three main groups of patients to whom this applies: edentulous (or partially edentulous) patients who require a construction of partial or full removable denture prostheses; patients who need full-mouth reconstruction, with or without implants; and patients who need full-mouth orthodontic and/or orthognathic therapy.

In addition, it is necessary to know the kinetics of the mandibular masticatory system in function with the masticatory muscles and temporomandibular joints, which are coordinated and controlled by the central nervous system. The evaluation of the occlusion is important in prosthodontics and restorative dentistry because the occlusal surfaces of the teeth to be restored must be functional units of the patient's masticatory system. Specifically, the morphology of the cusps, fossae, grooves, and marginal ridges should support the mandible in the intercuspal position and where appropriate, during eccentric jaw movements and in functional activities such as mastication. Restored teeth should not interfere with mandibular function in mastication, speech, and swallowing nor should they transmit excessive force to the attachment apparatus or the temporomandibular joint either in the intercuspal or eccentric jaw positions or during movement [17].

Occlusal contacts that occur between the teeth of the upper and lower dental arches during mandibular movements under the influence of the function of the masticatory muscles and the anatomy of the temporomandibular joints in the natural dentition can take place through the so-called occlusal concepts or schemes. The "ideal occlusion" for eccentric movements can be classified by three schemes according to the tooth contact condition: mutually protected occlusion (canine guidance), group function, and balanced occlusion. The balanced occlusion concept is applied to complete denture patients while mutually protected occlusion and group function are applied for natural dentition and prosthetic restorations [18–21].

The partial or complete loss of natural teeth and the various types of prosthetic restorations (complete and partial dentures) that compensate for this loss alter the normal regulation of oral functions such as movements and position of the lower jaw or functions such as biting or chewing. In a completely dentate individual, sensory organs (receptors in the dental pulp and periodontal mechanoreceptors) perform fine proprioceptive control of jaw function and influence the control of magnitude, direction and force of the bite (e.g. adaptation of masticatory muscle activity to the hardness of the food). In completely edentulous and partially edentulous individuals, the number of periodontal mechanoreceptors is completely and significantly reduced, so that an important source of tactile sensory input through the central nervous system is lost.

Wearers of complete and partial dentures often have impaired masticatory function because the remaining mechanoreceptors from the alveolar bone, gingiva and palatal mucosa cannot compensate for the loss of periodontal mechanoreceptors. This leads to a change in the basic forms of movement during mastication, i.e., an impairment of masticatory function in wearers of removable prostheses [22–26].

Fuentes et al. [27] come to a similar conclusion in their study. They claim that subjects with prosthetic restoration, regardless of the type of restoration, show a reduction in mandibular range of motion (border and functional) compared to the movements of a fully dentate subject. Rivera et al. [28] compared the characteristics of masticatory cycles in wearers of removable complete and partial dentures and overdentures retained with two dental implants. They concluded that wearers of mandibular overdentures supported by dental implants showed masticatory performance very similar to that described in the literature in younger, fully dentate subjects. They emphasize that this type of prosthetic restoration improves sensory perception and provides wearers with a greater sense of comfort, retention, and stability. They also point out that the wearers of removable implant prosthetic restoration (overdenture) showed increased chewing frequency and speed with smaller and faster chewing cycles.

Shiga et al. [29] studied the stability of mandibular movements during mastication in complete dentures, overdentures supported by dental implants and adult dentate subjects. The least differences in mandibular movements during mastication were found between adult dentate subjects and wearers of overdenture anchored by two implants. However, it was necessary to wear implant overdentures for 9 months to one year to adapt to the new chewing function.

These scientific findings gave rise to the term "osseoperception", which is used to describe the sensations evoked by the mechanical stimulation of dental implants or a prosthetic restoration supported by dental implants. Osseoperception is defined as mechanoreception in the absence of a functional periodontal mechanoreceptive input but derived from temporomandibular joint (TMJ), muscle, cutaneous, mucosal, and/or periosteal mechanoreceptors, and which provides mechanosensory information for oral kinesthetic sensibility in relation to jaw function and artificial tooth contacts. Patients with implant supported prostheses have improved tactile discriminative capabilities and report improved motor function [23, 24, 30, 31].

It has been found that the masticatory efficiency with implant prosthetic restorations is very close to that of the natural dentition and that the maximum bite force with such a prosthetic restoration is equally high, if not higher [24]. However, in subjects with implant-retained prosthetic restorations, regulation of masticatory muscle activity was arranged in response to gradual changes in food consistency that occur during mastication [32].

In this context, there are differences between osseointegrated dental implants and natural teeth. Individuals with implant prosthetic restoration retain a good sense of dynamic loads (such as tapping on the tooth or contact between the tooth and a hard object) but an impaired sensitivity to static loads (e.g. spatial aspects) because the sensory signals underlying osseoperception are qualitatively different from those evoked when the load is directed at the natural tooth [24, 32].

2.1 Occlusal concepts in the fabrication of conventional complete dentures

Being a toothless person entails almost continuous morphological, functional, and behavioral changes in the tissues that support the complete dentures. This is particularly true of the oral mucosa and alveolar ridges of the edentulous maxilla and mandible, but the changes also affect the masticatory muscles, the temporomandibular joints and the lower third of the face. The above changes normally occur as part of the physiological aging process, but they can also be the result of inadequately performed therapy with complete dentures, where the changes can accelerate and aggravate the already difficult condition of edentulism. It should not be forgotten that chronologically, edentulous patients are mostly elderly people who are prone to polypharmacy due to their compromised health status, which translates into a decrease in their overall quality of life [33].

The existence and definition of the concept of centric relation shares a similar confused and controversial fate as the selection of occlusal schemes in the fabrication of complete dentures in edentulous patients. There are various classifications of occlusal schemes for complete dentures mentioned in dental textbooks, professional and scientific journals. For this chapter, it seems appropriate to use the classification of balanced and non-balanced occlusal schemes/concepts in the fabrication of complete dentures. Regardless of the type of occlusal scheme (balanced and non-balanced) for complete dentures, the goals of the chosen scheme are generally the same: maximum retention/support and stability of complete dentures, adequate esthetics, adequate masticatory efficiency, sense of comfort and long-term preservation of oral tissues (gingiva and alveolar bone) due to optimal transfer of masticatory load to the loaded tissues [34–38]. In his 25-year study, Tallgren [39] found that the alveolar bone is continuously resorbed (resorption of the anterior mandibular ridge is four times higher than in the maxilla), which is the biggest problem for wearers of complete dentures.

In the latest edition (2017), "the Glossary of Prosthodontic Terms" [8] defines "bilateral balanced articulation" as: the bilateral simultaneous posterior occlusal contact of the teeth in maximum intercuspal position and eccentric positions. The concept of balanced or bilaterally balanced occlusion/articulation was created "under the pretext" of improving the stabilizing effect of complete dentures by achieving tooth contact on the non-working side of the dental arches during excursive movements of the mandible [40, 41]. In individuals with normal dentition (Angle Class I), occlusal contact on the non-working side of the dental arches represents premature/interferential tooth contact, which has a potential and pathological effect on the function of the masticatory system (association with temporomandibular disorder).

In contrast to the non-balanced occlusal schemes, whose representatives are the mutually protected occlusal schemes (canine guidance) and the group function of natural teeth [8, 42] and the concept of monoplane (neurocentric) occlusion without balance (with no compensating curves or posterior balancing ramp) [43, 44], the examples of balanced occlusal schemes are much more numerous. The confusion is compounded by the fact that different designs of the occlusal surfaces of artificial, predominantly acrylic posterior teeth (anatomical, semi-anatomical and non-anatomical (flat) posterior teeth) are used to realize balanced and non-balanced occlusal schemes for complete dentures. However, the most mentioned types of balanced occlusion schemes for complete dentures are bilateral balanced occlusion, lingualized occlusion, anatomical occlusion, monoplane occlusion with balance (with mediolateral and antero-posterior compensating curves or posterior balancing ramp), buccalized occlusion and others [45–52].

Scientific studies comparing balanced and non-balanced occlusion schemes of complete dentures regarding variables such as retention, stabilization, masticatory efficiency of complete dentures, comfort of wearing complete dentures, resorption of the alveolar ridge (load transfer) and quality of life of edentulous patients also come to contradictory results. Some studies [53, 54] favor non-balanced occlusal concepts (e.g., canine guidance) over balanced occlusions and vice versa [55], others emphasize the advantages of balanced occlusions per se and over non-balanced schemes [38, 47, 56–63], and third studies record no differences between balanced and non-balanced occlusal concepts for complete dentures [64–69] and consider them clinically equivalent in the context studied.

However, the general conclusion based on the current state of scientific evidence shows that edentulous patients function well with complete dentures regardless of the type of occlusal scheme chosen (it seems to the authors of this article that two occlusal schemes can be distinguished in the literature: lingualized occlusion (e.g. in clinical situations with severe resorption of the edentulous alveolar ridges [63]) and canine guidance (patients with good alveolar ridges without neuromuscular problems [46]), in which anatomical posterior teeth are used. Thus, the choice of occlusal scheme for complete dentures is mainly based on a well-determined vertical dimension of the occlusion, on the clinical assessment of the occlusal scheme and the experience of the clinician, and on the expected good relationship between patient and clinician [38, 46, 70]. Until the contrary is proved.

Figures 1–8 show the occlusion scheme (unbalanced occlusion—canine guidance/group function) for complete dentures used by the so-called "Zagreb School", Croatia. The occlusion scheme is based on compliance with the lingual or neutral space; correctly determined vertical and horizontal relationships using the physiological rest position and centric relation; and Gerber's reduced occlusion (the

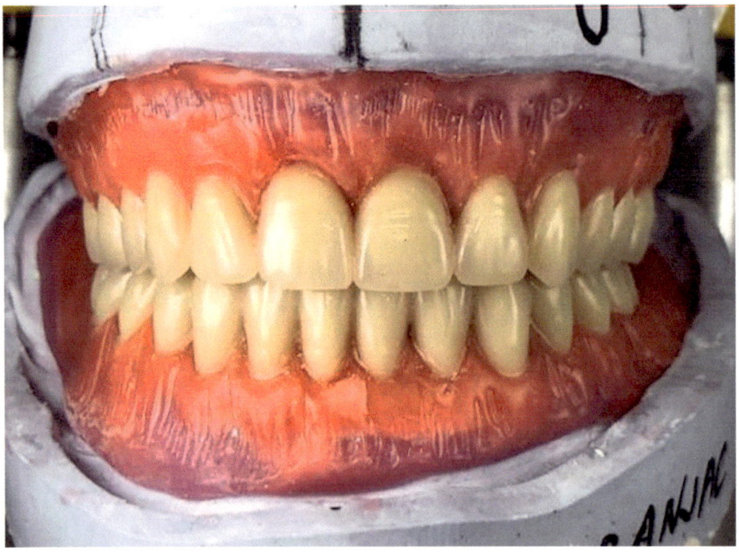

Figure 1.
Setting teeth for complete dentures in the dental articulator: position of maximum intercuspation—"Zagreb School" (unbalanced occlusion—canine guidance/group function).

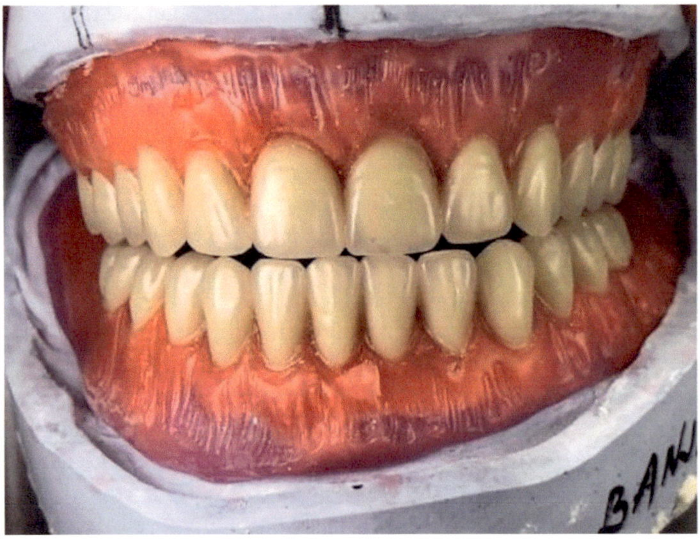

Figure 2.
Setting teeth for complete dentures in the dental articulator: protrusion movement—"Zagreb School" (unbalanced occlusion—canine guidance/group function).

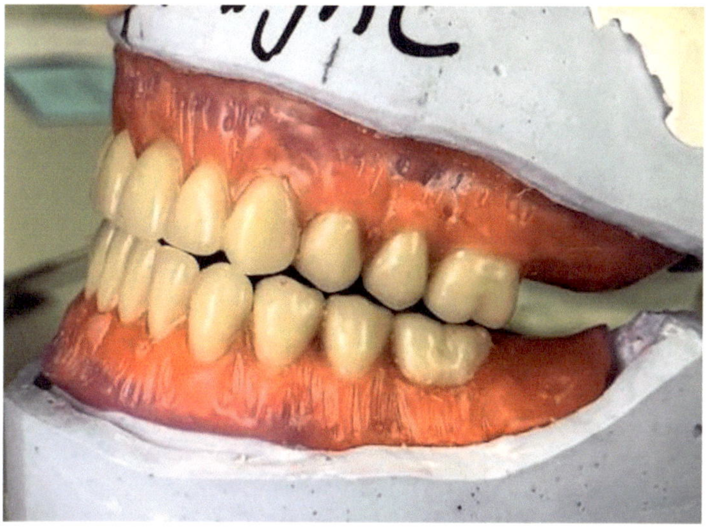

Figure 3.
Setting teeth for complete dentures in the dental articulator: left laterotrusion movement—"Zagreb School" (unbalanced occlusion—canine guidance/group function).

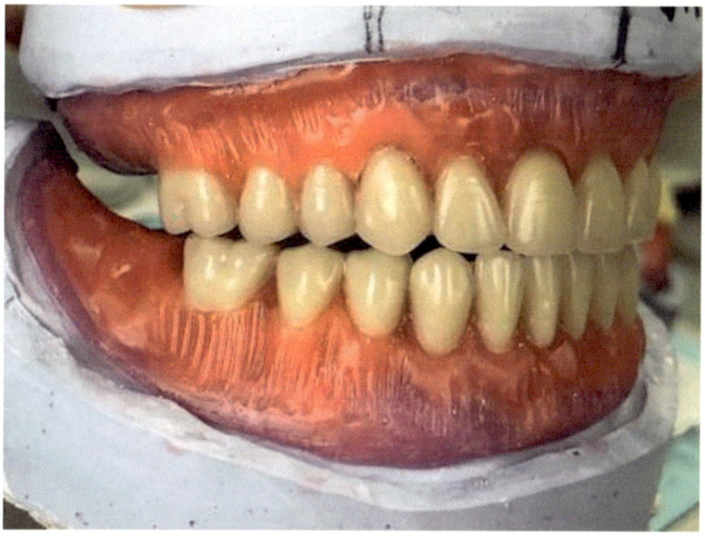

Figure 4.
Setting teeth for complete dentures in the dental articulator: right laterotrusion movement—"Zagreb School" (unbalanced occlusion—canine guidance/group function).

supporting cusps of the upper (palatal) and lower (buccal) artificial (anatomical) posterior teeth are projected to the centre of the edentulous alveolar ridges—the load generated by the function of such a posterior dentition is shifted to the centre of the edentulous ridge and lingually/palatinally, thus acting as a stabilizer in full dentures). One of the peculiarities of this scheme is that the upper and lower second molars in complete dentures are not placed at the expense of the lingual or neutral space.

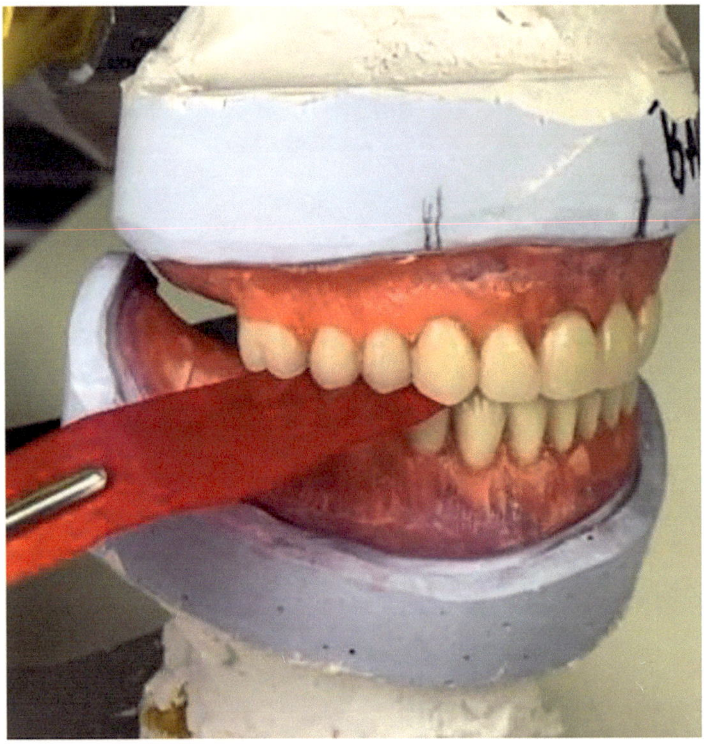

Figure 5.
Marking of the occlusal contacts during maximum intercuspation (red articulation paper) in the dental articulator.

2.2 Occlusal concepts in the fabrication of conventional fixed (single crowns, fixed bridges) restorations and removable partial dentures

Historically, occlusion and articulation are concepts/schemes that are constantly evolving. Bilaterally balanced occlusion, unilaterally balanced occlusion (group function) and mutually protected occlusion (canine guidance) are basic concepts of occlusion used daily in clinical practise. In the fabrication of fixed prosthetic restorations and removable partial dentures, but generally in restorative dentistry, the concept of mutually protected occlusion (alternatively group function) is recommended by the education of doctors and various specialists in dental medicine. As restorative treatment requirements may vary, the clinician should be aware of the combinations of occlusion schemes, their advantages and disadvantages and indications [71].

Otherwise, during the anamnesis and clinical examination of a patient who is to be treated with fixed restorations (single crowns, bridges), removable partial dentures (cast metal) or so called., combined removable-fixed prosthetic restorations (e.g., surveyed crowns + various types of dental attachments + removable partial dentures), an analysis of the state of the existing occlusion is inevitable. Those types of prosthetic restorations usually involve a partially edentulous patient in whom the vertical dimension of the occlusion (the height of the lower third of the face) may or may not be preserved. Regardless of the clinical situation, the

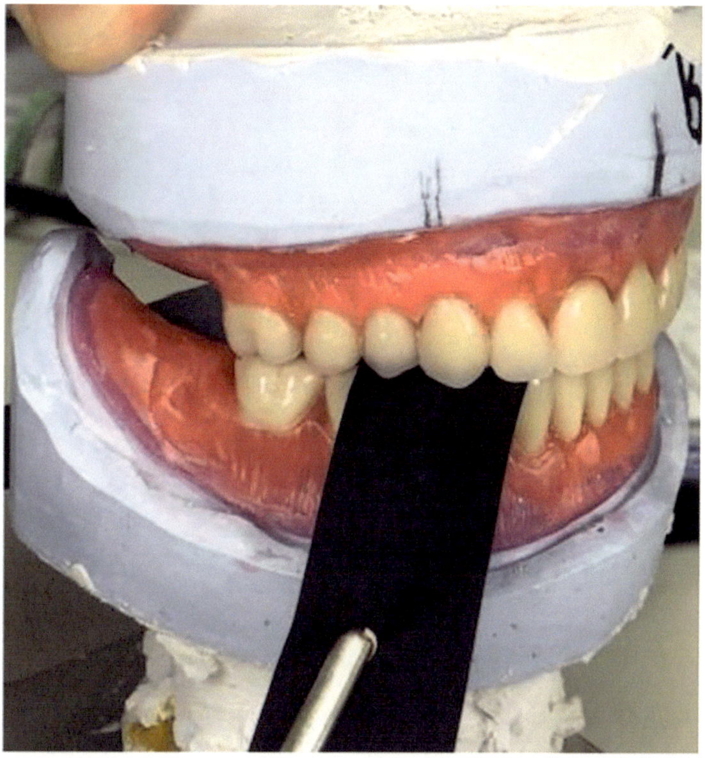

Figure 6.
Marking of the occlusal contacts during lateral and protrusive movements (blue articulation paper) in the dental articulator.

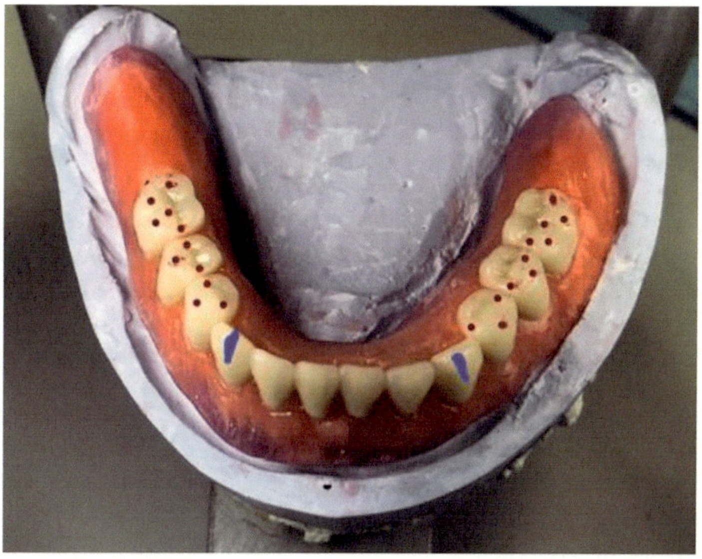

Figure 7.
Distribution of static (red) and dynamic (blue) occlusal contacts in lower occlusal rim/full denture.

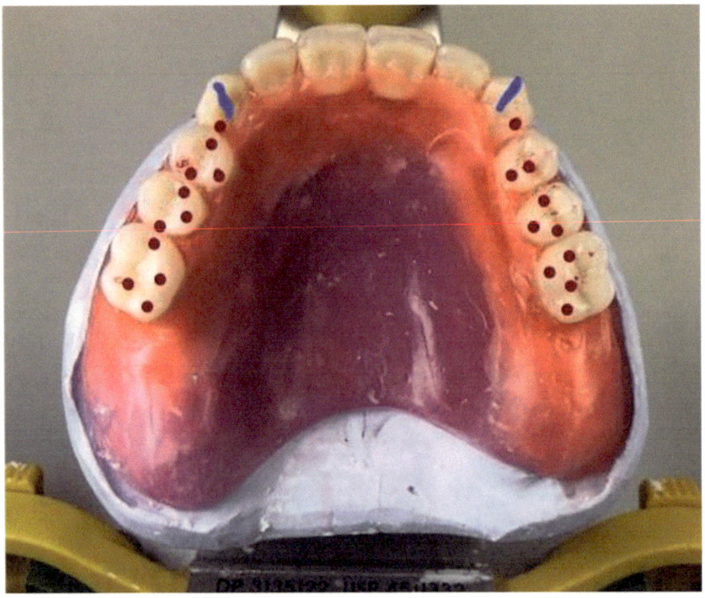

Figure 8.
Distribution of static (red) and dynamic (blue) occlusal contacts in upper occlusal rim/full denture.

static and dynamic occlusal relationships between the existing upper and lower dental arches (occlusogram) should be considered and recorded within the normal function (masticatory muscles and temporomandibular joints) of the masticatory system [24, 72]. Static occlusal contacts are the position and distribution of contacts/dots/marks in the tooth position of maximum intercuspation on the occlusal surfaces of the dental arches, and dynamic occlusal contacts are the position of the path line that occurs during excursion movements of the mandible (in the canine guidance concept, the line is on the palatal surface of the upper canine and the labial surface of the lower canine on the working side). Static and dynamic analysis of occlusion may reveal the presence of premature occlusal contacts in the centric (shift between the retruded contact position and maximum intercuspation) and on the non-working sides of the dental arches (most commonly the maxillary and mandibular second molars) during laterotrusion and protrusion movements of the mandible. In the literature, the causal relationship between premature occlusal contacts and the prevalence of clinical signs and symptoms of TMJ disorders is controversial, but in any case, they should be excluded when performing any form of prosthetic rehabilitation [7, 24, 71].

Occlusal analysis can be performed in the patient's mouth and on study plaster models in an adjustable dental articulator (with facebow transfer). As part of the occlusal analysis of the study models in the adjustable dental articulator, the vertical dimension of the occlusion (which may be adequate or reduced) can be assessed as part of the diagnostic procedure; resulting changes in the horizontal jaw relation (joint position of the centric relation); the degree of wear of the tooth structure on the remaining natural teeth as a result of parafunctional activity of the masticatory system in the form of bruxism; the change in the alignment of the occlusal/prosthetic plane; the presence of centric and eccentric premature occlusal contacts; the position of the teeth of the potential abutments of the prosthetic restorations (which may be inclined in all directions; in supraocclusion; in infraocclusion), etc. The above parameters may also influence the prosthetic treatment plan (selection of the type of prosthetic restoration) and the choice of occlusal scheme. The clinician has the option of choosing all three types, from canine guidance to group function to bilateral balanced articulation (e.g. when one jaw is edentulous and a complete prosthesis needs to be made and there is partial tooth loss in the opposing jaw that can be treated with a fixed restoration or removable partial denture) [24].

Static and dynamic occlusal contacts can be marked on the biting surfaces with thin articulating papers (in different colors) or with a shimstock foil. The resulting marks on the tooth surfaces provide information about their position, distribution, and intensity. When articulating paper is used to identify and verify static and dynamic occlusal contacts in the patient's mouth, the procedure is not entirely reliable in terms of leaving an adequate trace of each mark on the biting surface of the teeth due to the moist environment in the oral cavity. Even if articulating papers with a thickness of 200 μm are used, they leave a trace of a larger and unclear surface, which makes them additionally "positively false". Therefore, for testing and marking occlusal contacts, shimstock thickness (8–10 μm) or articulating paper thickness (12–40 μm) are more reliable, provided that a "dry working field" has been achieved on the occlusal surfaces [73–75]. Kernstein and Radke [76] also claim that the use of articulating paper in occlusal analysis is influenced by the clinician's subjective and inaccurate assessment, as it is not possible to distinguish how the differences between heavy and light loads/bites on the articulating paper affect the intensity of occlusal contact marks. They propose a more objective method of occlusal analysis where it is possible to determine the relative occlusal force and duration of static and dynamic occlusal contacts. Devices (e.g., the T-scan device) for so-called computer-assisted occlusal analysis [77, 78] are capable of this type of occlusal analysis. It should be noted that these devices are currently used more in scientific research and are slowly finding their way into clinical practise.

When fabricating fixed crowns and bridges supported by natural teeth, the occlusal settings are adapted to the static and dynamic occlusal contacts of the remaining natural teeth. If no natural canines are present (due to extraction or, for example, in individuals with Angle class anomalies II and III, where canine guidance is not realistic), the group function concept can be used for the above types of prosthetic restorations. In this case, with the group function concept, the central and lateral incisors take over the anterior guidance instead of the canines and "protect" the posterior teeth from overloading. In addition, a detailed occlusal analysis is important as it shows all tooth contacts and anterior guidance that should be maintained when preparing the teeth. It also allows clinicians to decide whether to maintain the existing occlusal scheme (conformative approach) or to modify the existing occlusal scheme (reorganized approach) [79].

After analyzing the occlusion of the existing condition of the masticatory system according to the same principles, the selection of an occlusal scheme (canine guidance and group function) for a removable partial denture can be recommended. Therefore, in a partially edentulous patient in whom, the fabrication of a removable (metal) partial denture (RPD) or a combined removable fixed prosthetic restoration (metal RPD + surveyed crowns + dental attachments) is indicated, the objectives of establishing an appropriate occlusal scheme should be combined in such a way that the supporting teeth, gingiva and alveolar bone are functionally appropriately loaded (within the adaptability of the aforementioned tissues) and that they are harmonized with the morphology and function of the masticatory muscles and temporomandibular joints [80].

3. Occlusal concepts in the fabrication of implant prosthetic restorations

In the recent literature, numerous factors are mentioned (e.g. bone quality and density; type, surface and design of dental implant; type of prosthetic restoration and its passive fit on dental implants; skills and experience of the clinician; health status of the patient (influence of systemic diseases); oral parafunctions; smoking, oral hygiene and others) that can affect the long-term clinical success of implant prosthetic therapy. Among other factors, the biomechanical environment in which implant prosthetic restoration's function is cited as a major cause (along with infectious factors) of initial and long-term bone loss around dental implants. Biomechanical loading refers to the way in which occlusal/masticatory forces are transferred to dental implants by different types of prosthetic restoration, which occurs during static and dynamic occlusal contacts, i.e. mandibular movements [5, 6, 81, 82]. Scientific studies [83–85] have shown that mandibular movements and function (by form and velocity), chewing efficiency (habitual chewing) are very similar in natural dentition and implant prosthetic restorations compared to wearers of removable conventional complete dentures thanks to the phenomenon of osseoperception.

In general and historically, the parameters of a therapeutically functionally optimal occlusion to ensure longevity (test of time) and success of prosthetic restoration are as follows: (a) simultaneous bilateral occlusal contacts and even distribution of occlusal forces; (b) absence of premature occlusal contacts in the centric; (c) smooth, even, symmetrical lateral and protrusive movements of the mandible without premature occlusal contacts on the non-working sides of the dental arches [1, 4, 86–88]. In other words, the concepts of occlusion in natural teeth were transferred, with certain modifications, to the concepts of occlusion in implant prosthetic restoration. The clinician's responsibility in selecting an occlusal scheme for implant prosthetic restoration is to minimize occlusal overload at the interface between the alveolar bone and the surface of the dental implant. In addition, an accurate diagnosis and treatment plan provides for an appropriate number and placement of dental implants inserted for removable and fixed implant prosthetic restoration, a passive fit of the prosthetic restoration supported by the implants, and progressive loading to increase the amount and density of bone around the implant to reduce the risk of loading beyond physiological limits [89].

Following the same principle, Misch [81, 89] introduced an occlusal scheme for implant prosthetic restoration at the end of the last century, called implant-protective occlusion (IPO) or medially positioned lingualized occlusion. The IPO concept considers several conditions to reduce the stress on the implant-bone interface. These include: the timing of occlusal contacts, the influence of the implant surface, the mutually protected articulation, the angle of the implant or crown body to the occlusal load, the cusp angle of the crowns, the cantilever (offset) distances, the crown height, the crown contour, the protection of the weakest component and the occlusal material of the implant crowns. The following modifications must be considered when creating the "IPO scheme" for implant prosthetic restoration: The occlusal morphology of the prosthetic restoration must direct the occlusal load in the axial direction, use a narrower occlusal plane, reduce the inclination of the tooth cusps, reduce the length of the cantilever in the mesiodistal and buccolingual directions and use crossbite occlusion in certain situations. In short, the principles of the "IPO scheme" are bilateral stability in the central (habitual) occlusion, even distribution of occlusal contact and load, no premature contacts between the centric relation and central occlusion, realization of the "freedom in centric" (relation between functional/supportive cusps and central fossae of the posterior teeth based on a flat plateau of 1–1.5 mm), anterior guidance whenever possible; and free lateral and protrusive excursion of the mandible without non-working premature occlusal contacts [89].

Teeth with associated periodontium and surrounding bone are inherently biomechanically designed to absorb excessive occlusal/masticatory forces, which is not the case with osseointegrated dental implants without periodontium, which are therefore more sensitive to occlusal overload. The mobility of a natural tooth under axial masticatory loading is in the range of 25–100 μm, that of a dental implant is 3–5 μm. Under axial loading, the periodontium of a natural tooth allows the tooth to move according to the indicated range and the functional adaptation of the alveolar bone ("shock or stress absorber"). The dental implant "reacts" to the axial load with a linear deflection to the extent allowed by the elastic deformation of the alveolar bone. Therefore, the compressibility and deformability of the periodontium of a natural tooth under axial loading may make a difference in adapting to static and dynamic occlusal forces compared to osseointegrated implants. A natural tooth subjected to transverse or lateral loading can move rapidly by 56–108 μm, rotating in the apical third of the root, and the periodontium of the tooth tends to immediately reduce this direction of force from the crestal bone towards the root. On the other hand, the movement of the implant under the same lateral load is gradual, reaching a displacement of 10–50 μm, and the concentration of the force is greater at the level of the crestal bone around the shoulder of the implant, without the possibility of implant rotation. In addition to direction, occlusal forces may show differences in duration and intensity during normal or parafunctional activity of the masticatory system. Conditions of increased muscle activity due to bruxism, unbalanced static and dynamic occlusal contacts on different types of implant prosthetic restorations, lack of passive fit between prosthetic components (mesostructure), prosthetic restorations and dental implants are potential factors for prolonged and more intense occlusal overload, which consequently increases the occurrence of biological and technical complications in implant prosthetic restorations [82, 89–93].

It must be said, however, that there are studies and opinions of authors [94, 95] who believe that the role of occlusal overload as a causative factor in the occurrence of peri-implant bone loss has not yet been scientifically proven and that this should be done by future studies. The relationship between mechanical loading and biological consequences (increase in peri-implantitis rate) on bone response has been established, but specific thresholds have not been correlated with prosthetic design (crown-implant length ratio, cantilever implant prostheses and splinting) and occlusal scheme guidelines.

3.1 Occlusal concepts in the fabrication of removable implant prosthetic restorations (overdentures)

The fabrication of removable implant overdentures in edentulous maxillae and mandibles represents a higher standard of treatment for completely edentulous patients. According to the ITI classification, there are three loading protocols for dental implants [96]: conventional, early and immediate loading. When fabricating lower overdentures in the edentulous mandible, the standard protocols are the conventional and immediate loading protocols (2–4 dental implants placed) and for upper overdentures the conventional protocol (4 dental implants placed, mainly due to the lower quality and density of the upper edentulous alveolar ridge). From a biomechanical point of view, the choice of an occlusal scheme for removable implant prosthetic restoration has the effect of reducing the magnitude of the load and the mechanical stress/strain force, especially in the crestal bone area around osseointegrated dental implants. As in the fabrication of conventional full dentures, occlusal schemes of bilateral balance occlusion and its variants are most frequently mentioned in the published papers. The lingualized occlusion is recommended in clinical situations when the edentulous alveolar ridges are preserved, while the so-called monoplane (flat plane) occlusion is recommended in extremely atrophic ridges [97–100]. On the other hand, Aarts et al. [101] compared patient satisfaction with the physiological occlusion (in terms of features most similar to the concept of canine guidance (unbalanced occlusion)—used for complete dentures with anatomical posterior acrylic teeth [8]) and lingualized occlusion over a 3-year period of wearing a lower overdenture supported by two implants and an upper conventional complete denture. Most patients rated prosthetic restorations with physiological occlusion better, which justifies the authors in concluding that, in addition to the expected balanced occlusion schemes, unbalanced occlusion concepts can also be used for implant-supported overdentures.

There are no clinical scientific studies that favor balanced occlusal schemes over nonbalanced ones and vice versa. The reasons for this are that there are not many published studies comparing different occlusal schemes for overdentures supported by dental implants. The most common are comparisons of occlusal schemes between the lower overdenture supported by two dental implants and the upper conventional complete denture (no comparison in relation to the natural dentition, different types of fixed, removable prosthetic or implant prosthetic restorations and possible combinations); the small number of subjects participating in this type of research (shortcomings in the study design); and the objective finding that this type of clinical research on this topic is very difficult to conduct for ethical, financial and other reasons.

3.2 Occlusal concepts in the fabrication of fixed implant prosthetic restorations (single crowns, fixed bridges and full arch implant-supported restorations)

When fabricating fixed implant prosthetic restoration such as single crowns or bridges with a shorter span (maximum five to six units), the application of the occlusal scheme is mainly determined by the loading protocol (conventional, early and immediate) of the dental implants and the status of the remaining natural occlusion. In such situations, the existing status of the natural occlusion (static and dynamic occlusal contacts) should be checked and recorded before starting implant-prosthetic therapy. If there is an indication for conventional loading, the occlusal characteristics registered on the natural dentition are applied after a phase of osseointegration. This would mean that the static occlusal contacts on the crown or bridge in the position of maximum intercuspation should be spatially distributed as close as possible to the axial axis of the inserted dental implant (reduced inclination of the tooth cusps in this type of prosthetic restoration to reduce the generation of damaging lateral forces due to leverage). Then, the distance between the single crown or fixed bridge in relation to the antagonist tooth/teeth is 30 μm (due to the lack of tooth periodontium), which avoids subjecting the prosthetic restoration to excessive forces during higher loading (e.g. bruxism). Adequate clearance between the centric relation position and the maximum intercuspation (1–1.5 mm) provides more favorable vertical load lines, reducing the possibility of premature contacts in the centric during function. The anterior and lateral guidance of the mandible (dynamic occlusal contacts) is provided by the patient's natural teeth (canine guidance or group function) and there are no premature working and non-working contacts during excursion movements of the mandible. In summary, the occlusal parameters mentioned are parameters that describe the occlusal scheme of implant-protective occlusion [5, 81, 89, 98, 102, 103].

However, in the indication for immediate loading of dental implants and the fabrication of single crowns and fixed bridges with shorter spans with preserved natural dentition, there are some changes regarding occlusion. First, this form of implant prosthetic therapy consists of two prosthetic phases. In the first phase, a provisional or temporary prosthetic restoration is fabricated on newly placed dental implants due to one or more lost teeth, using a "softer" prosthetic material such as polymethyl methacrylic (PMMA), composite, metal-acrylic or metal-composite. The aim of this approach with the use of "softer prosthetic material" is to protect newly placed dental implants from excessive functional and parafunctional forces (e.g. bruxism, severe clenching) in a partially edentulous patient [104, 105]. Second, for the same reasons, two occlusal loading protocols for dental implants have been proposed and applied: the immediate functional loading protocol and the immediate non-functional loading protocol. Immediate functional loading assumes that, in this case, a temporary single crown or a fixed bridge supported by a newly placed dental implant is in occlusal contact with the opposite part of the dental arch. The protocol of immediate non-functional loading implies that there is no occlusal contact (neither static nor dynamic contacts) between the provisional prosthetic restoration and the opposing dental arch [106]. The published scientific studies on this subject provide contradictory results and interpretations. There are studies [107, 108] that report lower survival rates of dental implants after immediate functional loading than after immediate non-functional and even delayed conventional loading. Other authors observe no differences between immediate functional and immediate non-functional loading in terms of implant survival, peri-implant bone loss or soft tissue healing, especially in the mandible, at short or medium follow-up periods [109–113]. In contrast to the mandible, the immediate loading protocol for single crowns and fixed bridges is carried out more cautiously in the maxilla. Zarrabi et al. [114], in their randomized clinical trial comparing a non-functional immediate protocol and a conventional protocol in the posterior maxilla, found no differences in the follow-up results (1 year) of clinical and radiological parameters evaluating implant prosthetic therapy. Once the period of osseointegration has elapsed (3–6 months), provisional fixed implant prosthetic restorations can be replaced by definitive single crowns and fixed bridges made of different prosthetic materials, using occlusal schemes based on the principles described in the first paragraph of this chapter.

In modern digital dentistry, the so-called "digital workflow" allows the virtual design of different types of prosthetic restorations supported by dental implants in accordance with the principles of the chosen occlusal scheme. **Figures 9–17** show the possibilities of virtual occlusal design in the fabrication of single crowns supported by dental implants in the partially edentulous mandible. In the specialized software, there is the possibility to perform mandibular movements in the virtual articulator and to adjust static and dynamic occlusal contacts to definitive prosthetic/implant prosthetic restorations. This reduces the need to grind the occlusal contacts in the patient's mouth and compromise the mechanical integrity of the finished prosthetic restoration.

The fabrication of full-arch fixed implant prosthetic restorations in completely edentulous patients also has its own peculiarities and occlusal characteristics, especially since they can be made of different prosthetic materials (e.g. metal-acrylic, metal-ceramic and all-ceramic prostheses). Between biological and technical complications

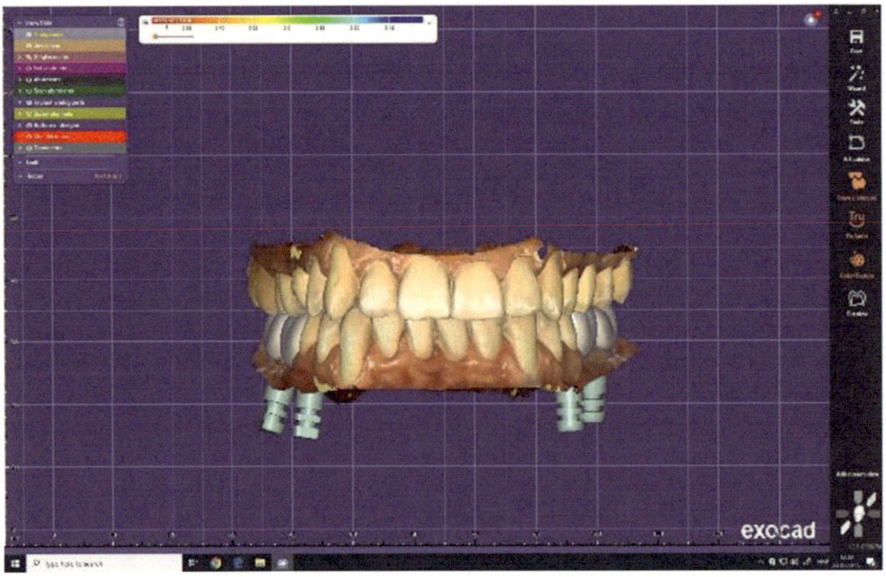

Figure 9.
Fabrication of fixed implant prosthetic restorations—virtual design of single crowns supported by dental implants.

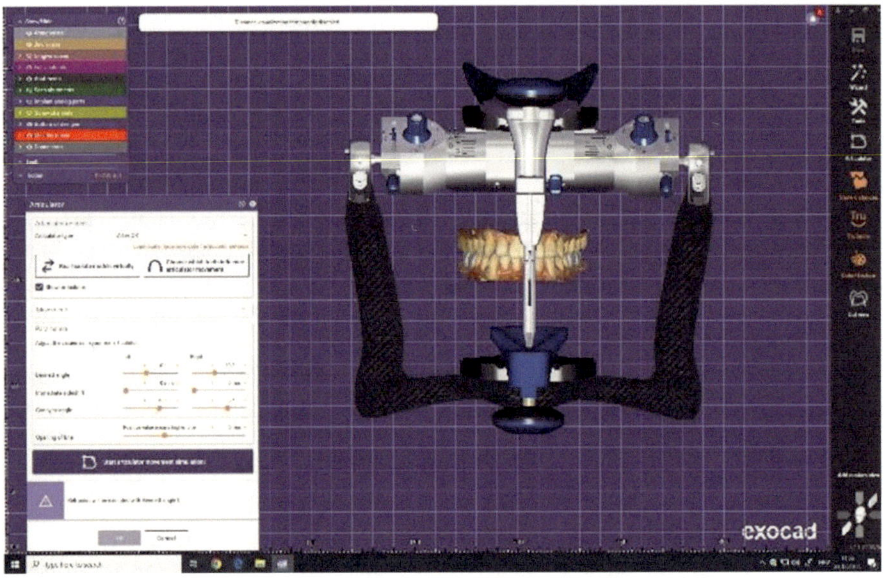

Figure 10.
Fabrication of fixed implant prosthetic restorations—virtual design of single crowns supported by dental implants. Display of modeling in the position of maximum intercuspation in the virtual articulator.

in implant prosthetic restoration, technical or prosthetic complications are more common, some of which can lead to occlusal instability and occlusal overload. This type of implant prosthetic restoration is most often applicable through a conventional and immediate protocol (e.g. All-on-4 concept) of loading dental implants with a high survival rate (95–100%) at a follow-up of at least five years or more [115–117].

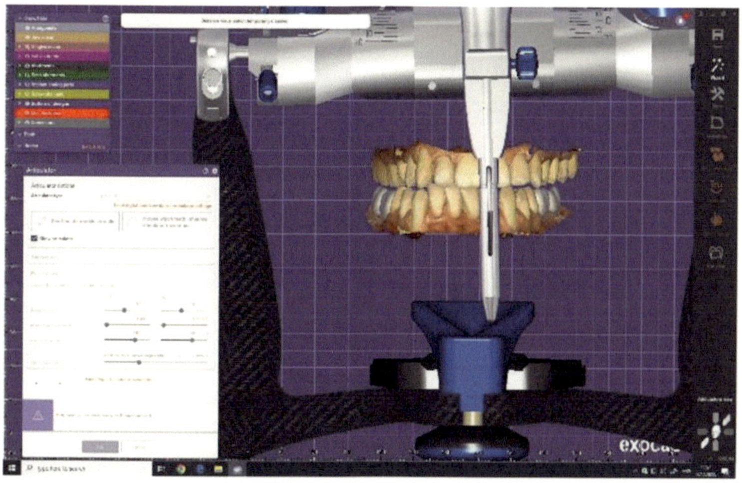

Figure 11.
Movement of the mandible in the virtual articulator (right laterotrusion).

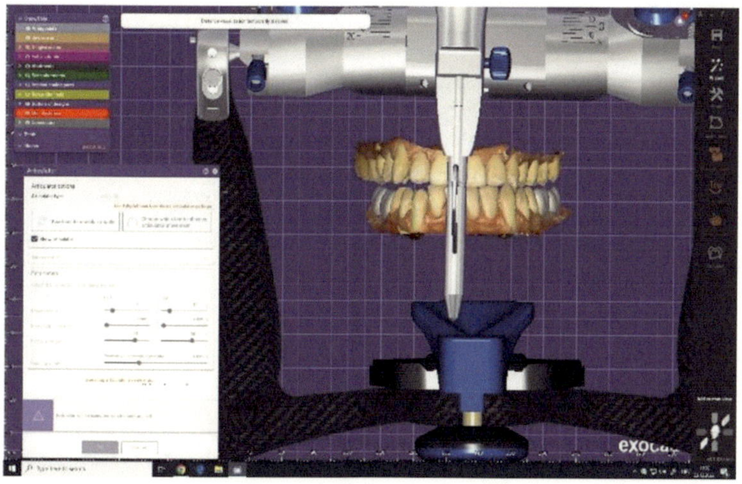

Figure 12.
Movement of the mandible in the virtual articulator (left laterotrusion).

For conventional loading, the characteristics of the occlusion scheme for definitive full-arch fixed implant prosthetic restoration are closest to those of implant-protective occlusion [81, 89], which combines the characteristics of natural dentition and lingualized occlusion.

Immediate loading is a little different. Temporary fixed implant prosthetic restorations are made of a "softer" prosthetic material (e.g. PMMA or metal acrylic) and the occlusion requirements fall under the protocol for immediate loading. This means that the temporary fixed implant prosthetic restoration has established stable centric static occlusal contacts with the opposing dental arch. If cantilevers are present on fixed prosthetic restorations, the cantilevers should be left out of contact (10–30 μm). As this is a one-piece fixed prosthetic construction (in other words, the dental implants are connected or splinted by a prosthetic restoration), the distribution of static and

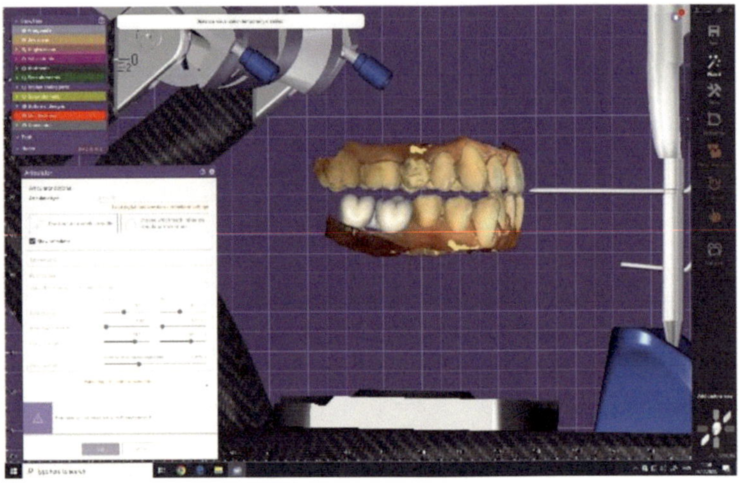

Figure 13.
Movement of the mandible in the virtual articulator (protrusion).

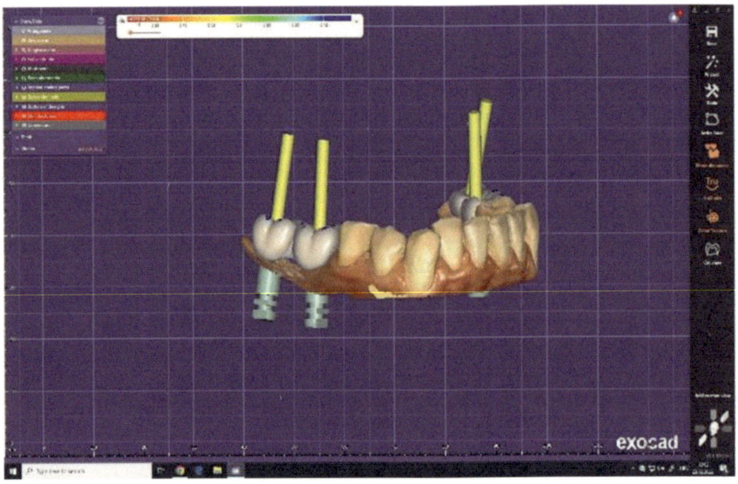

Figure 14.
Point occlusal contacts on the supporting cusps of modeled single implant crowns to achieve optimal load transfer along the longitudinal axes of the inserted dental implants (right lateral view).

dynamic occlusal contacts is within the area (polygon) created by the connection of the installed dental implants. Splinting of dental implants with fixed restorations, passive fit of a fixed structure supported by implants, harmonized static and dynamic occlusal contacts (without premature interference) ensure successful completion of osseointegration of dental implants. After six months of osseointegration of dental implants, the provisional prosthetic restorations can be replaced and final fixed implant prosthetic restorations with different prosthetic materials can be started and completed.

Recently, Yoon et al. [118] in their review suggested occlusal considerations or guidelines based on the prosthetic material chosen when performing a full-arch fixed implant prosthetic restoration in both edentulous jaws. They presented five

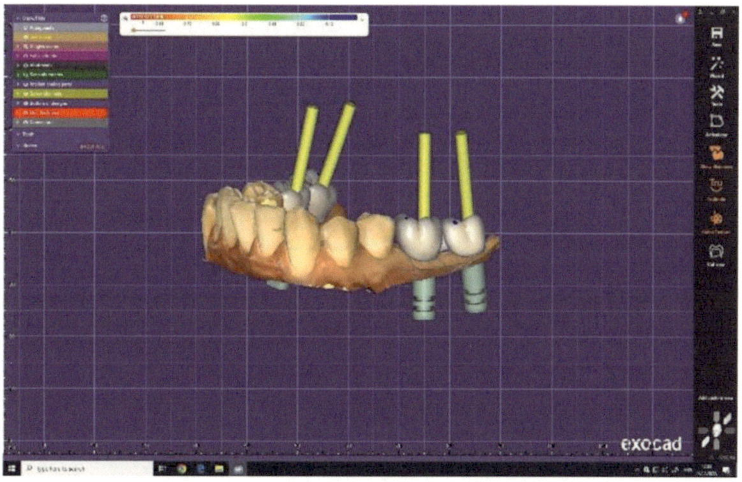

Figure 15.
Point occlusal contacts on the supporting cusps of modeled single implant crowns to achieve optimal load transfer along the longitudinal axes of the inserted dental implants (left lateral view).

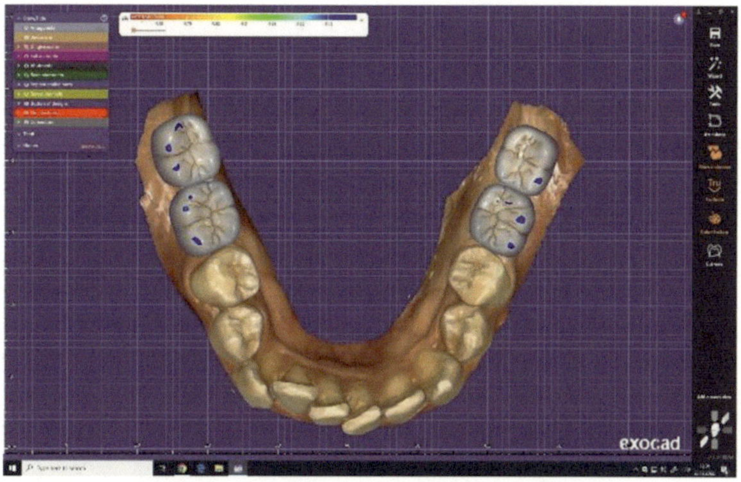

Figure 16.
Point occlusal contacts on the supporting cusps of modeled single implant crowns to achieve optimal load transfer along the longitudinal axes of the inserted dental implants (occlusal view).

combinations of prosthetic materials and natural dentition, focusing on centric and eccentric occlusal relationships. In the situation where a fixed *metal acrylic* implant prosthetic restoration is fabricated in a centric relationship compared to a *metal acrylic* restoration, the following should be achieved: simultaneous bilateral contacts; "freedom in centric" (1–1.5 mm) and on the cantilevers and anteriors a distance of 10 μm (thickness of the Schimstock foil); and during excursion movements: in laterátrusion (group function); in protrusion (shallow anterior guidance) and without occlusal contacts on the cantilevers. Combination of full-arch fixed *metal-acrylic* bridge and *natural dentition* in centric relationship: simultaneous bilateral contacts; "freedom in centric" (1–1.5 mm) and on the cantilevers and anteriors 10 μm; in

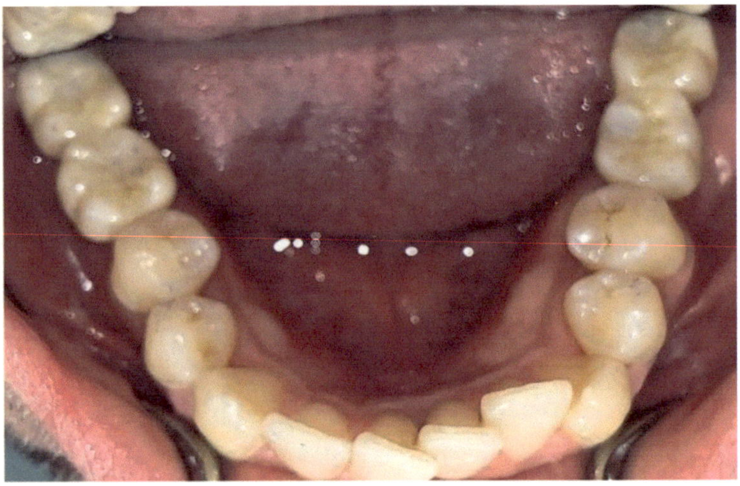

Figure 17.
Final appearance of single implant crowns (screw retained) in the mandible (occlusal view).

excursion movements: in laterotrusion canine guidance if canines are preserved/group function if canines are compromised; in protrusion shallow anterior guidance and no occlusal contacts on the cantilevers. The third combination of prosthetic materials shows centric and eccentric occlusal relationships when fixed implant prosthetic restorations are made of *all-ceramic* systems such as zirconia ceramics (zirconia framework design: fully contoured monolithic and veneering ceramics limited to facial/buccal). In centric relationship, simultaneous bilateral contacts are present; 10 μm distance at cantilevers; equal intensity of contact at anterior and posterior teeth; and "freedom in centric". In case of excursion movements of the mandible, i.e. laterotrusion, group function; in case of protrusion, shallow anterior guidance and no occlusal contact at the cantilever. Combination of *all-ceramics* and fixed *metal-acrylic* prosthetics in centric: simultaneous bilateral contacts; 10 μm spacing on cantilevers and anterior teeth; and "freedom in centric". In lateratrusion, group function; narrow anterior guidance of protrusions and without the presence of occlusal contacts on the cantilever. The last combination shows an *all-ceramic* and *natural dentition*. In a centric relationship: simultaneous bilateral contacts; 10 μm spacing on cantilevers and anterior teeth; and "freedom in centric". In an eccentric relationship: in laterotrusion, canine guidance when canines are preserved/group function when canines are compromised; in protrusion, shallow anterior guidance, and no occlusal contacts on cantilevers. Türker et al. [119] compared three occlusal schemes (canine guidance, group function and lingualized occlusion) using finite element analysis (in vitro study) in full-arch All-on-4 restorations. The stress/strain forces transmitted from the occlusal contacts to the prosthetic restorations, the abutments, and the screws during maximum intercuspation and excursion movements were measured. The most even distribution of forces on the mentioned prosthetic components was recorded with the group function. Different prosthetic materials have different mechanical properties in function and should be considered when designing an occlusal scheme for full-arch fixed implant prosthetic restoration. The article is transparent and very useful for clinicians working on this topic, although the authors themselves point out that the recommendations are "experience-based" and not "evidence-based".

4. Conclusion

The selection of an occlusal scheme for prosthetic restoration supported by natural teeth or osseointegrated dental implants in fully and partially edentulous patients must be adapted to each individual patient, with the aim of ensuring the longevity of the prosthetic restoration in accordance with the health and function of all components of the masticatory system. The characteristics of the natural dentition in relation to the characteristics of the "prosthetic/implant prosthetic occlusion" in the fabrication of fixed and removable prosthetic or implant prosthetic restorations are presented and compared. The design of an artificial prosthetic occlusion or occlusal scheme aims to ensure optimal load transfer to supports such as natural teeth or dental implants, mucosa, and alveolar bone and to reduce the occurrence of biological and mechanical complications.

Unfortunately, it is almost traditional that when this topic is written about, the most common conclusion of published papers is that occlusion is an important and contradictory factor in prosthetic therapy, but that there is no significant scientific evidence recommending one type of occlusal scheme over another [95]. The main argument for this is that it is difficult to design and conduct well-controlled long-term studies with large sample sizes to compare different occlusal schemes. Clinicians are on their own and must apply their knowledge of occlusion acquired in training, through experience and practise, and their common sense in their daily work. The good thing is that with this knowledge and skills about occlusion, the prosthetic restorations will work in the patient's mouth. It will stay that way until future research proves otherwise.

Author details

Robert Ćelić[1]*, Hrvoje Pezo[2], Stanislava Senzel[2] and Gracia Ćelić[2]

1 Department of Removable Prosthodontics, School of Dentistry University of Zagreb, Zagreb, Croatia

2 Dental Polyclinic Ars Salutaris, Zagreb, Croatia

*Address all correspondence to: celic@sfzg.hr

References

[1] Ash MM, Ramfjord S. Occlusion. Philadelphia: WB Saunders Company; 1995. pp. 50-84

[2] McNeill C. Fundamentals Treatment Goals. Science and Practice of Occlusion. Chicago: Quintessence Publishing; 1997. pp. 306-322

[3] Knezović Zlatarić D, Ćelić R. Koncepcije okluzije. Sonda. 2002;**6**:62-64

[4] Schillingburg HT, Hobo S, Whitsett LD, Jacobi R, Brackett SE. Fundamentals of Fixed Prosthodontics. Chicago: Quintessence; 1997

[5] Ćelić R, Pandurić J, Klaić B. Understanding of occlusion—key for osseointegration success. Medix. 2005;**60**(61):180-184

[6] Carlsson GE. Dental occlusion: Modern concepts and their application in implant prosthodontics. Odontology. 2009;**97**:8-17

[7] Okeson JP. Management of Tempormandibular Disorders and Occlusion. St Louis: Mosby; 2003. pp. 1-127

[8] The Glossary of Prosthodontic Terms: Ninth Edition. The Journal of Prosthetic Dentistry. 2017;**117**(5S):e1-e105

[9] Lucia VO. Centric relation—Theory and practice. The Journal of Prosthetic Dentistry. 1960;**10**:849-856

[10] Keshvad A, Winstanley RB. An appraisal of the literature on centric relation. Part I. Journal of Oral Rehabilitation. 2000;**27**:823-833

[11] Jasinevicius TR, Yellowitz JA, Vaughan GG, Brooks ES, Baughan LW, Cline N, et al. Centric relation definitions taught in 7 dental schools: Results of faculty and student surveys. Journal of Prosthodontics. 2000;**9**:87-94

[12] Truitt J, Strauss RA, Best A. Centric relation: A survey study to determine whether consensus exists between oral and maxillofacial surgeons and orthodontists. Journal of Oral and Maxillofacial Surgery. 2009;**67**:1058-1061

[13] Gottsegen R. Centric relation: The periodontist's viewpoint. The Journal of Prosthetic Dentistry. 1996;**16**:1034-1038

[14] Goldstein GR, Andrawis M, Choi M, Wiens JP, Janal MN. A survey to determine agreement regarding the definition of centric relation. The Journal of Prosthetic Dentistry. 2017;**3**:426-429

[15] Wiens JP, Goldstein GR, Andrawis M, Choi M, Priebe JW. Defining centric relation. The Journal of Prosthetic Dentistry. 2018;**120**:114-122

[16] Zonnenberg AJJ, Türp JC, Greene CS. Centric relation critically revisited—What are the clinical implications? Journal of Oral Rehabilitation. 2021;**48**:1050-1055

[17] Weiner S. Biomechanics of occlusion and the articulator. Dental Clinics of North America. 1995;**39**:257-284

[18] McCollum BB. Fundamentals involved in prescribing restorative dental remedies. Dental Items of Interest. 1939;**61**:522, 641, 724, 852, 942

[19] Schuyler CH. Principles employed in full denture prosthesis which may be applied to other fields of dentistry. Journal of the American Dental Association (1939). 1929;**16**:2045-2054

[20] D'Amico A. Canine teeth-normal functional relation of the natural teeth of man. Journal—Southern California Dental Association. 1958;**26**:6-23, 49-60, 127-142, 175-182, 194-208, 239-241

[21] Tiwari B, Ladha K, Aaruti Lalit A, Naik BD. Occlusal concepts in full mouth rehabilitation: An overview. Journal of Indian Prosthodontic Society. 2014;**14**:344-351

[22] Bejoymony CM, Hemasathya B, Mitthra S. Perception and proprioception in relation to masticatory act. Biomedical and Pharmacology Journal. 2015;**8**:149-154

[23] Klineberg I, Murray G. Osseoperception: Sensory function and proprioception. Advances in Dental Research. 1999;**13**:120-129

[24] Klineberg I, Eckert SE. Functional Occlusion in Restorative Dentistry and Prosthodontics. Edinburgh: Elsevier Mosby; 2015. pp. 131-151

[25] Jemt T, Hedegard B, Wickberg K. Chewing patterns before and after treatment with complete maxillary and bilateral distalextension mandibular removable partial dentures. The Journal of Prosthetic Dentistry. 1983;**50**:566-569

[26] Gonçalves TMSV, Vilanova LS, Goncalves LM, Rodrigues Garcia RC. Effect of complete and partial removable dentures on chewing movements. Journal of Oral Rehabilitation. 2014;**41**:177-183

[27] Fuentes R, Arias A, Lezcano MF, Saravia D, Kuramochi G, Dias FJ. Systematic standardized and individualized assessment of masticatory cycles using electromagnetic 3D articulography and computer scripts. BioMed Research International. 2017;**3**:1-9

[28] Rivera P, Farfan C, Arias A, Lezcano MF, Dias FJ, Navarro P, et al. Characteristics of mandibular movement and mastication in older adults with removable dental prostheses: Three-dimensional analysis. International Journal of Odontostomatology. 2020;**14**:81-88

[29] Shiga H, Shin Ogura S, Yasushi Hiraga Y, Hitoshi Takamori H, Nerihisa Namba N, Kobayashi Y. Stability of masticatory movements after placement of implant-supported denture. Odontology. 2022;**110**:216-222

[30] Klineberg I. Introduction: From osseointegration to osseoperception. The functional translation. Clinical and Experimental Pharmacology & Physiology. 2005;**32**:97-99

[31] Klineberg I, Calford MB, Dreher B, Henry P, Macefield V, Miles T, et al. A consensus statement on osseoperception. Clinical and Experimental Pharmacology & Physiology. 2005;**32**:145-146

[32] Grigoriadis A, Johansson RS, Trulsson M. Adaptability of mastication in people with implant-supported bridges. Journal of Clinical Periodontology. 2011;**38**:395-411

[33] Zarb GA, Hobkirk J, Eckert S, Jacob R. Prosthodontic Treatment for Edentulous Patients: Complete Dentures and Implant-Supported Prostheses. 13th ed. London: Elsevier Mosby; 2012

[34] Jacobson T, Krol A. A contemporary review of the factors involved in complete denture retention, stability, and support. Part I: Retention. Journal of Prosthetic Dentistry. 1983;**49**:5-15

[35] Jacobson T, Krol A. A contemporary review of the factors involved in complete dentures. Part III: Support.

Journal of Prosthetic Dentistry. 1983;**49**:306-313

[36] Jacobson T, Krol A. A contemporary review of the factors involved in complete dentures. Part II: Stability. Journal of Prosthetic Dentistry. 1983;**49**:165-172

[37] Adam RZ. Do Complete Dentures Improve the Quality of Life of Patients? (Master Thesis). Cape Town: University of the Western Cape; 2006

[38] Shirani M, Mosharraf R, Shirany M. Comparisons of patient satisfaction levels with complete dentures of different occlusions: A randomized clinical trial. Journal of Prosthodontics. 2014;**23**:259-266

[39] Tallgren A. The continuing reduction of the residual alveolar ridges in complete denture wearers: A mixed-longitudinal study covering 25 years. The Journal of Prosthetic Dentistry. 1972;**27**:120-132

[40] Hobo S, Hisao T. Oral Rehabilitation: Clinical Determination of Occlusion. Tokyo: Quintessence; 1997

[41] Mann AW, Pankey LD. Oral rehabilitation: Part I. Use of the P-M instrument in treatment planning and in restoring the lower posterior teeth. Journal of Prosthetic Dentistry. 1960;**10**:135-150

[42] Nagao M. Comparative studies on the curve of spee in mammals, with a discussion of its relation to the form of the fossa mandibularis. Journal of Dental Research. 1919;**1**:159-202

[43] DeVan M. The concept of neutrocentric occlusion as related to denture stability. Journal of the American Dental Association (1939). 1954;**48**:165-169

[44] Engelmeier RL. The development of nonanatomic denture occlusion: Part IV: Nonanatomic denture occlusion development. Journal of Prosthodontics. 2019;**28**:e159-e171

[45] Becker CM, Swoope CC, Guckes AD. Lingualized occlusion for removable prosthodontics. The Journal of Prosthetic Dentistry. 1977;**38**:601-608

[46] Goldstein G, Kapadia Y, Campbell S. Complete denture occlusion: Best evidence consensus statement. Journal of Prosthodontics. 2021;**30**:72-77

[47] Moradpoor H, Arabzade Hoseini M, Savabi O, et al. Patient satisfaction with occlusal scheme of conventional complete dentures: A randomised clinical trial (part I). Journal of Oral Rehabilitation. 2018;**45**:41-49

[48] Abe J, Kokubo K, Sato K. Mandibular Suction-Effective Denture and BPS: A Complete Guide. Tokyo: Quintessence; 2012

[49] Huff KD, Benting DG. The Art of Complete Denture Therapy for the General Practitioner. Berlin: Quintessence; 2022. pp. 120-133

[50] Rangarajan V, Gajapathi B, Yogesh PB, Mohamed Ibrahim M, Ganesh Kumar R, Karthik P. Concepts of occlusion in prosthodontics: A literature review, part I. Journal of Indian Prosthodontic Society. 2015;**15**:200-205

[51] Rangarajan V, Yogesh PB, Gajapathi B, Ibrahim MM, Kumar RG, Karthik M. Concepts of occlusion in prosthodontics: A literature review, part II. Journal of Indian Prosthodontic Society. 2016;**16**:8-14

[52] Raghavan R, Shajahan PA, Purushothaman P, et al. Occlusal concepts in complete denture

prosthodontics: A literature review. International Journal of Health Sciences and Research. 2020;**5**:96-100

[53] Peroz I, Leuenberg A, Haustein I, et al. Comparison between balanced occlusion and canine guidance in complete denture wearers—A clinical, randomized trial. Quintessence International. 2003;**34**:607-612

[54] Brandt S, Danielczak R, Kunzmann A, et al. Prospective clinical study of bilateral balanced occlusion (BBO) versus canine-guided occlusion (CGO) in complete denture wearers. Clinical Oral Investigations. 2019;**23**:4181-4188

[55] Rehmann P, Balkenhol M, Ferger P, et al. Influence of the occlusal concept of complete dentures on patient satisfaction in the initial phase after fitting: Bilateral balanced occlusion vs canine guidance. The International Journal of Prosthodontics. 2008;**21**:60-61

[56] Winter CM, Woelfel JB, Igarashi T. Five-year changes in the edentulous mandible as determined on oblique cephalometric radiographs. Journal of Dental Research. 1974;**53**:1455-1467

[57] Kimoto S, Gunji A, Yamakawa A, et al. Prospective clinical trial comparing lingualized occlusion to bilateral balanced occlusion in complete dentures: A pilot study. The International Journal of Prosthodontics. 2006;**19**:103-109

[58] Sutton AF, McCord JF. A randomized clinical trial comparing anatomic, lingualized, and zero-degree posterior occlusal forms for complete dentures. Journal of Prosthetic Dentistry. 2007;**97**:292-298. Erratum in: Journal of Prosthetic Dentistry 2007;**98**:16-19

[59] Moradpoor H, Salari F, Ebadian B, et al. Patient satisfaction with occlusal scheme of conventional complete dentures: A randomised clinical trial (part II). Journal of Oral Rehabilitation. 2018;**45**:702-709

[60] Zhao K, Mai QQ, Wang XD, et al. Occlusal designs on masticatory ability and patient satisfaction with complete denture: A systematic review. Journal of Dentistry. 2013;**41**:1036-1042

[61] Niwatcharoenchaikul W, Tumrasvin W, Arksornnukit M. Effect of complete denture occlusal schemes on masticatory performance and maximum occlusal force. The Journal of Prosthetic Dentistry. 2014;**112**:1337-1342

[62] Klineberg I, Kingston D, Murray G. The bases for using a particular occlusal design in tooth and implant-borne reconstructions and complete dentures. Clinical Oral Implants Research. 2007;**18**(Suppl. 3):151-167

[63] Abduo J. Occlusal schemes for complete dentures: A systematic review. The International Journal of Prosthodontics. 2013;**26**:26-33

[64] Paleari AG, Marra J, Rodriguez LS, et al. A cross-over randomised clinical trial of eccentric occlusion in complete dentures. Journal of Oral Rehabilitation. 2012;**39**:615-622 16

[65] Schierz O, Reissmann D. Influence of guidance concept in complete dentures on oral health related quality of life - canine guidance vs. bilateral balanced occlusion. Journal of Prosthodontic Research. 2016;**60**:315-332

[66] Farias-Neto A, Carreiro AF. Complete denture occlusion: An evidence-based approach. Journal of Prosthodontics. 2013;**22**:94-97

[67] Patel J, Granger C, Morrow L. The effect of complete denture occlusion on function and patient quality of life: Systematic review. The European Journal

of Prosthodontics and Restorative Dentistry. 2018;**26**:24-30

[68] Lemos CAA, Verri FR, Gomes JML, et al. Bilateral balanced occlusion compared to other occlusal schemes in complete dentures: A systematic review. Journal of Oral Rehabilitation. 2018;**45**:344-354

[69] Sabir S, Regragui A, Merzouk N. Maintaining occlusal stability by selecting the most appropriate occlusal scheme in complete removable prosthesis. Japanese Dental Science Review. 2019;**55**:145-150

[70] Carlsson GE. Clinical morbidity and sequelae of treatment with complete dentures. The Journal of Prosthetic Dentistry. 1998;**79**:17-23

[71] Rosenstiel SF, Land MF, Walter RD. Contemporary Fixed Prosthodontics. Phaladephia: Elsevier; 2022. pp. 107-146

[72] Zarina R, Jaini JL, Raj RS. A systematic approach for rehabilitation of occlusion in fixed partial denture. International Journal of Clinical Preventive Dentistry. 2017;**4**:136-141

[73] Carey JP, Craig M, Kernstein RB, Radke J. Determining a relationship between applied occlusal load and articulating paper mar karea. The Open Dentistry Journal. 2007;**1**:1-7

[74] Brizuela-Velasco A, Álvarez-Arenal A, Ellakuria-Echevarria J, del Río-Highsmith J, Santamaría-Arrieta G, Martín-Blanco N. Influence of articulating paper thickness on occlusal contacts registration: A preliminary report. International Journal of Prosthodontics. 2015;**28**:360-362

[75] Malta Barbosa J, Urtula AB, Hirata R, Caramês J. Thickness evaluation of articulating papers and foils. Journal of Esthetic and Restorative Dentistry. 2018;**30**:70-72

[76] Kerstein RB, Radke J. Clinician accuracy when subjectively interpreting articulating paper markings. Cranio. 2014;**32**:13-23

[77] Maness WL, Benjamin M, Podoloff R, Bobick A, Golden RF. Computerized occlusal analysis: A new technology. Quintessence International. 1987;**1**:287-292

[78] Afrashtehfar KI, Qadeer S. Computerized occlusal analysis as an alternative occlusal indicator. Cranio. 2016;**34**:52-57

[79] Amin K, Vere J, Thanabalan N, Elmougy A. Occlusal concepts and considerations in fixed prosthodontics. Primary Dental Journal. 2019;**8**:20-27

[80] Chang T-L, Orellana D, Beumer J III. Kratochvil's Fundamentals of Removable Partial Dentures. Berlin: Quintessence; 2019. pp. 138-148

[81] Misch CE. Dental Implant Prosthetics. Occlusal Considerations for Implant-Supported Prostheses: Implant-Protective Occlusion. St Louis: Elsevier Mosby; 2005. pp. 472-510

[82] Parekh RB, Shetty O, Tabassum R. Occlusion in implant prosthodontics. Journal of Dental Implants. 2013;**3**:153-156

[83] Jemt T, Lindquist L, Hedegard B. Changes in chewing patterns of patients with complete dentures after placement of osseointegrated implants in the mandible. Journal of Prosthetic Dentistry. 1985;**53**:578-583

[84] Gartner JL, Mushimoto K, Weber HP, et al. Effect of osseointegrated implants

in the coordination of the masticatory muscles: A pilot study. The Journal of Prosthetic Dentistry. 2000;**84**:185-193

[85] Gonzalez-Gil D, Dib-Zaitum I, Flores-FraileJ,Lopez-MarcosJ.Importance of Osseoperception and tactile sensibility durin masticatory function in different prosthetic rehabilitations: A review. Medicina. 2022;**58**:92

[86] Beyron HL. Characteristics of functionally optimal occlusion and principles of occlusal rehabilitation. Journal of the American Dental Association (1939). 1954;**48**:648-656

[87] Beyron HL. Optimal occlusion. Dental Clinics of North America. 1969;**13**:537-554

[88] Dawson PE. Evaluation, Diagnosis and Treatment of occlusal Problems: A Textbook of Occlusion. St Louis: Mosby; 1989

[89] Misch CE, Bidez MW. Implant-protected occlusion: A biomechanical rationale. Compendium. 1994;**15**: 1330-1344

[90] Parfitt GJ. Measurement of the physiological mobility of individual teeth in axial direction. Journal of Dental Research. 1960;**39**:608-618

[91] Richter EJ. In vivo horizontal bending moment on implants. International Journal of Oral & Maxillofacial Implants. 1998;**13**:232-244

[92] Gross MD. Occlusion in implant dentistry. A review of the literature of prosthetic determinants and current concepts. Australian Dental Journal. 2008;**53**:60-68

[93] D'Amico C, Bocchieri S, Sambataro S, Surace G, Stumpo C, Fiorillo L. Occlusal load considerations in implant-supported fixed restorations. PRO. 2020;**2**:252-260

[94] Sadowsky SJ. Occlusal overload with dental implants: A review. International Journal of Implant Dentistry. 2019;**5**:29

[95] Goldstein G, Goodacre C, Taylor T. Occlusal schemes for implant restorations: Best evidence consensus statement. Journal of Prosthodontics. 2021;**30**:84-90

[96] Gallucci G, Benic G, Eckert S, Papaspyridakos P, Schimmel M, Schrott A, et al. Consensus statements and clinical recommendations for implant loading protocols. The International Journal of Oral & Maxillofacial Implants. 2014;**29**(Supplement):287-290

[97] Reitz J. Lingualized occlusion in implant dentistry. Quintessence International. 1994;**25**:177-180

[98] Kim Y, Oh TJ, Misch CE, Wang HL. Occlusal considerations in implant therapy: Clinical guidelines with biomechanical rationale. Clinical Oral Implants Research. 2005;**16**:26-35

[99] Nikolopoulou F, Ktena-Agapitou P. Rationale for choices of occlusal schemes for complete dentures supported by implants. The Journal of Oral Implantology. 2006;**32**:200-203

[100] Ismail HA, Yousief SA, Mahrous AI, Shaban AA, Azzeghaiby SN, Aljehani D. Clinical and radiographic evaluation of median lingualized occlusion in implant retained mandibular complete overdenture. Journal of International Oral Health. 2015;7(Suppl. 1):5-8

[101] Aarts JM, Payne AGT, Thomson WM. Patients' evaluation of two occlusal schemes for implant overdentures. Clinical Implant Dentistry and Related Research. 2008;**10**:140-156

[102] Davies SJ, Gray RJM, Young MPJ. Good occlusal practice in the provision

of implantat borne prostheses. British Dental Journal. 2002;**192**:79-88

[103] Al-Ani Z, Maghaireh H. Occlusion on a single implant-supported crown: Any differences? Primary Dental Journal. 2022;**11**:32-38

[104] Romanos GE. Wound healing in immediately loaded implants. Periodontology 2000. 2015;**68**(1):153-167

[105] Glauser R, Rée A, Lundgren A, Gottlow J, Hämmerle CH, Schärer P. Immediate occlusal loading of Brånemark implants applied in various jawbone regions: A prospective, 1-year clinical study. Clinical Implant Dentistry and Related Research. 2001;**3**:204-213

[106] Degidi M, Piattelli A. Immediate functional and nonfunctional loading of dental implants: A 2 to 60-month follow-up study of 646 titanium implants. Journal of Periodontology. 2003;**74**:225-241

[107] Zembic A, Glauser R, Khraisat A, Hämmerle CH. Immediate vs early loading of dental implants: 3-year results of a randomized controlled clinical trial. Clinical Oral Implants Research. 2010;**21**:481-489

[108] Margossian P, Mariani P, Stephan G, Margerit J, Jorgensen C. Immediate loading of mandibular dental implants in partially edentulous patients: A prospective randomized comparative study. International Journal of Periodontics & Restorative Dentistry. 2012;**32**:e51-e58

[109] Degidi M, Nardi D, Piattelli A. A comparison between immediate loading and immediate restoration in cases of partial posterior mandibular edentulism: A 3-year randomized clinical trial. Clinical Oral Implants Research. 2010;**21**:682-687

[110] Donati M, Botticelli D, La Scala V, Tomasi C, Berglundh T. Effect of immediate functional loading on osseointegration of implants used for single tooth replacement. A human histological study. Clinical Oral Implants Research. 2013;**24**:738-745

[111] Östman PO, Wennerberg A, Ekestubbe A, Albrektsson T. Immediate occlusal loading of NanoTite tapered implants: A prospective 1-year clinical and radiographic study. Clinical Implant Dentistry and Related Research. 2013;**15**:809-818

[112] Singh JP, Gupta AK, Dhiman RK, Chowdhury SKR. Comparative study of immediate functional loading and immediate non-functional loading of monocortical implants. Medical Journal, Armed Forces India. 2015;**71**(Suppl. 2): S333-S339

[113] Vogl S, Stopper M, Hof M, Theisen K, Wegscheider WA, Lorenzoni M. Immediate occlusal vs nonocclusal loading of implants: A randomized prospective clinical pilot study and patient centered outcome after 36 months. Clinical Implant Dentistry and Related Research. 2019;**21**:766-774

[114] Zarrabi MJ, Radvar M, Shiezadeh F, Mokhtari MR, Nejat A. Immediate nonfunctional loading of a single implant in the posterior maxillary area: A randomized clinical trial. Journal of Long-Term Effects of Medical Implants. 2018;**28**:145-153

[115] Malo P, Araujo Nobre MD, Lopes A, Rodrigues R. Diuble full-arch versus single-arch four implant-supported restorations: A retrospective, 5-year cohort study. Journal of Prosthodontics. 2015;**24**:263-270

[116] Malo P, Araujo Nobre MD, Lopes A, Ferro A, Botto J. The all-on-4 treatment

concept for the rehabilitation of the completely edentulous mandible: A longitudinal study with 10 to 18 years of follow-up. Clinical Implant Dentistry and Related Research. 2019;**21**:565-577

[117] Gallardo YNR, Rodrigues da Silva-Olivio I, Gonzaga L, Sesma N, Martin W. A systematic review of clinical outcomes on patients rehabilitated with complete-arch fixed implant-supported prostheses according to the time of loading. Journal of Prosthodontics. 2019;**28**:958-968

[118] Yoon D, Pannu D, Hunt M, Londono J. Occlusal considerations for full-arch implant-supported prostheses: A guideline. Dentistry Review. 2021;**2**:100042

[119] Türker N, Alkiş HT, Sadowsky SJ, Büyükkaplan US. Effects of occlusal scheme on all-on-four abutments, screws, and prostheses: A three-dimensional finite element study. The Journal of Oral Implantology. 2021;**47**:18-24

Chapter 15

Elimination of Retainers in Anterior Zone in Removable Partial Dentures

Diego Muñoz, Christian Rochefort, Nicole Manzur and Sergio Cáceres

Abstract

Partial edentulism, one or more teeth missing, is one of the most common dental problems that can be found in the population. It has multiple resolution possibilities depending on the individual characteristics of the patient. One of these options is the removable partial denture (RPD), which allows for solving complex clinic problems in a faster way and with a considerably lower operational cost compared to other treatment modalities. However, important discontentment has been reported among patients who have opted for this treatment. One of the discomforting factors for users of removable partial prostheses is the presence of visible retainers in the anterior area. In some cases, it could be possible to eliminate the retainers in the anterior sector, providing retention by mechanical friction and the choice of a removal axis.

Keywords: esthetic, retainers, metallic elements, retention, visibility

1. Introduction

Partial edentulism is one of the most common dental problems in the population, which can be solved in many ways, depending on the individual characteristics of each patient. Therefore, the options can be using prostheses on implants, fixed prostheses, and removable prostheses. Restoring with removable partial dentures allows the resolution of complex clinical problems, including extensive edentulous spaces and distal extensions, relatively quickly and at a substantially lower operational cost than other treatment modalities. However, it has been reported that many patients are not satisfied with this type of dental prosthesis [1].

Patients report that one of the most important features is the cosmetic aspect, so if direct retainers are visible when opening the mouth, they are commonly rejected [2].

All direct retainers must provide the following functions in order to be effective and not harm the abutment teeth or tissues of the denture foundation area: support, retention, cross-tooth reciprocation, fixation, and passivity [1, 2]. The literature makes it clear that dissatisfaction can be avoided if the basic requirements of minimal dislodgement during masticatory function are taken into account when making the prosthetic element and if the esthetic demands of the patients can be met. On the other hand, it is not feasible to assume that by achieving an acceptable masticatory capacity with the prosthetic appliance, the esthetic shortcomings of the appliance can be tolerated [3].

On another point, retainers play a key role in removable partial dentures, providing direct retention. However, their esthetic component has been widely recognized as an obstacle to patient acceptance [4, 5]. Also, using a metal clasp in RPD has many disadvantages. They provide sites for bacterial plaque accumulation, can abrade tooth surfaces, and become deformed over time [6]. Therefore, circumferential metal retainers or clasp can challenge esthetic restoration when anterior teeth are present, helping to support and retain removable partial dentures [6].

There are numerous options for the complication caused by visible anterior clasp [7]; however, many of these alternatives do not offer similar retentive characteristics to those provided by circumferential clasps.

Despite the existence of the implant option to replace missing teeth, many patients are not candidates for these solutions. In this group are patients whose economic position does not allow the choice of treatment or whose general systemic health is too deteriorated to undergo dental surgery; and others with psychological or anatomical limitations. Patients in this category may receive a removable partial denture to replace edentulous areas [8].

The removable partial denture (RPD) is an orthopedic appliance intended to restore lost oral structures, which, without causing damage to the remaining components, is installed and dislodged from the oral cavity at the patient's will, which gives it the character of being removable. It must withstand the forces produced during the physiological acts of chewing, swallowing, and phonation that tend to dislodge it and restore the lost function and esthetic appearance [9–11]. RPD, for reasons of cost, time of fabrication, and because it is a nonaggressive and reversible treatment, is still widely used in patients who need partial tooth replacement [9–12].

Because this prosthesis is not rigidly fixed, it is subject to movement in response to loads in function. Since these movements induce stress and displacement of the prosthetic base, retainers are designed to control these possible movements [12]. Many prostheses are fabricated without the essential requirements for proper function to produce a better esthetic appearance. However, they need more stability [13, 14].

Properly fabricating parallel guide planes, well-defined rest seats, and a tangentiograph (surveyor) are important factors to consider for the retention, support and stability of RPD [14]. Although RPDs are a favorable option for restoring many clinical situations, many patients are dissatisfied with this type of prosthesis, especially when they are unstable during mastication [15]. Masticatory function and esthetics should be considered when planning prosthetic-based treatment without overriding one over the other. When anterior teeth must be used to support an RDP, or when they are visible when speaking or smiling, extracoronal retainers are not well accepted by patients for their esthetic demands and can challenge esthetic restoration [16].

With increasing esthetic requirements, more patients ask their dentists to position retainers closer to the gingiva, where the retentive areas tend to be more profound. The rigidity of cobalt chrome retainers makes them impossible to place in these larger retentive areas due to the unacceptable stress that would be placed on the abutments; therefore, their placement cannot be hidden, and they are likely to fit in an area that is still visible when the mouth is opened [17, 18].

2. Classification of removable partial dentures

In oral rehabilitation, classifications have been proposed to allow the application of basic principles in the design of each partial denture so that not only mechanical aspects are taken into account but also the biological conditions of the oral tissues that will be in contact with the denture.

Partially edentulous arches have been classified by various methods. The primary objective of the classification is to facilitate communication about the combination of missing teeth to edentulous ridges among students, dental practitioners, and laboratory technicians. The most widely accepted classification for partially edentulous patients is that of Kennedy in 1925, who organized the different types of patients into classes numbered from I to IV [9].

Kennedy's classification of Partially Edentulous Arch

- **CLASS I**: Bilateral edentulous areas located posterior to the remaining natural teeth.
- **CLASS II**: A unilateral edentulous area located posterior to the remaining natural teeth.
- **CLASS III**: A unilateral edentulous area with natural teeth remaining both anterior and posterior to it.
- **CLASS IV**: A single, but bilateral (crossing the midline), edentulous area located anterior to the remaining natural teeth.

Considering this classification, patients with unilateral edentulous areas and teeth remaining anterior and posterior to them are established as class III [9, 11, 19].

According to the support and transmission of loads to the remaining oral tissues, RDP can be classified as follows:

2.1 Tooth-supported prostheses

These are prosthetic devices whose support or load path depends exclusively on the abutment teeth. For example, this is the case when there are edentulous spans delimited on both sides by teeth [20].

2.2 Tooth mucosa-supported prosthesis

Also called mixed-loaded track prostheses, these are prostheses in which the teeth and the surrounding soft tissues provide support. In this case, there is at least one free end [20].

2.3 Mucosa-supported prosthesis

In this type of prosthesis, the forces exerted by occlusion are transmitted directly to the bone through the alveolar mucosa. Total prostheses and acrylic prostheses generally have this type of support, since they do not have a rest seat for support and are seated directly on the mucosa [20].

3. Components of the removable partial denture (RPD)

The components of an RDP are diverse. The selection and variation of each will depend on the classification of the edentulous arch and the particular characteristics of each clinical case [21].

3.1 Denture base

It is the structure primarily responsible for supporting artificial teeth. This component is located on the soft tissues and provides stability and retention, allowing the transmission of occlusal loads to the biological support structures. In Kennedy class III its function is only to support the artificial teeth, whereas, in a tooth mucosa-supported prosthesis, it must also transmit the loads, provide stability, retention, and, in some cases, lip contouring [22].

3.2 Minor connector

The structure joins the major connector to the anti-rotational elements and the denture base to the clasp assembly. This mechanical element provides mesiodistal stability, since it is attached to the guide plane made in the abutment tooth. It also provides secondary retention to the removable partial prosthesis by means of friction between the guide plane and the minor connector [20, 22].

3.3 Artificial teeth

These elements replace the anatomy, esthetics, and function of missing natural teeth. The most commonly used teeth are acrylic; however, porcelain teeth can also be used [23].

3.4 Major connector

This is the prosthetic structure responsible for joining the RDP components from one side of the arch to the other. All prosthesis components are attached to this element, directly or indirectly. Its function in the tooth-supported prosthesis is only to connect, and in the tooth mucosa-supported prosthesis, it also provides support [24]. The central connector provides support, stability, and retention to the prosthetic appliance. To function correctly, they must meet specific structural requirements, such as rigidity, to prevent torsion, leverage forces on the abutments, and ensure better distribution of forces on the supporting tissues [24].

3.5 Circumferential clasps assembly

This is the RDP unit responsible for resisting the displacement of the prosthesis from its final seating in the supporting tissues on which it rests. This complex involves an abutment tooth on which it will seat, mechanically preventing the prosthesis from shifting out of place in the patient's functional movements, primarily in the vertical direction [25]. When the prosthesis is at rest, and there are no active vertical displacement forces, the retentive arm is in a passive state relative to the abutment. Therefore, when adjusting a retentive arm for more retention, the tip of the active arm of the circumferential clasp should be positioned as cervically as possible to exert a greater degree of retention [25].

4. Component elements of the clasp assembly

In a conventional RDP, the clasp assembly consists of the retentive arm, a reciprocating arm connected through a minor connector, and a rest (**Figure 1**) [26].

4.1 Retentive arm

It is the main actor in the retention of an RDP. The shape of the retentive arm must be such as to allow it to be flexible, so the initial two-thirds of the arm is rigid,

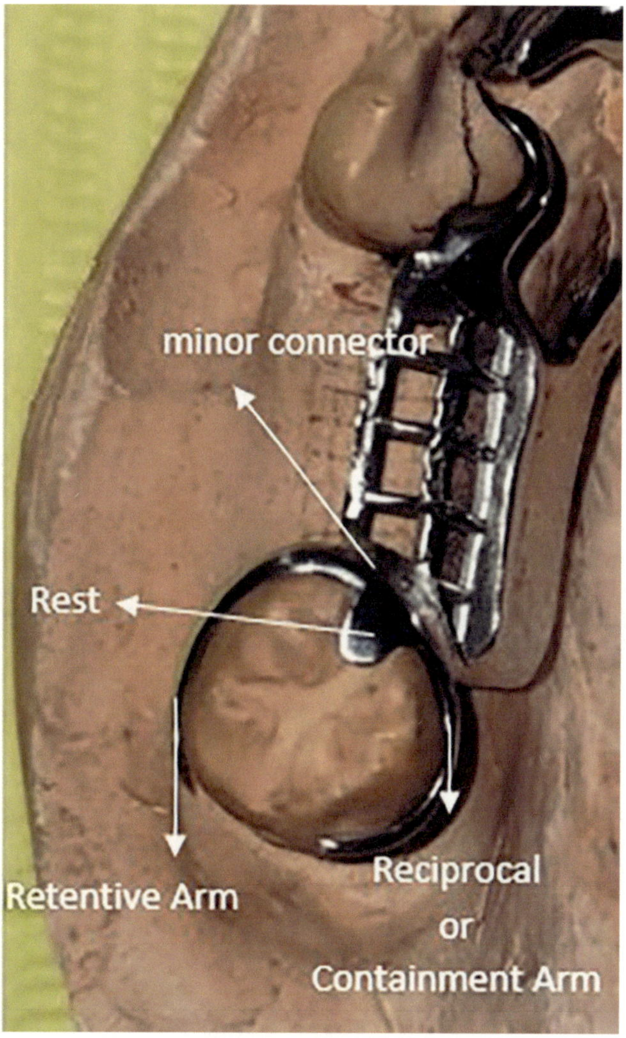

Figure 1.
Components of clasp assembly.

part of which is located over the prosthetic equator, and a thinner terminal end, which sits in the retentive area under the prosthetic equator as it decreases in diameter toward the tip, its flexibility increases. As a result, it can deform as it passes through the prosthetic equator, offering resistance to prosthesis displacement along its path of insertion and removal, which determines the functional forces [9, 11, 20].

The location of the retentive arm generally corresponds to the visible side of the abutment, that is, in the vestibular area. However, this may vary depending on multiple factors, mainly esthetics, which may be detrimental to its function if necessary [9, 11, 20].

4.2 Reciprocal or containment arm

This element is responsible for opposing the forces of the retentive arm on the abutment to prevent damage to the abutment. It exerts an opposing force on the retentive arm when the prosthesis moves out of place and neutralizes it. It is generally located on the opposite side of the retentive arm and toward the occlusal of the prosthetic equator line. Its thickness is uniform throughout and thicker than the retentive arm. In all load-bearing track cases, this element is rigid due to its construction's shape and volume, allowing it to fulfill its primary function. The rigidity also allows it to contribute to prosthetic stability [9, 11, 20].

4.3 Rest

Its rigid structure comes from the minor connector and rest seat milled in the abutment teeth. In posterior teeth, it will be on the occlusal face, while in anterior teeth, it will be in the cingular or incisal area. The rest seat is a preparation on the tooth or restoration created to receive occlusal, cingular, incisal, or root support [5, 21]. It is responsible for transmitting functional forces to the abutment teeth acting on the prosthesis along the axial axis of the prosthesis, preventing movement of the prosthesis into the soft tissues (intrusion), maintaining the positional relationship of the retentive arms concerning the prosthetic equator and also contributes to stability [20, 22]. This structure must be rigid; therefore, it must have a minimum volume that fits into the milled support and does not interfere with the antagonist's occlusion [20, 22]. The rest is considered one of the most important components because, in addition to providing support and controlling the prosthetic position of the oral tissues, the support also helps to restore the occlusal plane [27].

5. Retention in RPD

In RDP, retention of prosthetic appliances is generally not a factor of concern. Through direct retentive arms, the parallelism of guide planes and indirect retention concepts, adequate retention is achieved, which is demanded by patients [28].

5.1 Direct retainers

There are two classes of direct retainers:

1. Intracoronal: These are mechanical devices placed in the casting of a full crown. They are usually reserved for RPD therapies that require exceptional effort to

produce ideal esthetics. There are precision attachments on the one hand and semi-precision attachments on the other.

2. Extracoronal: They engage an external surface of an abutment in a natural undercut or in a prepared depression. There are two categories of extracoronal: 1) Clasp and 2) extracoronal attachment. Clasp are classified as circumferential or ackers (suprabulge) and bar clasp (infrabulge).

Retention is derived by placing a clasp arm into an undercut area so that it is forced to deform upon vertical dislodgement. Resistance of the clasp to deformation generates retention. Resistance is proportionate to the flexibility of the clasp arm. Nonflexible portions of clasp arms must be placed occlusal to the height of contour (suprabulge area) [29].

5.2 Indirect retainers

They are those elements of a removable partial denture, which prevent the rotational displacement of the prosthesis on the supports of the main abutment teeth. Indirect retainers usually take the form of rests, on the opposite side of a fulcrum line.

The retention and stability of an RDP depend on the retentive arms, the mesial bedding component for support, the rest seat within its bedding, and the contact of the minor connector against the guide planes. The tooth-to-base metal relationship must be maintained to allow positive contact.

Noted authors agree that the factors that determine the amount of retention, or if preferred, the tensile strength of an RDP, are given by the seating of the base, the number and distribution of teeth, the design, the material and quality of the retainer, the point of force application, the cervical convergence angle or depth of the retentive area, the location of the active arm tip at the convergence angle, its diameter, cross-sectional shape and the texture of the surfaces on which the active arm of the retainer is placed [28].

The total resistance of the indirect retainer to removal reaches its most significant magnitude before the retentive arm reaches the largest diameter of the tooth, that is, the prosthetic equator, and decreases as it moves toward the occlusal surface of the abutment after reaching this position. The maximum force value reached corresponds to the retention force of the retainer, which in some studies has been determined to be around 500 gf (4.903 Newton) [30].

6. In vitro study for retention of RDP without anterior retainer

Rochefort et al. evaluated the retention and settlement of the RPD in an in vitro study assessing the possibility of removing anterior retainers in patients classified as Kennedy class III subdivision 1 (**Figures 2** and **3**).

In the study, they used ivory artificial teeth provided by Nissin dental products Inc. as prosthetic abutments in maxillary dental models in plaster type IV, where the main characteristic of these is to include in the design retainers in anterior abutment teeth that compromise esthetics (canine teeth or first premolars), which were subjected to tensile tests to evaluate retentive strength (**Figure 4**).

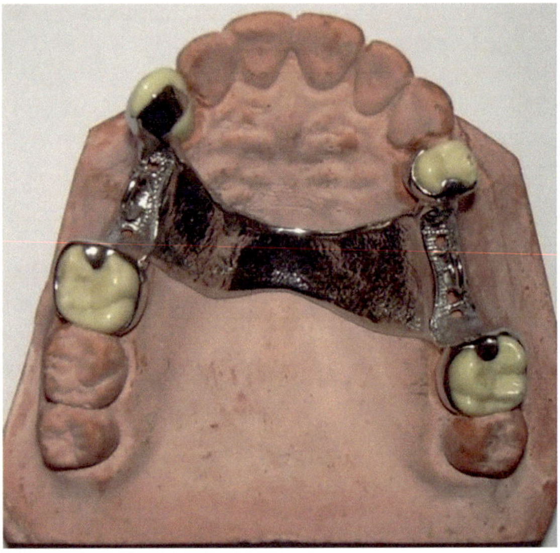

Figure 2.
Removal of clasp from the anterior sector.

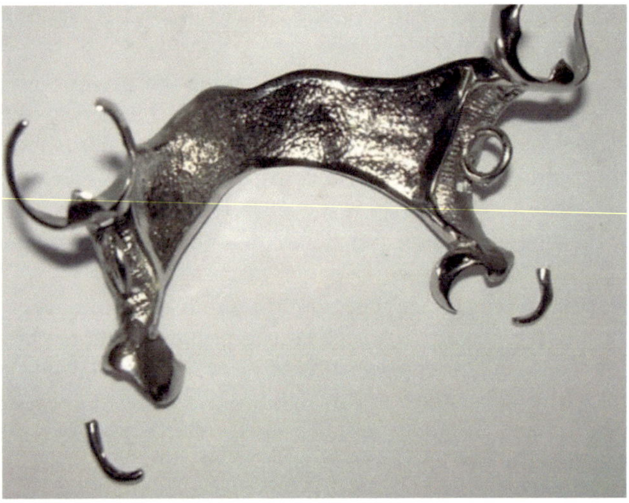

Figure 3.
Front view of metal bases positioned on the plaster model without the anterior clasp.

7. Conclusions

Given the tensile tests performed in the study, it was determined that it is possible to remove the anterior retainers without detriment to the retention of the prosthetic appliance while maintaining its functionality in the in vitro study. However, it may be a therapeutic alternative for patients showing abutments in anterior teeth that represent an esthetic requirement (**Figures** 5 and **6**).

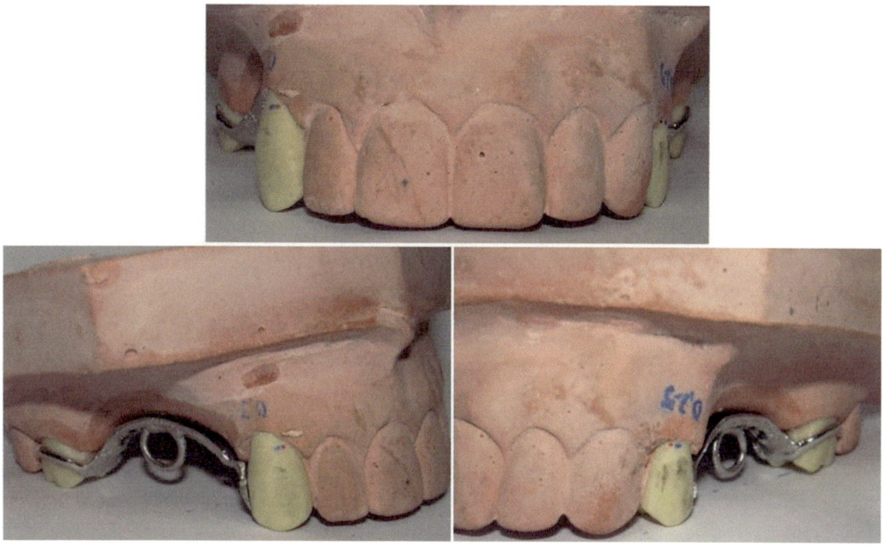

Figure 4.
Side view of metal bases positioned on the plaster model without the anterior clasp.

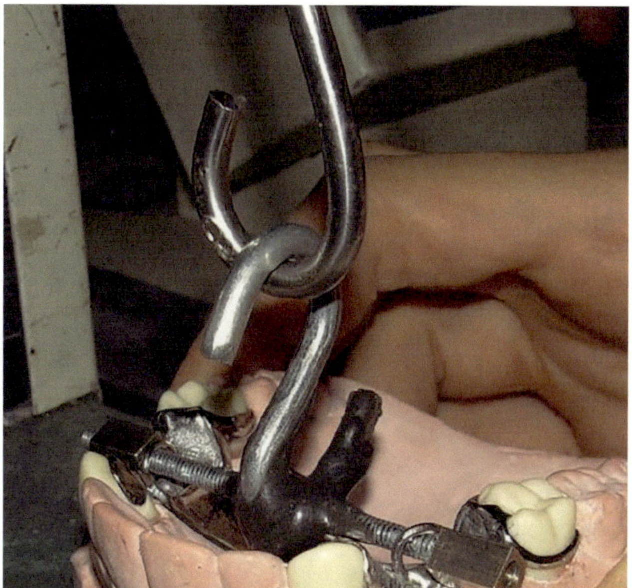

Figure 5.
Pulling device installed on the metal bases for testing.

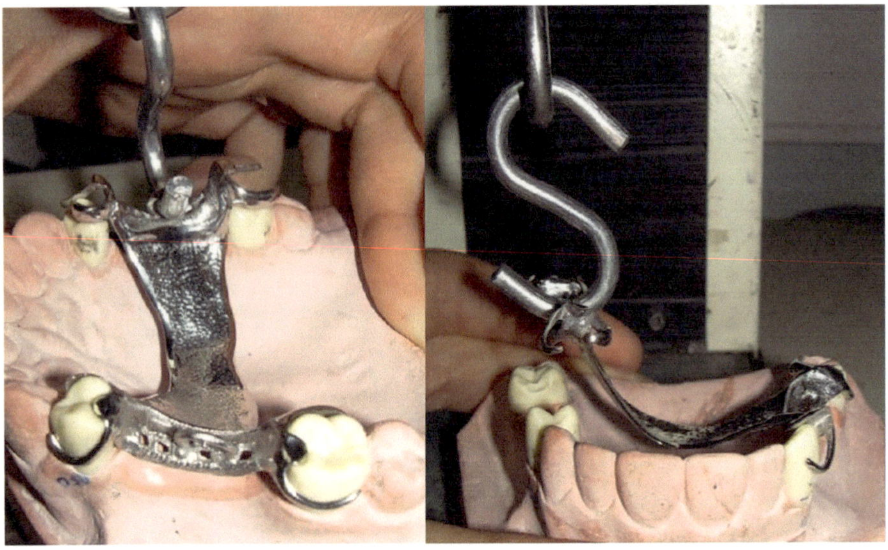

Figure 6.
Pulling device installed on the metal bases for test execution.

Author details

Diego Muñoz*, Christian Rochefort, Nicole Manzur and Sergio Cáceres
Faculty of Dentistry, University of Chile, Chile

*Address all correspondence to: dmunozr@odontologia.uchile.cl

References

[1] Santana-Penín U, Mora MJ. An esthetically attractive twin-flex clasp for removable partial dentures. The Journal of Prosthetic Dentistry. 1998;**80**:367-370

[2] Bezzon OL, Mattos MGC, Ribero RF. Surveying removable partial dentures: The importance of guiding planes and path of insertion for stability. The Journal of Prosthetic Dentistry. 1997;**78**:412-418

[3] de Rossi A, Junior RFA, Bezzon OL. Esthetic options for the fabrication of removable partial dentures: A clinical report. The Journal of Prosthetic Dentistry. 2001;**86**:465-467

[4] Arda T, Arikan A. An in vitro comparison of retentive force and deformation of acetal resin and cobalt- chromium claps. The Journal of Prosthetic Dentistry. 2005;**94**:267-274

[5] Brudvik J, Palacios R. Lingual retention and the elimination of the visible clasp arm. Journal of Esthetic and Restorative Dentistry. 2007;**19**:247-255

[6] Ahmad S, Waters. The effect of reducing the number of clasps on removable partial denture retention. The Journal of Prosthetic Dentistry. 1992;**68**:928-933

[7] Soo S, Leung T. Hidden clasps versus C clasps and I bars: A comparison of retention. The Journal of Prosthetic Dentistry. 1996;**75**:622-625

[8] Suh JS, Billy E. Rotational path removable partial denture (RPD): Conservative esthetic treatment option for the edentulous mandibular anterior region: A case report. Journal of Esthetic and Restorative Dentistry. 2008;**20**:98-107

[9] Loza D. Removable Partial Prosthesis. Caracas: Ed. 1° Actualidades Médico Odontológicas Latinoamericanas; 1992. p. 172

[10] Borel JC, Schittly J, Exbrayat J. Manual of Removable Partial Prosthetics. Vol. 335. Barcelona: Ed. 1° Masson; 1996. pp. 51-120 Chap. 7

[11] Craken M. Removable Partial Prosthesis. Buenos Aires: Ed. 10th Panamericana; 2004. p. 569

[12] Sandu L, Faur B. Finite element stress analysis and fatigue behaviour of cast circumferential clasps. The Journal of Prosthetic Dentistry. 2007;**97**:39-44

[13] Waghorn S, Kuzmanovic DV. Technique for preparation of parallel guiding planes for removable partial dentures. The Journal of Prosthetic Dentistry. 2004;**92**:201-202

[14] Shimizu H, Takahashi Y. Highly filled composite partial coverage restorations with lingual rest seats and guide planes for removable partial dentures. The Journal of Prosthetic Dentistry. 2008;**99**:73-74

[15] Saito M, Notani K, Miura Y, Kawasaki T. Complications and failures in removable partial dentures: A clinical evaluation. Journal of Oral Rehabilitation. 2002;**29**:627-633

[16] Jorge JH, Giampaolo E, Vergani C, Machado A, Pavarina A, Cardoso de Oliveira M. Clinical evaluation of abutment teeth of the removable partial denture by means of the Periotest method. Journal of Oral Rehabilitation. 2007;**34**:222-227

[17] Davenport JC, Basker R, Heath J, Ralph J, Glantz P-O, Hammond P. Clasp

Design. Bristish Journal Dental. 2001;**190**:71-81

[18] Bridgeman J, Marker V, Hummel S, Benson B, Pace L. Comparison of titanium and cobalt-chromium removable partial denture clasps. The Journal of Prosthetic Dentistry. 1997;**78**:187-193

[19] Vanzeveren C, D'Hoore W, Bercy P, Leloup G. Treatment with removable partial dentures: A longitudinal study. Part I. Journal of Oral Rehabilitation. 2003;**30**:447-458

[20] Loza D, Valverde HR. Diseño de Prótesis Parcial Removable. Vol. 239. Madrid: Ed. Ripano; 2007. pp. 45-66 Chap. 2

[21] Dunham D, Brudvik J, Morris J, Plummer K, Cameron S. A clinical investigation of the fit of removable partial denture prosthesis clasp assemblies. The Journal of Prosthetic Dentistry. 2006;**95**:323-326

[22] Díaz A. In Vitro Study of the Retention Capacity of a New Aesthetic Clasp Assembly: Palatal Circumferential in Removable Prostheses". Undergraduate thesis. Santiago: Faculty of Dentistry, University of Chile; 1994

[23] Miller E. Removable Partial Prosthesis. Mexico: Ed. Interamericana; 1975. p. 352

[24] García J, Olavarría L. Prótesis parcial removible: secuencia lógica y practica para su diseño U. de Valparaíso; 2003. p. 187. Available from: https://www.iberlibro.com/9789806574403/DISE%C3%91O-PROTESIS-PARCIAL-REMOVIBLE-SECUENCIA-9806574400/plp

[25] Zarb G. Prosthodontic Treatment for the Partially Edentulous. Buenos Aires: Ed. Mundi; 1985

[26] Sato Y, Shimodaira O, Kitagawa N. Systematic clinical evaluation and correction procedures for support of removable partial dentures. Journal of Prosthodontics. 2007;**2008**:228-232

[27] Henderson D, Seward T. Design and force distribution with removable partial dentures: A progress report. The Journal of Prosthetic Dentistry. 1967;**17**:350-364

[28] Romo F, Contreras C. In vitro measurement of different types of retainers. Review Society Stomatological Prosthodontics. 1988;**IV**:39-43

[29] Robert W. Removable Partial Denture Manual. DMD, MS: Loney; 2011

[30] Sato Y. Clinical methods for adjusting retention force of cast clasps. The Journal of Prosthetic Dentistry. 1999;**82**:557-561

Chapter 16

Novel Prosthetic Solutions for High-Quality Aesthetics

Lavinia Cosmina Ardelean, Laura-Cristina Rusu, Daniela Elisabeta Jumanca and Codruta Victoria Tigmeanu

Abstract

Human teeth play an important role in facial aesthetics. The modern society trends for more and more demanding aesthetics strongly reflect in teeth appearance. Perfect shaped, aligned, and white teeth are considered the business card of an appealing smile. Fixed prosthetics not only aim to restore the lost function of human teeth but also their aesthetics. The ideal in this matter is to provide an indirect restoration that perfectly matches the neighboring teeth or aims to improve the overall appearance of the patient's teeth, depending on the situation. However, the choice of aesthetic materials and technologies has experienced a significant development in recent years. By far, dental ceramics are the state-of-the-art material when a high-quality indirect fixed prosthetic restoration is the goal. This chapter will provide information on this class of materials and their indications in fixed prosthodontics, focusing on novel manufacturing technologies, as well.

Keywords: fixed prosthetic restoration, dental ceramics, CAD/CAM milling, 3D printing, prosthetic solutions

1. Introduction

Human teeth play an important role in facial aesthetics. The modern society trends for more and more demanding aesthetics strongly reflect in teeth appearance. Perfect shaped, aligned, and white teeth are considered the business card of an appealing smile.

Humans attempted to replace the lost teeth or to improve their look through teeth ornamentation since immemorial times. For instance, in ancient Egypt, "donated" teeth were tied to the patient's remaining teeth with golden wires. Mayans used to adorn their teeth with precious stones or even painted them [1]. In the Middle Ages, the main procedure to treat dental issues was extraction, so, prosthodontics made of ivory, precious metals, and other materials were imagined. However, only rich people could afford them, and played mostly an aesthetic role, because they were quite uncomfortable [2].

In the eighteenth century, Pierre Fouchard, the founder of modern dentistry, attempted to use posts, manufactured removable dentures retained by arches, and proposed covering damaged teeth with gold crowns, coated with porcelain, for aesthetic reasons. He founded the first workshop for dental prostheses, and trained jewelers as the first dental technicians. His vision opened the way for the first dental schools and for recognizing dental surgery as a separate medical profession [3].

In 1746, Claude Mouton used a post, retained in the root canal, covered by a gold crown [4], and in the late 1700s, Dubois de Chemant patented the "mineral paste" porcelain artificial teeth [5].

In 1825, in Philadelphia, commercial porcelain teeth were produced [6]. Later on, screw joint retention between the pontic and abutment was introduced by Winder, opening the way to bridge manufacturing [7].

Starting with the twentieth century, the evolution of fixed prosthodontics became fulminant. In 1903, Charles Land first introduced the porcelain jacket crown. In 1907, William Taggart described the lost wax technique, which was a cornerstone for prosthetic dentistry [8]. In 1926, Ante's law, postulated in 1926, still represents the standard principle for abutment selection [9]. The introduction of the high-speed by John Borden in 1957, operating at 300,000 rotations/minute, enabled much easier abutment preparation for fixed prosthodontics [10]. In 1962, porcelain fused to metal was introduced by Weinstein et al. [7]. In 1973, the Rochette bridge concept lead to the idea of minimal tooth reduction [11]. Starting in 1985, with the introduction of CAD/CAM systems, prosthetic dentistry experienced a tremendous transformation, both for patients and practitioners [12].

Fixed prosthetics not only aim to restore the lost function of human teeth but also their aesthetics, aiming to regain the patient's overall comfort and satisfaction. For a successful result, every detail should be considered, starting with the patient's interview, diagnosis, treatment plan, and subsequent phases, as well as follow-up once the restoration is placed.

The ideal in this matter is to provide an indirect restoration that perfectly matches the neighboring teeth or aims to improve the overall appearance of the patient's teeth, depending on the situation. In the quest for the best result, various dental materials for prosthetic restorations have been attempted over time, including rubber, porcelain, aluminum, and later plastic.

However, the choice of aesthetic materials and technologies has experienced significant development in recent years. By far, dental ceramics are the state-of-the-art contemporary material, when a high-quality fixed prosthetic restoration is the goal. This chapter aims to provide information on this class of materials and their indications in fixed prosthodontics, focusing on novel materials and manufacturing technologies.

2. Brief history of dental ceramics

Following Dubois de Chemant's "mineral paste" porcelain teeth, in the beginning of the nineteenth century, Giuseppangelo Fonzi developed single-tooth dental prostheses, held in place thru an embedded platinum pin. The porcelain material, "earth-metal" as he called it, to distinguish it from Chemant's type, was made available in 28 different colors [6].

The era of modern ceramic restorations begins in the 1900s when Charles Land patented the all-ceramic jacket crown, which, despite its low strength, was extensively used until the 1950s, when porcelain-fused-to-metal was developed by Abraham Weinstein [13]. This type of restoration provided adequate strength and reasonable aesthetics, however, diminished by the presence of the opaque layer.

Early air-fired dental porcelain consisted of large-size particle powders, resulting in entrapped air bubbles and increased porosity. Upon the emergence of vacuum-fired porcelain in the 1960s, its appearance improved, due to reduced internal porosity.

Adding leucite filler to feldspathic porcelain intended to tailor their thermal expansion/contraction behavior, to match it with the alloys they were fired on. It also contributes to slow crack propagation [14].

All-ceramic restorations gained interest once again in 1965 when John W. McLean attempted adding alumina (40–50%) to feldspathic porcelain, developing a new version of the all-ceramic jacket crown. Because of the alumina core, it had twice the strength of the traditional one, but still could be used only for anterior teeth. Its major drawbacks were due to high opacity, brittleness, and poor marginal adaptation [15].

During the 1980–1990s, several new ceramic technologies were introduced. Some of them were later abandoned, and some of them were further developed, resulting in the modern all-ceramic pressed and computer-aided design/computer-aided manufacture (CAD/CAM) systems.

The introduction of the Cerestore (Johnson & Johnson, New Brunswick, NJ, USA) "shrink-free" all-ceramic crown system and the Dicor (Dentsply International, York, PA, USA) castable glass-ceramic crown system in the 1980s provided innovative processing methods, such as pressing and centrifugal casting, and further stimulated the renewed interest for all-ceramic restoration. Both were later on abandoned, because of the high fracture incidence, doubled, in case of Dicor, by the difficult and high processing cost [16].

The high-leucite pressed ceramic IPS Empress 1 was introduced by Ivoclar Vivadent (Schaan, Liechtenstein) in the late 1980s. In this case, a heated ceramic ingot is being pressed into a mold, by means of a special pressing furnace (**Figure 1**). Despite its increased strength, IPS Empress 1 was still not suitable for usage in the posterior area, due to frequent fractures, but became popular due to ease of use and high aesthetic appearance. The second generation, IPS Empress 2, released in 1998, is a lithium disilicate ceramic, suitable for single and multiple-unit restorations of the anterior teeth. The IPS e.max system was introduced in 2005. It includes lithium-disilicate IPS e.max Press (Ivoclar Vivadent, Schaan, Liechtenstein), with high strength, which enables fabricating single-tooth restorations and bridges in the anterior and

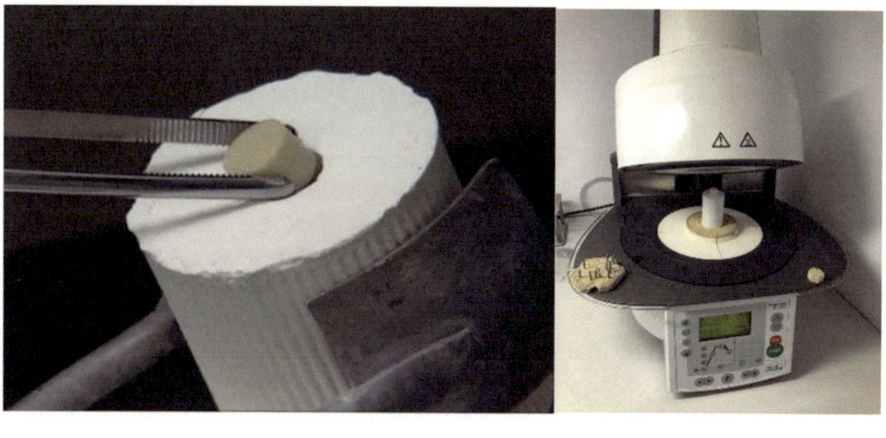

Figure 1.
Pressing of a ceramic ingot.

premolar area, as well as thin inlays, onlays, and veneers. The ingots are available in 4 levels of translucency and 2 sizes and are subsequently stained or layered for better aesthetic results [17].

Back to the year 1989, Vita developed the In-Ceram system (Vita Zahnfabrik, Bad Sackingen, Germany), initially consisting of a glass-infused porous alumina ceramic core, indicated for the anterior region. To increase its aesthetic appearance, the alumina core was later replaced with spinel, resulting in a restoration material suitable for the anterior teeth. The third variant of In-Ceram is zirconia, obtained by mixing alumina with zirconium oxide, gaining a doubled flexural strength, compared with the alumina type. Its indications include posterior crowns and bridges. Following sintering, the core is layered for gaining the final aesthetic result [18, 19].

Dental computer-aided design/computer-aided manufacture (CAD/CAM) applications also became available in the late 1980s-early 1990s. Since then, due to continuous developments, the CAD/CAM technology, either subtractive or additive, has metamorphosed itself not only in the greatest success of contemporary prosthodontics but also disseminated in other dental specialties, as well as in promising emerging domains, such as tissue engineering and regenerative medicine, with outstanding future applications in the dental field [20].

3. The CAD/CAM era

The godfather of digital dentistry and inventor of dental CAD/CAM is considered the French professor Francois Duret, who, in the 1970s, developed a CAD/CAM system capable to fabricate single-tooth restorations. The Duret system was later patented and marketed as the Sopha Bioconcept System, but it did not last long on the dental market, due to its cost and complexity [21].

The first viable dental CAD/CAM system was developed in 1987 by Werner Mormann and Marco Brandestini, under the name of Cerec (Sirona, Bensheim, Germany). It was initially chair-side, used to fabricate feldspathic ceramic inlays. The material used was Vita Mark I feldspathic ceramic (Vita Zahnfabrik, Bad Sackingen, Germany). The Cerec 1 milling unit used a grinding wheel. Short after, Cerec 2, equipped with an additional cylinder diamond, enabled manufacturing of partial and full crowns. Cerec 3 introduced the two-bur system, followed, in 2006, by the "step bur", gaining high precision. In 2000, inLab, its sibling for the dental laboratory, was released. The Cerec dedicated software developed, as well, the two-dimensionally displayed design of Cerec 1 and 2 became three-dimensional in 2003, accompanying Cerec 3. The current version of Cerec is currently used worldwide, with a high rate of clinical success, and various choices of prosthetic restorations, including inlays, onlays, veneers, anterior and posterior crowns and bridges, copings, bridge frameworks, telescope crowns, bars, attachments, provisionals, denture bases, as well as implant mesostructures and surgical guides [22–24].

Another pioneer CAD/CAM system is Procera AllCeram (Nobel Biocare, Zurich, Switzerland), developed by Matts Anderson and Agneta Oden in 1993. Introducing a new ceramic material, consisting mainly of alumina, subsequently layered with a feldspathic ceramic, enabled the fabrication of strong and aesthetic all-ceramic single-unit restorations. Later on, it evolved to multiple-unit restorations, and zirconia copings, as well. However, the technique is extremely sensitive, as it uses enlarged dies to precisely match the shrinkage of the sintered copings. Currently, the centralized production Procera milling centers offer the advantage of possibility of producing high-quality restorations [25, 26].

Top dental manufacturers developed their own CAD/CAM milling systems, such as DCS Precident (Dental Concept Systems, Wesertal, Germany), Cercon (Dentsply Sirona, Charlotte, NC, USA), Arctica and Everest (KaVo, Biberach, Germany), Lava (3 M ESPE, Seefeld, Germany), ZenoTec (Wieland Dental, Pforzheim, Germany), and Tizian (Schutz Dental, Rosbach, Germany). Some started as only CAM or synchronous copy milling systems and have developed later on, adding up the CAD part and different milling units.

3D printing in dentistry was first used in the 1990s, but, until recently, ceramic materials were not suitable for this particular technology. Mainly, different types of resin-based materials and alloys were used for fabricating prosthetic restorations. By means of 3D printing, not only various prosthetics can be fabricated, but tissue constructs, as well [20].

CAD/CAM systems have three major components. The data acquisition unit collects the data from the area of the preparation, adjacent, and opposing structures by means of intraoral scanners. The collected data, either through intraoral scanning or indirectly, by scanning a conventional impression or a model (**Figure 2**) obtained by means of a conventional impression, are converted into virtual models. The second component is the software, used for designing the virtual restorations on the virtual model (**Figure 3**), then sending the data as an STL file and computing the milling parameters (**Figure 4**). Third, the milling machine is used for manufacturing the restoration from a solid block of restorative material (**Figure 5**). In case of CAD/CAM additive manufacturing, the third component is a 3D printing device.

Since its beginning, more than 30 years ago, CAD/CAM subtractive systems have undergone substantial changes [27]. The data acquisition methods evolved from optical cameras to contact digitization and laser scanning. Various intraoral and extraoral digitizing systems are currently available. Complex designing software and the 5-axis modern milling machines, the possibility of dry or wet milling choice, enable the fabrication of complex shapes, including fixed and removable prosthetic devices. The development of novel ceramic materials, such as alumina and zirconia, with

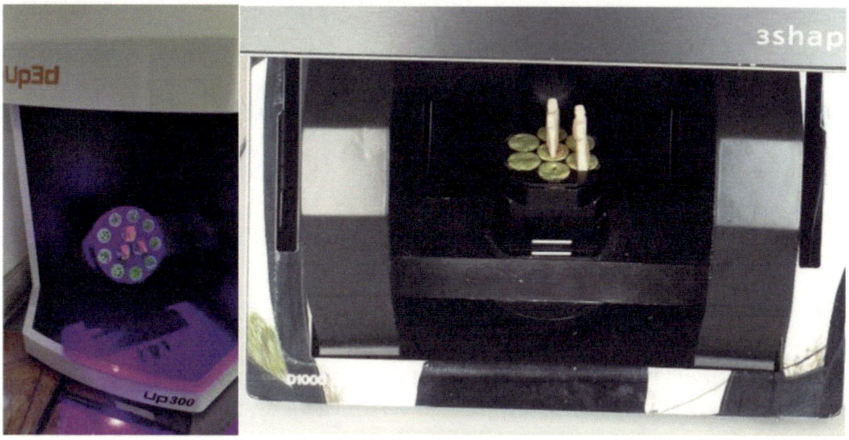

Figure 2.
Scanning of a model.

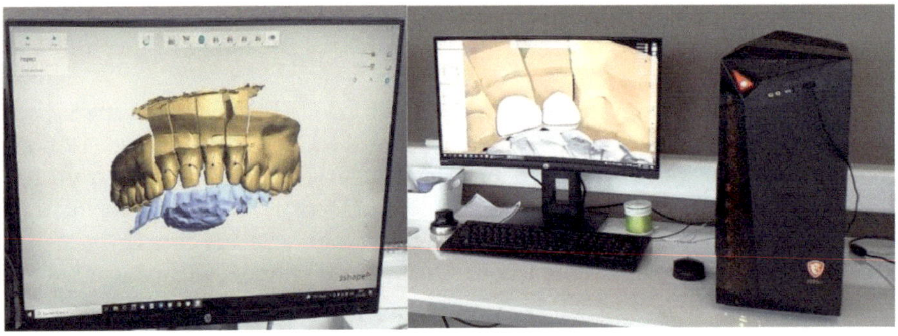

Figure 3.
Designing the restorations on the virtual model.

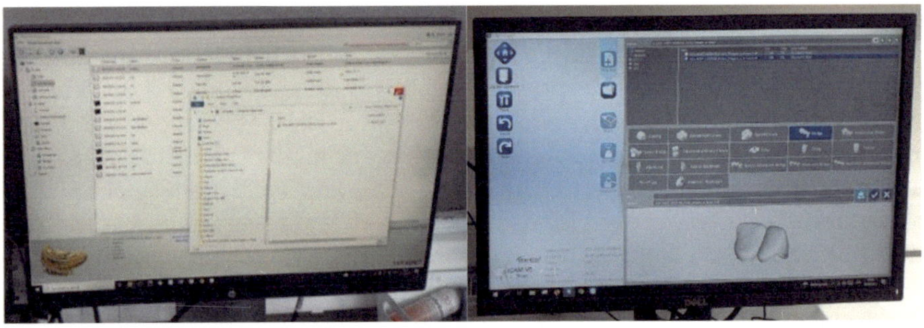

Figure 4.
Sending data, saved as an STL file, and choosing the milling parameters.

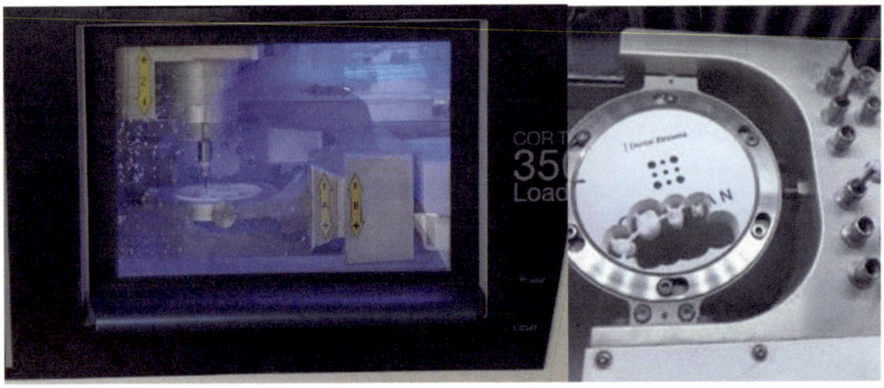

Figure 5.
Milling of a ceramic block.

high strength and excellent machinability, as well as extending the materials range to different types of alloys, resins, and composite materials, currently offers a wide range of possibilities. Integration of the technological advances resulted in the three categories of currently available CAD/CAM systems: chairside, laboratory (**Figure 6**), and centralized production milling centers.

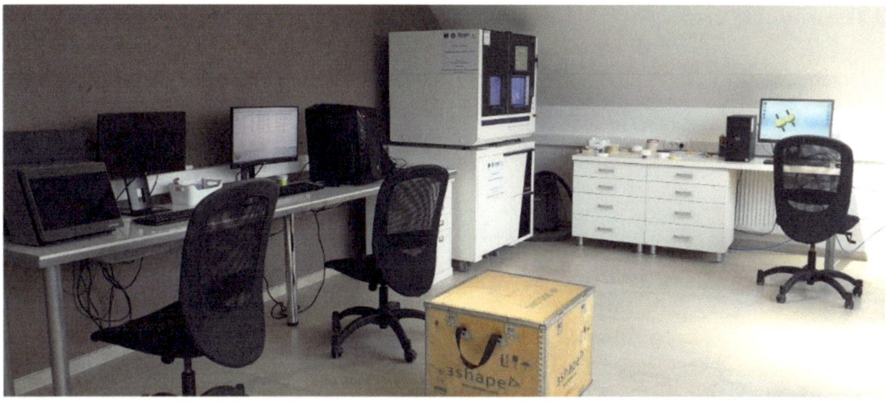

Figure 6.
Components of a laboratory CAD/CAM system.

CAD/CAM can be classified into open and closed systems, according to data sharing. The workflow of closed systems does not permit interchangeability with other manufacturers' systems, the data acquisition scanner, virtual design software, and restoration manufacturing unit are provided by the same company. In case of an open system, the digital data generated by the scanner, transferred in a standard format, allows reading by different manufacturer's milling units, offering more versatility [28].

Automation of fabrication, increased quality, short manufacturing time, reduced infectious cross-contamination hazards, and minimized inaccuracies are some advantages offered by the CAD/CAM technology. One major disadvantage of the chairside and laboratory systems is the high initial cost, a large-scale production being necessary for the investment to pay off. The solution stands in the acquisition of a scanner, scanning and designing the restoration in the laboratory, and then transferring data to a centralized production milling center for fabrication. In case it is needed, subsequent layering is accomplished in the laboratory [28].

4. Ceramic materials for CAD/CAM systems

All-ceramic restorations may be classified as dual restorations (bicomponent or ceramo-ceramic), which consist of two chemically different types of ceramics, which are mechanically and aesthetically compatible, and monobloc restorations (single-component or monolithic), consisting of one type of ceramic. In case of dual ones, the milled ceramic framework is subsequently layered with the second type of ceramic, to finalize its form and aesthetic appearance. In case of monolithic restorations, only a surface makeup is used to ensure the aesthetics [29].

The large choice and structural variety of all-ceramic materials for CAD/CAM systems may represent a tough dilemma, as no material is ideal for all clinical situations, and the selection of the best option is of most importance for accomplishing the best prosthetic restoration [30].

According to their chemical nature, milled ceramics are classified into four categories: vitreous ceramics, glass-infiltrated ceramics, polycrystalline ceramics, and resin matrix ceramics.

4.1 Vitreous ceramics

Vitreous ceramics are further divided into feldspathic ceramics and reinforced glass ceramics.

4.1.1 Feldspathic ceramics

Feldspathic ceramics, consisting of vitreous and crystalline phases, were the first used for CAD/CAM, as Vita Mark I blocks (Vita Zahnfabrik, Bad Sackingen, Germany) for Cerec (Sirona, Bensheim, Germany), but their unsatisfactory clinical performance leads to the development of monochromatic, multiple shade Vita Mark II blocks and the multi-layered Triluxe, Triluxe Forte, and Real Life (Vita Zahnfabrik, Bad Sackingen, Germany), with better mechanical properties, but still only indicated for inlays, onlays, veneers, and anterior crowns. The main plus of feldspathic ceramics is their remarkable aesthetics, due to a blend of luminosity, shade, and saturation [30].

4.1.2 Reinforced glass ceramics

Reinforced glass ceramics contain different minerals: leucite, lithium-disilicate, or zirconia, as fillers, in their crystalline phase.

Empress ProCAD (Ivoclar Vivadent, Schaan, Liechtenstein) was the first block of leucite-reinforced glass ceramic, introduced in 1998, and substituted in 2006 by Empress CAD (Ivoclar Vivadent, Schaan, Liechtenstein), which contains 35–45% fine size leucite particles. It has better resistance to machining, but due to low resistance to flexion, its indications are limited to veneers and anterior crowns. Three types of blocks are available: low and high-translucency and multi.

The lithium disilicate-reinforced glass ceramic, provided in a pre-crystallized stage, is easily milled, and subsequently crystallized in the oven. The blue-colored IPS e.max CAD (Ivoclar Vivadent, Schaan, Liechtenstein) blocks thus gain the shade and translucency of choice. The resistance to flexion is three times higher than that of the leucite-reinforced type. Its indications include veneers, inlays, onlays, anterior and posterior crowns, and even three-unit anterior bridges, extending to the premolar area [31].

In 2013, lithium silicate-reinforced glass-ceramic doped with zirconium dioxide was released, aiming to combine the properties of the components. It has proven to display a higher mechanical resistance to the propagation of fissures in comparison with the lithium disilicate-reinforced glass ceramic, because of the inclusion of 8–10% zirconium dioxide particles. Two commercial products are Celtra Duo (Dentsply Sirona, Charlotte, NC, USA), in a completely crystallized form and Vita Suprinity (Vita Zahnfabrik, Bad Sackingen, Germany), in a partially crystallized phase, requiring an additional crystallization treatment after milling. It has better translucency compared to lithium disilicate-reinforced glass ceramics and is indicated for veneers, inlays, onlays, and anterior and posterior crowns [32].

4.2 Glass-infiltrated ceramics

The blocks for CAD/CAM milling, belonging to this category, were introduced in 1993, based on the original In-Ceram Alumina, Spinel, and Zirconia ceramics (Vita Zahnfabrik, Bad Sackingen, Germany), with indications for crowns and bridges frameworks, up to three elements, depending on the type. Subsequently, layering with a feldspathic ceramic is needed to obtain the final form and aesthetics [33].

4.3 Polycrystalline ceramics

Due to the fact that this type of ceramic is integrally composed of oxides, and does not include a vitreous matrix, they are characterized by excellent mechanical properties, as their dense crystal network opposes the propagation of fissures. It comprises of two types: alumina-based and zirconia-based.

4.3.1 Alumina-based polycrystalline ceramics

Alumina-based polycrystalline ceramics, introduced by Procera AllCeram system (Nobel Biocare, Zurich, Switzerland), consist of more than 99.5% aluminum oxide.

Vita In-Ceram AL for inLab (Vita Zahnfabrik, Bad Sackingen, Germany) pre-sintered blocks consist of the same type of ceramic. They are easily processed, resulting in enlarged bridge and crown frameworks. Shrinkage occurs during the subsequent dense sintering process in a high-temperature furnace. The highly stable and precision-fit frameworks are indicated for anterior and posterior crowns and short anterior bridges, and are to be layered with corresponding ceramic [26].

4.3.2 Zirconia-based polycrystalline ceramics

Zirconia is a polymorphic material, showing three temperature-depending crystallographic phases: monoclinic (from room temperature up to 1170°C), tetragonal (from 1170 to 2370°C), and cubic (from 2370 to 2716°C). Noticeable volumetric changes are associated with these transformations. When sintered above 1170°C, during the monoclinic to tetragonal transformation, a 5% decrease in volume occurs. At cooling, the reverse phenomenon takes place, the change from tetragonal to monoclinic phase is accompanied by a 3–4% volume expansion, leading to a subsequent high level of stress within the sintered material, which often leads to fracture [34] (**Figure 7**).

In order to force it to maintain its tetragonal and/or cubic phases until reaching room temperature, stabilizing oxides: CeO_2, MgO, CaO, or Y_2O_3 have been added.

This allows the generation of the multiphase partially-stabilized zirconia (PSZ), consisting, at room temperature, of cubic zirconia as the major phase, with monoclinic and tetragonal zirconia precipitates as the minor phase [35].

The presence of a small amount of stabilizing oxides, such as 2–3% mol of Y_2O_3, at room temperature, enables obtaining a monophasic material, which consists of tetragonal structured crystals only, known as tetragonal zirconia polycrystal (TZP) [30]. The Y-TZP zirconia blocks are frequently used materials for CAD/CAM dental applications, because of their high mechanical properties and thermodynamically metastable tetragonal phase [36].

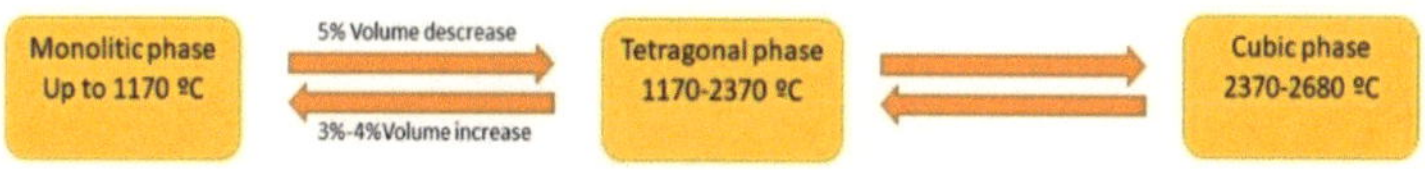

Figure 7.
Temperature-related phase transformation of zirconia.

Zirconia has the capacity of auto-reparation, which makes it resistant to crack propagation. The mechanism is as follows: a propagating crack applies stress, which determines the transformation of the tetragonal particles into monoclinic ones. The phase transformation is accompanied by a 3–5% increase in volume, which results in compression at the crack tip, stopping and squeezing the fracture, and shielding it from its surroundings. This capacity to transform in response to stress makes zirconia a smart material [37].

The high flexural strength of Y-TZP and its fracture resistance recommend it as an alternative to metallic frameworks, with indications for long-range prosthetic pieces [36].

Because of its lower translucency, compared to glass ceramics, initially, the zirconia infrastructure was designed to be layered with feldspathic ceramic, but a relatively high percentage of fracture of the layering ceramic was detected (**Figure 8**). In order to solve the problem, monolithic restorations were attempted (**Figure 9**). Monochromatic

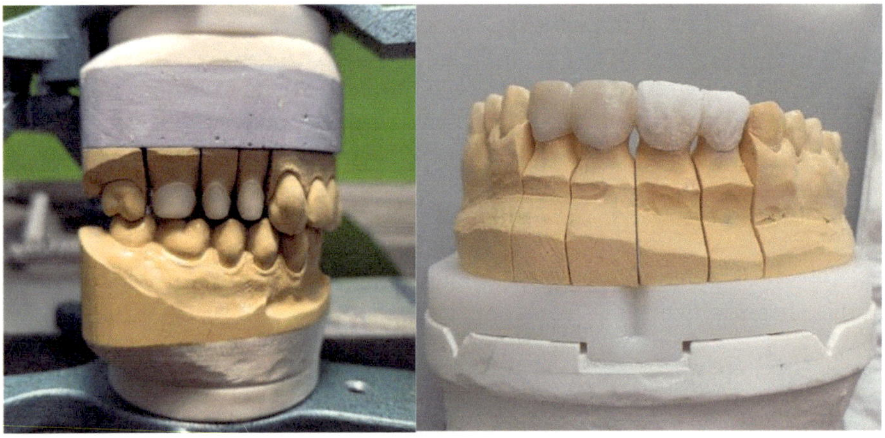

Figure 8.
Layering of a zirconia framework.

Figure 9.
Monolithic zirconia crown (1.6) and zirconia framework to be layered with feldspathic ceramic (2.6).

blocks, which can be colored if needed, and polychromatic zirconia blocks, which imitate the variation in color from dentin to enamel were put on the market. The first polychromatic zirconia block, Katana Zirconia ML, was released in 2013 by Kuraray Noritake (Tokyo, Japan). Much more, different types of zirconia blocks with increased translucency, such as Lava Plus (3 M ESPE, Seefeld, Germany), Cercon ht. KL (Dentsply Sirona, Charlotte, NC, USA), Zenostar T (Wieland Dental, Pforzheim, Germany), and Vita YZ HT, ST, XT (Vita Zahnfabrick, Bad Sachingen, Germany) are currently available. Another possibility is pressing fluorapatite glass-ceramic ingots onto CAD/CAM milled zirconia frameworks: IPS e.max ZirPress (Ivoclar Vivadent, Schaan, Liechtenstein), which allows pressing onto IPS e.max ZirCAD (Ivoclar Vivadent, Schaan, Liechtenstein), Zenostar Wieland Dental (Pforzheim, Germany), and other zirconia frameworks with a CTE of 10.5–11.0. The press-on technique is indicated for multi-unit bridges and crowns and implant superstructures [38–40].

Three chemically identical types of zirconia are available for dental application, characterized by slightly different physical properties: green stage, pre-sintered, or the completely sintered blocks. The zirconia powder is shaped into ceramic preforms using cold isostatic pressing, producing chalk-like non-sintered, green-stage Y-TZP blocks. Further stabilizing and condensing by sintering in special furnaces, results in pre-sintered blocks. Additional compression by hot isostatic pressing (carried out at 1000 bar and 50°C below the sintering temperature) removes residual porosity, resulting in dense, fully-sintered zirconia blocks, usually of gray-black shade, requiring subsequent oxidizing to restore their whiteness [41].

Green-stage zirconia blocks can be milled using dry milling and carbide burs, pre-sintered zirconia blocks require wet milling and carbide burs, and fully-sintered zirconia blocks need diamond burs and wet milling.

Green and pre-sintered zirconia frameworks need milling in an enlarged form to compensate for the sintering shrinkage, and can be individualized by pigmentation, in the green-stage faze (**Figure 10**). The sintering takes place in special furnaces (**Figure 11**) [42].

Each type of zirconia has its advantages and disadvantages. Fully-sintered zirconia is denser, with less porosity and increased resistance to fracture. On the other hand, milling of fully-sintered zirconia is time-consuming, causes greater wear of the milling device, and is more expensive [43].

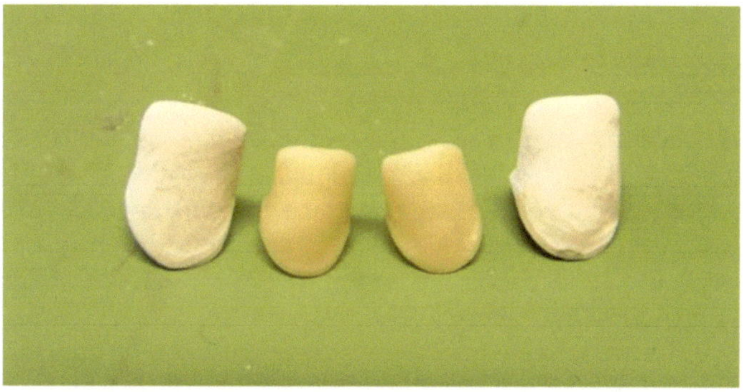

Figure 10.
Zirconia blocks before and after staining and sintering. The shrinkage is easily observed.

Figure 11.
Sintering zirconia frameworks.

4.4 Resin matrix ceramics

Resin matrix ceramics, specially designed for CAD/CAM milling, were intended to match the elastic modulus of dentin. They are easily milled and adjusted, compared to CAD/CAM ceramics. Resin matrix ceramics consist of an organic matrix reinforced by inorganic filler particles [44, 45].

The three types of resin matrix ceramics are resin nano ceramic, glass-ceramic in a resin interpenetrating matrix, and zirconia-silica ceramic in a resin interpenetrating matrix.

Resin nanoceramics: Lava Ultimate (3 M ESPE, Seefeld, Germany) is indicated only for inlays, onlays, and veneers. Glass-ceramic in a resin interpenetrating matrix: Vita Enamic (Vita Zahnfabrik, Bad Sackingen, Germany) is described as a hybrid ceramic, consisting of a paired polymer network and feldspathic ceramic, in 86:14% ratio. It is indicated for crowns, inlays/onlays, veneers, and contraindicated for bridges, in cases of para-functional habits. Zirconia-silica ceramic in a resin interpenetrating matrix: Paradigm MZ100 (3 M ESPE, Seefeld, Germany) was introduced in 2000. It has 85% inorganic content, consisting of ultrafine zirconia-silica ceramic particles and 15% organic polymer matrix, and a patented ternary initiator system. In fact, it is a factory-processed version of the Z100 restorative resin, indicated for inlays, onlays, veneers, and crowns [46–48].

4.5 3D printed ceramics

The subtractive method has the advantage of using homogenous materials, but its major drawbacks result from material loss and high costs. Additive manufacturing or 3D printing is an alternative technology that addresses the drawbacks of subtractive manufacturing in the CAM step. In this case, the prosthetic devices are fabricated by layering materials, and it enables manufacturing of ceramic, as well as polymeric materials and alloys, with lower costs [49].

Zirconia ceramics are suitable for 3D printing by different technologies: vat photopolymerization (stereo-lithography and direct light processing), selective laser sintering, material jetting, fused deposition modeling, enabling fabricating crowns, bridges, and implants [50].

Vat Photo-polymerization uses a slurry of fine ceramic particles incorporated in a photo-curable solution. Because ceramic particles are inert to light emission, polymerization occurs exclusively in the organic phase, resulting in uniformly dispersed ceramic particles in the organic network. Each layer is photo-polymerized until the full 3D ceramic prosthetic is constructed. Following, de-binding is carried out to remove the remaining organic resins and sintering, which removes the pores between the particles, resulting in fully dense, high-performance zirconia prosthodontics [51].

Selective laser sintering (SLS) uses powder beds containing loose ceramic particles to construct the prosthetic restoration. The ceramic powders are either combined with a polymer binder, or no polymer binder is employed. In the first case, the laser melts the binder, bonding the ceramic particles together, followed by de-binding and sintering. In the second case, the laser beam directly sinters the ceramic particles, no further de-binding or sintering is required. Because of its high melting point, zirconia is difficult to process by SLS, and thermal stress might result in cracks development [49].

Material jetting technology allows manufacturing of ceramic restorations with full density, low porosity, high accuracy, complicated shapes, and minimal material usage at a low cost. It uses a suspension of ceramic particles, in form of droplets, which are selectively deposited onto a substrate by the print nozzle. When the droplets come in contact they experience a phase transition and solidify. The low porosity results in the formation of a solid portion [52].

Fused deposition modeling is a material extrusion-based technique, in which the ceramic material, in a filament form, is heated and extruded through a nozzle. It needs de-binding and sintering for densification. The flexible ceramic filament is difficult to obtain, because of the brittle nature of polymer binder and ceramic powder mixture. Surface roughness is the main concern for this technique [49].

5. Conclusion

Duret's prediction regarding CAD/CAM, made in 1991: "The systems will continue to improve in versatility, accuracy, and cost-effectiveness, and will be a part of routine dental practice by the beginning of the twenty-first century," proved to be accurate [53].

The evolution of CAD/CAM systems provided increasing user-friendliness, improved quality and complexity, a wide range of applications, and extended capabilities. The new classes of ceramic materials, specially designed for usage with CAD/CAM systems, enable fabrication of aesthetic and functional crowns and bridges [54].

The modern demands for aesthetic and function have channeled the evolution of fixed prosthodontics in seeking new paths of restoring patients' comfort and health, doubtlessly, will reflect in the future perspectives of this field.

Conflict of interest

The authors declare no conflict of interest.

Author details

Lavinia Cosmina Ardelean[1]*, Laura-Cristina Rusu[2], Daniela Elisabeta Jumanca[3] and Codruta Victoria Tigmeanu[1]

1 Department of Technology of Materials and Devices in Dental Medicine, Multidisciplinary Center for Research, Evaluation, Diagnosis and Therapies in Oral Medicine, "Victor Babeş" University of Medicine and Pharmacy Timisoara, Timisoara, Romania

2 Department of Oral Pathology, Multidisciplinary Center for Research, Evaluation, Diagnosis and Therapies in Oral Medicine, "Victor Babeş" University of Medicine and Pharmacy Timisoara, Timisoara, Romania

3 Department of Preventive Dentistry, Community and Oral Health, Translational and Experimental Clinical Research Center in Oral Health, "Victor Babeş" University of Medicine and Pharmacy Timisoara, Timisoara, Romania

*Address all correspondence to: lavinia_ardelean@umft.ro

References

[1] Hussain A, Khan FA. History of dentistry. Archieves Medicine & Health. 2014;**2**:106-110. DOI: 10.4103/2321-4848.133850

[2] Anderson T. Dental treatment in Medieval England. British Dental Journal. 2004;**197**:419-425. DOI: 10.1038/sj.bdj.4811723

[3] Lynch C, O'Sullivan V, McGillycuddy C. Pierre Fauchard: The 'Father of Modern Dentistry'. British Dental Journal. 2006;**201**:779-781. DOI: 10.1038/sj.bdj.4814350

[4] Mahant R, Agrawal SV, Kapoor S, Agrawal I. Milestones of dental history. CHRISMED Journal of Health Research. 2017;**4**:229-234. DOI: 10.4103/cjhr.cjhr_37_17

[5] Kurdyk B. Nicholas Dubois de Chémant and the use of porcelain in dentistry. Le Chirurgien-Dentiste de France. 1991;**61**(49-50):53-54

[6] Anusavice KJ, Phillips RW, Shen C, Rawls HR. Phillips' Science of Dental Materials. 12th ed. St. Louis, MO: Elsevier/Saunders; 2013

[7] Kadavakolanu VP, Shetty J, Reddy C, Nair C. Prosthodontic rehabilitation: Back to future. International Dental Journal Studies Research. 2016;**4**:35-39. DOI: -10.18231

[8] Donaldson JA. The use of gold in dentistry: An historical overview. Journal of the History of Dentistry. 2012;**60**:134-147

[9] Lulic M, Bragger U, Lang NP, Zwahlen M, Salvi GE. Ante's (1926) law revisited: A systematic review on survival rates and complications of fixed dental prostheses (FDPs) on severely reduced periodontal tissue support. Clinical Oral Implants Research. 2007;**18**:63-72. DOI: 10.1111/j.1600-0501.2007.01438.x

[10] Schulein TM. The era of high speed development in dentistry. Journal of the History of Dentistry. 2002;**50**:131-137

[11] Rochette AL. Attachment of a splint to enamel of lower anterior teeth. The Journal of Prosthetic Dentistry. 1973;**30**:418-423

[12] Susic I, Travar M, Susic M. The application of CAD/CAM technology in dentistry. Materials Science and Engineering Conference Series. 2017;**200**:012020. DOI: 10.1088/1757-899X/200/1/012020

[13] Asgar K. Casting metals in dentistry: Past-present-future. Advances in Dental Research. 1998;**2**:33-43

[14] Kelly JR, Benetti P. Ceramic materials in dentistry: Historical evolution and current practice. Australian Dental Journal. 2011;**56**:84-89. DOI: 10.1111/j.1834-7819.2010.01299.x

[15] McLean JW, Hughes TH. The reinforcement of dental porcelain with ceramic oxides. British Dental Journal. 1965;**119**:251-267

[16] Kelly JR, Nishimura I, Campbell SD. Ceramics in dentistry: Historical roots and current perspectives. The Journal of Prosthetic Dentistry. 1996;**75**:18-32. DOI: 10.1016/S0022-3913(96)90413-8

[17] Marquardt P, Strub JR. Survival rates of IPS Empress 2 all-ceramic crowns and fixed partial dentures: Results of a 5-year prospective clinical study. Quintessence International. 2006;**37**:253-259

[18] Wagner WC, Chu TM. Biaxial flexural strength and indentation fracture toughtness of three new dental core ceramics. The Journal of Prosthetic Dentistry. 1996;**76**:140-144

[19] Wassermann A, Kaiser M, Strub J. Clinical long-term results of VITA In-Ceram classic crowns and fixed partial dentures: A systematic literature review. The International Journal of Prosthodontics. 2006;**19**:355-363

[20] Tigmeanu CV, Ardelean LC, Rusu LC, Negrutiu ML. Additive manufactured polymers in dentistry, current state-of-the-art and future perspectives-a review. Polymers. 2022;**14**:3658. DOI: 10.3390/polym14173658

[21] Duret F, Blouin JL, Duret B. CAD-CAM in dentistry. Journal of the American Dental Association (1939). 1988;**117**:715-720. DOI: 10.14219/jada.archive.1988.0096

[22] Mormann WH, Brandestini M. Cerec-system: Computerized inlays, onlays and shell veneers. Zahnärztliche Mitteilungen. 1987;**77**:2400-2405

[23] Mormann WH, Brandestini M, Lutz F. The Cerec-system: Computer-assisted preparation of direct ceramic inlays in 1 setting. Die Quintessenz. 1987;**38**:457-470

[24] Mormann WH. The evolution of the Cerec system. Journal of the American Dental Association (Chicago, IL). 2006;**137**:7S-13S

[25] Andersson M, Oden A. A new all-ceramic crown. A dense-sintered, high-purity alumina coping with porcelain. Acta Odontologica Scandinavica. 1993;**51**:151-161

[26] Ottl P, Piwowarczyk A, Lauer HC, Hegenbarth EA. The Procera AllCeram system. The International Journal of Periodontics & Restorative Dentistry. 2000;**20**:151-161

[27] Reclaru L, Ardelean LC. Alternative processing techniques for CoCr dental alloys. In: Narayan R, editor. Encyclopedia of Biomedical Engineering. Vol. 1. Amsterdam: Elsevier; 2019. pp. 1-15. DOI: 10.1016/B978-0-12-801238-3.11100-6

[28] Abdullah AO, Muhammed FK, Zheng B, Liu Y. An overview of computer aided design/ computer aided manufacturing (CAD/CAM) in restorative dentistry. Journal of Dental Materials Technology. 2018;**7**:1-10. DOI: 10.22038/JDMT.2017.26351.1213

[29] Sen N, Us YO. Mechanical and optical properties of monolithic CAD-CAM restorative materials. The Journal of Prosthetic Dentistry. 2018;**119**:593-599. DOI: 10.1016/j.prosdent.2017.06.-12

[30] Nasr E, Makhlouf AC, Zebouni E, Makzoume J. All-ceramic computer-aided design and computer-aided manufacturing restorations: Evolution of structures and criteria for clinical application. The Journal of Contemporary Dental Practice. 2019;**20**:516-523

[31] Culo L, McLaren EA. Lithium disilicate: The restorative material of multiple options. The Compendium of Continuing Education in Dentistry. 2010;**31**:716-720

[32] Elsaka SE, Elnaghy AM. Mechanical properties of zirconia reinforced lithium silicate glass-ceramic. Dental Materials. 2016;**32**:908-914. DOI: 10.1016/j.dental.2016.03.013

[33] Chaar MS, Passia N, Kern M. Ten-year clinical outcome of three-unit posterior FDPs made from a

glass-infiltrated zirconia reinforced alumina ceramic (In-Ceram Zirconia). Journal of Dentistry. 2015;**43**:512-517. DOI: 10.1016/j.jdent.2015.02.016

[34] Saridag S, Tak O, Alniacik G. Basi properties and types of zirconia: An overview. World Journal of Stomatology. 2013;**2**:40-47. DOI: 10.5321/wjs.v2.i3.40

[35] Hannink HJ, Kelly PM, Muddle BC. Transformation Tougheningin zirconia-containing ceramics. Journal of the American Ceramic Society. 2000;**83**:461-487. DOI: 10.1111/j.1151-2916.2000.tb01221.x

[36] Garvie R, Hannink R, Pascoe r. Ceramic steel? Nature. 1975;**258**:703-704. DOI: 10.1038/258703a0

[37] Badami V, Ahuja B. Biosmart materials: Breaking new ground in dentistry. Scientific World Journal. 2014;**2014**:986912. DOI: 10.1155/2014/986912

[38] Ziyad TA, Abu-Naba'a LA, Almohammed SN. Optical properties of CAD-CAM monolithic systems compared: Three multi-layered zirconia and one lithium disilicate system. Heliyon. 2021;7:e08151. DOI: 10.1016/j.heliyon.2021.e08151

[39] Ban S. Classification and properties of dental zirconia as implant fixtures and superstructures. Materials. 2021;**14**:4879. DOI: 10.3390/ma14174879

[40] Johansson C, Kmet G, Rivera J, Larsson C, Vult Von Steyern P. Fracture strength of monolithic all-ceramic crowns made of high translucent yttrium oxide-stabilized zirconium dioxide compared to porcelain-veneered crowns and lithium disilicate crowns. Acta Odontologica Scandinavica. 2014;**72**:145-153. DOI: 10.3109/00016357.2013.822098

[41] Gracis S, Thompson VP, Ferencz JL, Silva NRFA, Bonfante EA. A new classification system for all-ceramic and ceramic-like restorative materials. The International Journal of Prosthodontics. 2015;**28**:227-235. DOI: 10.11607/ijp.4244

[42] Rusu LC, Ardelean L. CAD/CAM technology concerning biocompatibility in zirconia all-ceramic Rrestorations. Revista de Chimie. 2012;**63**:513-515

[43] Luthardt RG, Holzhüter MS, Rudolph H, Herold V, Walter MH. CAD/CAM-machining effects on Y-TZP zirconia. Dental Materials. 2004;**20**:655-662. DOI: 10.1016/j.dental.2003.08.007

[44] Chen C, Zardo Trindade F, de Jager N, Kleverlaan CJ, Feilzer AJ. The fracture resistance of a CAD/CAM resin Nano ceramic (RNC) and a CAD ceramic at different thicknesses. Dental Materials. 2014;**30**:954-962. DOI: 10.1016/j.dental.2014.05.018

[45] Coldea A, Swain MV, Thiel N. Mechanical properties of polymer-infiltrated-ceramic network materials. Dental Materials. 2013;**29**:419-426. DOI: 10.1016/j.dental.2013.01.002

[46] Della Bona A, Corazza PH, Zhang Y. Characterization of a polymer-infiltrated ceramic network material. Dental Materials. 2014;**30**:564-569. DOI: 10.1016/j.dental.2014.02.019

[47] Albero A, Pascual A, Camps I, Grau-Benitez M. Comparative characterization of a novel CAD-CAM polymer-infiltrated-ceramic-network. Journal of Clinical and Experimental Dentistry. 2015;7(4):e495-e500. DOI: 10.4317/jced.52521

[48] Shetty R, Shenoy K, Dandekeri S, Syedsuhaim K, Ragher M, Francis J. Resin-matrix ceramics – An overview.

International Journal of Recent Science and Research. 2015;**6**:7414-7417

[49] Khanlar LN, Salazar Rios A, Tahmaseb A, Zandinejad A. Additive manufacturing of zirconia ceramic and its application in clinical dentistry: A review. Dental Journal. 2021;**9**:104. DOI: 10.3390/dj9090104

[50] Galante R, Figueiredo-Pina CG, Serro AP. Additive manufacturing of ceramics for dental applications: A review. Dental Materials. 2019;**35**:825-846. DOI: 10.1016/j.dental.2019.02.026

[51] Della Bona A, Cantelli V, Britto VT, Collares KF, Stansbury JW. 3D printing restorative materials using a stereolithographic technique: A systematic review. Dental Materials. 2021;**37**:336-350. DOI: 10.1016/j.dental.2020.11.030

[52] Ozkol E, Zhang W, Ebert J, Telle R. Potentials of the "direct inkjet printing" method for manufacturing 3Y-TZP based dental restorations. Journal of the European Ceramic Society. 2012;**32**:2193-2201. DOI: 10.1016/j.jeurceramsoc.2012.03.006

[53] Duret F, Preston J. CAD/CAM imaging in dentistry. Current Opinion in Dentistry. 1991;**1**:150-154

[54] Vellingiri K, Nooji D, Suhas Rao K, Shetty B, Kumar M, Meghashri K. Current trends in fixed prosthodontics. Jounal of Advanced Clinical Research Insights. 2020;**7**:51-54. DOI: 10.15713/ins.jcri.303

Section 4

Orthodontics

Chapter 17

From the Orthodontic Smile to the Perfect Smile: A New Categorization

Maria do Rosário Dias, Valter Pedroso Alves, Gunel Mammadova Kizi and Ana Sintra Delgado

Abstract

The facial symmetry from an esthetic standpoint is crucial. However, there are not much research that examine the malocclusion issue or the significance of the mouth and smile in the mental representation of the face. In this study, 151 kids and teenagers, both genders, aged 8 to 24, were asked to sketch two self-portraits of their mouths or smiles—before (and during) the usage of the orthodontic appliance. Participants seek therapy mostly for functional problems rather than cosmetic ones. The findings of this study provide insight into the significance of the mouth and smile for an individual's sense of self and psychological well, where the maximization of the mental representation of the orthodontic smile emerges as a new categorization of the *perfect smile*.

Keywords: orthodontic smile, esthetic smile, mental representation, drawing content analysis, self-image

1. Introduction

It is still of great scientific curiosity to comprehend the significance and self-image that a smile psychologically conveys in people's daily lives [1–3]. It is essential that the smile contribute to the facial harmony. There aren't many empirical studies, nevertheless, that consider both the issue of malocclusion and the significance of the mouth and mouth in the mental representation of the self-image of the face. The face is the area of the body where actions and expressions are concentrated and can either internalize or externalize feelings and emotions [4]. Since a large portion of the person's affection is communicated at this level during interactions and bonds with others, the expression on the face is pragmatic. Among these sayings. Among these expressions is the smile, the first psychic organizer and a crucial element in the development of one's own self-image, is one of these expressions [5]. In fact, the ability to understand a patient's facial expression through non-verbal language is crucial in the therapeutic environment of dental medicine since the gift of compassion communicated through a smile is difficult to measure empirically [6].

In a clinical environment, the "art of the smile" resides in the dentist's capability to fete the positive beauty rudiments in each case and to produce a strategy that enhances the attributes that fall outside the parameters of the prevailing esthetic conception [7]. A panoply of logical principles regarding the "smile" order need to come incorporated in the field of dental medicine orthodontics [8], to gain enhanced clinical results. Nonetheless, in the conception of esthetics is underscored a private perception of the beauty conception operated by the subject [9], in which the harmony of the smile plays an important part in the perception of the beauty of the face [10].

The smile is seen as a fundamental cognitive-affective ability and a trait in the development of a person's personality, from birth to the conclusion of the life trajectory, according to Freitas-Magalhães [11]. Four different types of smiles have been identified by Freitas-Magalhães the broad smile, the neutral smile, the superior smile, and the closed smile. The superior grin and the closed smile appear to be most associated with inter-psychosocial relationships from an empirical standpoint. The broad smile and the neutral smile, on the other hand, are not seen as loving since they represent the extremes of the smiling spectrum. The wide smile reveals the dental arch in the upper and lower jaws, whilst the neutral smile conceals all facial emotions. The superior smile described by Freitas-Magalhães [11] appears to be equivalent to the smile arc described by Sarver [12] as a type of smile defined by the "relationship of the curvature of the incisal edges of the maxillary incisors and canines to the curvature of the lower lip in the posed smile". While in a nonconsonant smile, the maxillary incisal are line is flatter than the curvature of the lower lip, according to Sarver [12], an ideal smile arc "has the maxillary incisal edge curvature parallel to the curvature of the lower lip". According to Sabri [13], the balance between eight factors that make up the perfect smile should not be viewed as rigorous criteria, but rather as esthetic guidelines to aid orthodontists in treating specific individuals. Additional authors have offered definitions of the ideal and discordant smile [7, 12–15]. A younger-appearing smile is one in which the central incisors are longer than the lateral incisors, resulting in a larger interincisal space, while an older-appearing smile is one in which the top incisal edge looks straight during the smile. In the field of orthodontics, esthetic considerations and the attractiveness of the smile are crucial because they appear to motivate treatment more effectively than functional considerations [16, 17]. Nevertheless, the functional factor for orthodontic treatment motivation outweighed the esthetic factor, according to an empirical study [3] done at the Egas Moniz-Lisbon University Clinic by Do Rosário Dias et al. Indeed, in spite of the multitudinous references set up in the literature review that highlight the significance of the functional and biomechanical determinants of malocclusion as a distinguishing diagnosis, there's still a lack of empirical studies that combine the case of malocclusion to the subject's self- image perception of the face in terms of the existent's psychosocial environment [18, 19], insofar as the orthodontic treatment leads to notorious changes in the subject's intrapsychic experience.

The current chapter fills in this information vacuum by attempting to explain the significance of the self-perception of the mouth and grin in both the individual's mental representation of his or her own self-image and perception of the face as well as in his or her wellbeing. Additionally, this research has helped define the orthodontic smile as a novel categorization linked to a beautiful, functional smile.

2. Subjects and methods

A descriptive and exploratory empirical study was conducted to get access to how the individual's mouth and smile are seen in their own minds. The methodological approach used a combination of quantitative and qualitative criteria to analyze the content of a group of patient drawings. The participants of the sample accurately recognized how the task (centered on the teeth) was being carried out, which indicates that just one person drew the category teeth with the orthodontic device fitted (**Figure 1**) [1–3].

The convenience sample included 151 children and teenagers, aged 8 to 24, who had undergone orthodontic treatment at the Egas Moniz University Clinic in Portugal for a period ranging from 6 months to 1 year. The following questions guided patients in creating two self-portraits of their smiles and mouths: (1) "What was your mouth like before you had the orthodontic appliance?" (Moment 1-[**M1**]); and (2) "How do you think your mouth will be when you remove the orthodontic appliance?" (Moment 2-[**M2**]). In addition, all participants answered a socio-demographic inquiry and provided a written answer to the question: "Why do you use an orthodontic appliance?" According to the selected sample, 302 valid drawings were collected and analyzed, half representing **M1** and half representing **M2**. A qualitative content grid for the analysis of the 302 drawings was originally designed to study the pictorial representations found in the sample, with four categories: (1) "smile," (2) "drawing of the figure (3) "appearance," and (4) "teeth". To further detail the content analysis, 10 subcategories were created: (1) "absence of teeth," (2) "teeth without detail," (3) "fractured teeth," (4) "teeth with diastema," (5) "crowded teeth," (6) "crooked teeth," (7) "teeth in saw," (8) "misplaced teeth," (9) "teeth with spacing" and (10) "gingival deformation" (**Table 1**). Other subcategories were created such as "size", "contours/limits", "opening of the mouth", "lips", and" jaw".

It has been noted that research projects in the field of health sciences' studies frequently follow the methodological guidelines of using drawings as a graphical instrument for qualitative research goals [16, 20]. Patients were instructed to depict only the mouth/smile in their drawings as a criterion for objectivity, concentrating the smile as the subject of empirical study [3, 17].

Following the theoretical presumptions of Sawyer et al. [21] and Pikunas [22], the data were statistically analyzed, and the sample was broken down into four groups of children and teenagers: "children" (8–12 years); "preteens" (13–14 years); "adolescents" (15–17 years); and "emerging adults" (18–24 years) (**Table 2**).

The statistic treatment of the descriptive data was analyzed using the software SPSS–Statistical Package for the Social Sciences IBM SPSS Statistics, version 23 for Windows, Lisbon, Portugal.

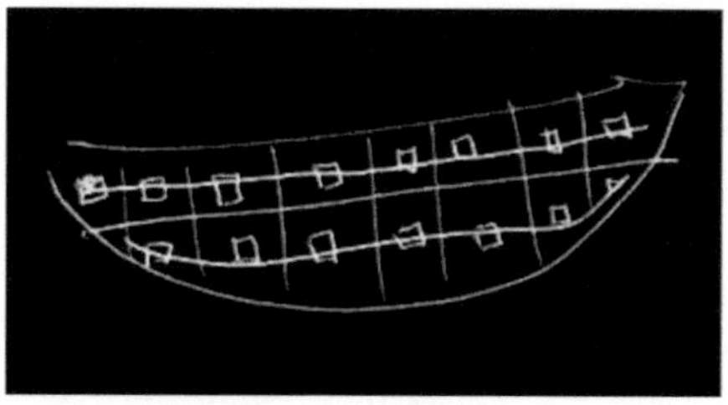

Figure 1.
Drawing of the teeth from a patient with orthodontic appliance.

	Children		**Preteens**		**Adolescents**		**Emerging adults**	
	M1	M2	M1	M2	M1	M2	M1	M2
Absence of teeth	0	0	0	0	0	2.3	8.6	14.3
Teeth without details	57.9	60.5	72.7	84.8	43.2	61.4	60	54.3
Fractured teeth	7.9	0	6.1	0	0	2.3	8.6	0
Teeth with diastema	10.5	0	24.2	6.1	27.3	0	8.6	0
Crowded teeth	55.3	2.6	45.5	0	40.9	6.8	40	0
Crooked teeth	89.5	21.1	84.8	9.1	88.6	20.5	88.6	20
In saw teeth	23.7	18.4	18.2	6.1	22.7	13.6	25.7	11.4
Misplaced teeth	89.5	21.1	90.9	18.2	95.5	15.9	88.6	14.3
Teeth with spacing	26.3	5.3	36.4	6.1	36.4	9.1	40	8.6
Gingival deformation	2.6	0	6.1	0	4.5	2.3	2.9	0

Table 1.
Frequency of the subcategories found in participants' drawings before (M1) and after (M2) use of an orthodontic appliance (%).

	Male, *n* (%)	**Female, *n* (%)**	**Total, *n* (%)**
Children (8–12 years)	22 (29.3)	16 (21.3)	38 (25.3)
Preteens (13–14 years)	12 (16)	21 (28)	33 (22)
Adolescents (15–17 years)	25 (33.3)	19 (25.3)	44 (29.3)
Emerging adults (18–24 years)	16 (21.3)	19 (25.3)	35 (23.3)
Total sample (8–24 years)	75 (100)	75 (100)	150 (100)

Table 2.
Frequency according to gender in all age groups (%).

3. Results and discussion

By comparing the self-portraits of patients' mouths and smiles before and after the placement of the orthodontic appliance, it was possible to understand the impact of the orthodontic appliance on the mental representation of the individual's self-image, particularly that of the mouth and smile (**Figure 2**).

The findings point to variations in the mental representations of the mouth and the smile, namely in the drawings made before **(M1)** and after **(M2)** the usage of the orthodontic device, as well as in the expressiveness and expression of emotions in these two situations. The results of the content analysis of the written responses to the question "Why do you use an orthodontic appliance?" show that the primary motivation for using an orthodontic appliance is thought to be related to the functional well-being of the oral cavity. A total of 103 written responses (68.67%) were collected. The content analysis of drawings in **M2** appears to support this, showing the presence of a wide smile (**M2** = 86.1%) and a sense of well-being (**Figure 3**). This places emphasis on the broad smile, which is seen to be the closest to the idea of happiness [12].

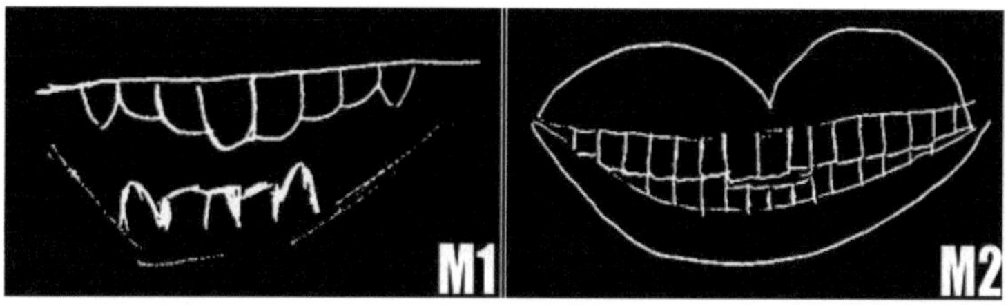

Figure 2.
Self-portraits drawn before (M1) and after (M2) the placement of an orthodontic appliance.

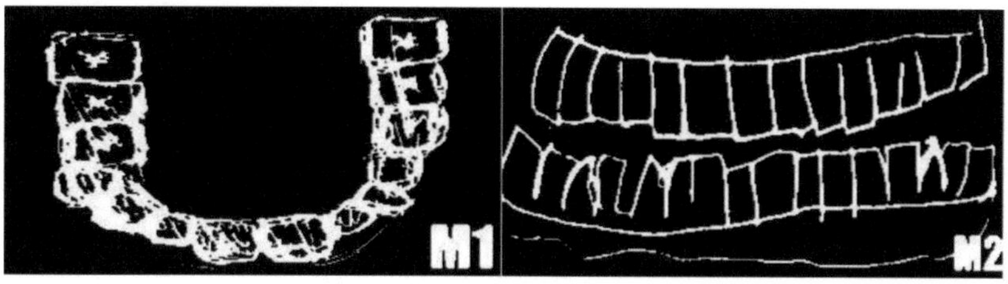

Figure 3.
Drawings of a patient which illustrates dental arcades converted to jaws and absence of lips in M1 and M2.

The findings also show alterations in how people see their own mouths after using the orthodontic appliance **(M2)**, as evidenced by more involved drawings, drawings that are more comprehensive and detailed, and drawings that have straighter mouths and better-aligned teeth. This fact may indicate that patients may have clinically induced discourse regarding the orthodontic smile given the time of the questionnaire in the treatment pathway (6 M - 1Y).

When comparing **M1** and **M2**, **Table 1** reveals three characteristics that show changes in the mental image of the mouth and smile both before and after the orthodontic device was positioned: crowded teeth, which were noticeable before the orthodontic equipment was used but completely disappeared after; crooked teeth, which are described in (**M1**); and misplaced teeth, which are also described in (ibidem). Apart from emerging adults, where there were no discernible changes, different scores were obtained in **M1** and **M2** for the graphical depiction of the category with a diastema. From **M1** to **M2**, there was a noticeable decrease in dental spacing across all age groups. A normal distribution was discovered in all age groups according to gender, per the statistical analysis carried out (**Table 2**), which appears to be consistent with the results of Rodrigues et al. [23].

In reference to the written response to the inquiry: "Why do you use an orthodontic appliance? Correction of crooked teeth (68.7%), followed by the closure of interdental gaps (12.7%), and malocclusion (12.7%), was the most important response across all age groups. Participants were least likely (0.7%) to cite improving breathing or loving braces (**Table 3**).

Overall, results point that the improvement of *oral health* is another justification for the use of the orthodontic appliance (9.3%), surpassing esthetic motivations such as to be*good looking* (8.7%) or to have *beautiful teeth* (7.3%). The findings indicate that the improvement of oral health (9.5%) is a more compelling reason to utilize an orthodontic appliance than cosmetic factors like having beautiful teeth (7.3%) or being attractive (8.7%). However, the investigation uncovered a few unique characteristics for each age group: The desire to have the ideal smile surfaced as a significant

	Children	Preteens	Adolescents	Emerging adults	Total (%)
Correction of crooked teeth	73.7	63.6	79.5	54.3	68.7
Interdental spaces	15.8	18.2	4.5	14.3	12.7
Correction of malocclusion	15.8	9.1	9.1	17.1	12.7
To have the perfect smile	5.3	15.2	2.3	22.9	10.7
Oral health	13.2	3	9.1	11.4	9.3
To be good looking	2.6	12.1	4.5	17.1	8.7
Beautiful teeth	15.8	6.1	0	8.6	7.3
Well-being	0	3	4.5	0	2
Improve self-esteem	0	0	2.3	2.9	1.3
Muscle pain	0	0	0	5.7	1.3
Improve breathing	2.6	0	0	0	0.7
Enjoying braces	0	0	0	2.9	0.7

Table 3.
Frequency of reasons to use an orthodontic appliance in all age groups (%).

issue for preteens (15.2%) and emerging adults (22.9%), but not as much for teenagers (2.3%), who appear to be less motivated by the esthetic motive for the ideal smile. Only emerging adults have acknowledged it as an incentive for improving physical difficulties, such as muscular soreness (5.7%), since it is uncomfortable. The findings also imply that in the children's age group, the primary justifications for using orthodontic appliances are connected to the treatment of crooked teeth (73.7%), poor occlusion (15.8%), interdental gaps (15.8%), and to have attractive teeth (15.8%). Braces were not highlighted for reasons like increased self-confidence, wellbeing, relief from muscular soreness, or enjoyment in this age range. In the preteen age group, straightening crooked teeth appears to be the primary motivation for using an orthodontic appliance (63.6%), followed by closing gaps between teeth (18.2%), achieving the ideal smile (6.1%), and being attractive (12.1%). In this age group, no mention was made of the categories of improved self-esteem, reduced muscle pain, better breathing, or enjoyment of braces. The repair of crooked teeth is once again the category most frequently cited as the primary justification for orthodontic treatment in the teenage age group, followed by poor occlusion (9.1%) and oral health (9.1%). Adolescents seldom bring up the topics of having perfect teeth, dealing with muscular discomfort, improving their respiration, or loving braces as being pertinent. Finally, the rising adult age group likewise cites the repair of crooked teeth as the primary reason for receiving orthodontic treatment (54.3%), followed by the desire to have the ideal smile (22.9%), poor occlusion (17.1%), and excellent looks (17.1%). In this age range, there was no mention of better breathing or wellbeing.

The results also show that the smile category (**Figure 3**) is primarily illustrated by the teeth category (in terms of dental arcades) in contrast to the lips subcategory's absence (**M1** = 70.9%; **M2** = 70.8%), with an exaggeration of the upper and lower dental arch. This shows that people are worried about the condition of their teeth, which are viewed as clinical objects and make up the dental arches.

Drawings in **M2** with more prominent labial commissures regarding the lips and complete visualizations of the top and lower dental arches demonstrate a reduction in the pictorial depiction of the partially open mouth, as shown by the appearance category's openness of the mouth subcategory. The drawing and teeth of the figure category categories (**Figure 2**) support this idea since both have more visible teeth.

Additionally, more detailed drawings were submitted in the smile category in **M2** compared to **M1**: These findings go against other findings of the two studies conducted by Gonçalves and Torres at the University Clinic Egas Moniz-Portugal, which highlighted the esthetic element and the pursuit of the "perfect smile" as the driving forces behind orthodontic treatment among patients. According to research by Dias et al. [1], children utilize orthodontic appliances mostly to address crooked and malocclusion teeth and interdental gaps, with adolescent patients being the only ones who care about having a flawless smile. The percentage of the upper and lower lips that are visually represented by the participants in the lip's subcategory has increased from **M1** to **M2** by 2.7%. Additionally, there was a notable increase in the representation of normal lips (**M1** = 16.5%, **M2** = 21.9%), which was offset by a drop in the number of thin lips (**M1** = 4%, **M2** = 2.6%) and thick lips (**M1** = 6%, **M2** = 2%).

The category with the most noticeable alterations between **M1** and **M2** is teeth. Indeed, the fragmented, diastema, crooked, straight, and well-positioned category show the most notable alterations. Additionally, the straight and cracked divisions (**Figure 4**). Drawings of shattered teeth (**M1** = 5.3%, **M2** = 0.7%) and crooked teeth (**M1** = 88.1%, **M2** = 17.9%) show a much lower percentage in **M2**.

As a representation of reinvented smiles following the use of the orthodontic appliance, there is a significant percentage decrease from **M1** to **M2** in the subcategories of teeth with diastema (**M1** = 17.9%, **M2** = 1.3%), crooked teeth (**M1** = 45.7%, **M2** = 2.6%), and well-positioned (**M1** = 91.4%, **M2** = 17.2%) (**Figure 5**).

In the current study, we discovered that age appeared to be a preponderant factor affecting subjects' perceptions of the esthetics of the smile, which seems to be consistent with the viewpoint of several authors, despite gender having no effect on

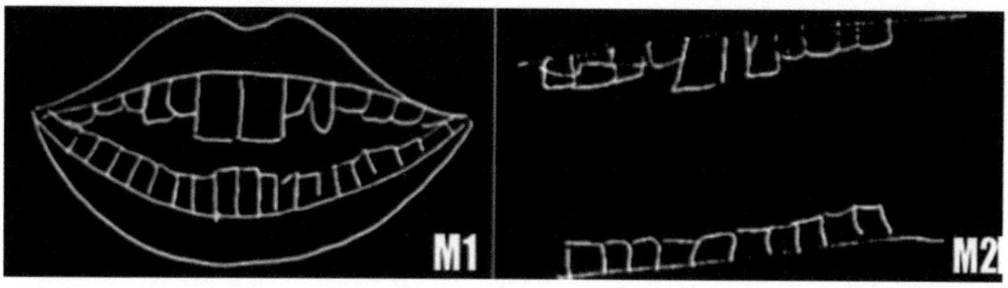

Figure 4.
Decrease in predominance of fractured and crooked teeth illustrated by the drawing from different Patients in M2.

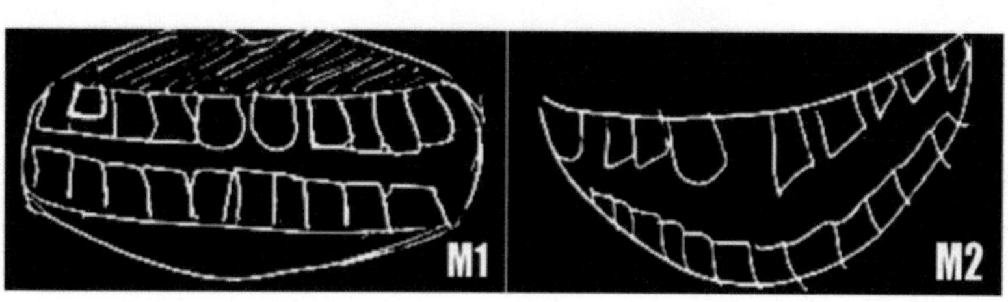

Figure 5.
Drawing from different Patients which illustrates badly positioned teeth with visible diastemas in M1.

results [23–27]. The correction of misaligned teeth, diastemas, and fractures are the main reasons for orthodontic treatment, according to results regarding participants' mental representations of their mouths and smiles before (**M1**) and after (**M2**) using orthodontic appliances. Drawings reveal that **M1's** smile was depicted by longer anterior incisors, which created a line that dipped in the middle and rose in the corners. In contrast, **M2's** smile was depicted by straight upper incisal edges and all teeth had the same vertical dimension. These factors support those mentioned by Goldstein [20] in relation to the younger-appearing smile and the older-appearing smile, respectively, distinguishing smiles based on age. According to the current study, the smile changes from appearing older in **M1** to appearing younger in **M2** after treatment (**Figure 6**).

Diastemas may be seen in drawings in **M1** (**Figure 5**) but not in **M2** across all age groups, as shown by **Table 1**. The exception to this rule is preteens. According to this line of reasoning, Rodrigues et al. [28] hypothesized that for children and adults, the presence of diastemas harmed the esthetic look and had a detrimental effect on how the smile develops. Almeida et al. [27] believe, on the other hand, that diastemas in children have a natural connotation, are present in deciduous teeth, and disappear naturally or with the help of a straightforward intervention.

Despite this paradox, children in the current study consistently depict the mouth and smile in **M2** without any diastemas (**Figure 7**), which is consistent with the finding of Almeida et al. [29].

Additionally, the superior smile as described by Freitas-Magalhães [11] appears to be the same as the representation of the smile arc in **M2** as described by Sarver [7]. Most age groups in **M1** exhibit the dental part fractures depicted in drawings (**Figure 8**), with emerging adults making up most of this subcategory (8.6%).

In general, the categories of crooked teeth rectification, malocclusion correction, and interdental gaps outline the primary motivating elements that decide the practical usage of an orthodontic device.

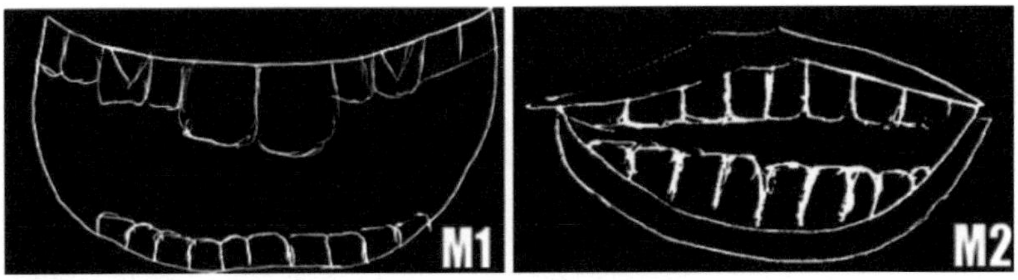

Figure 6.
Younger-appearing smiles (**M1**) *and older-appearing smile* (**M2**).

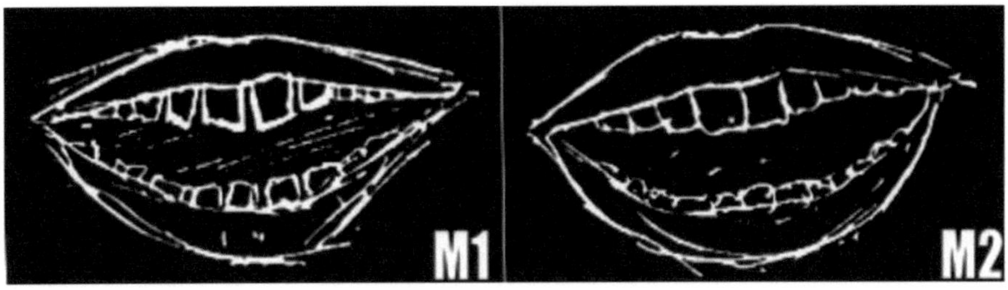

Figure 7.
Presence of diastema in M1 and absence of diastema in **M2**.

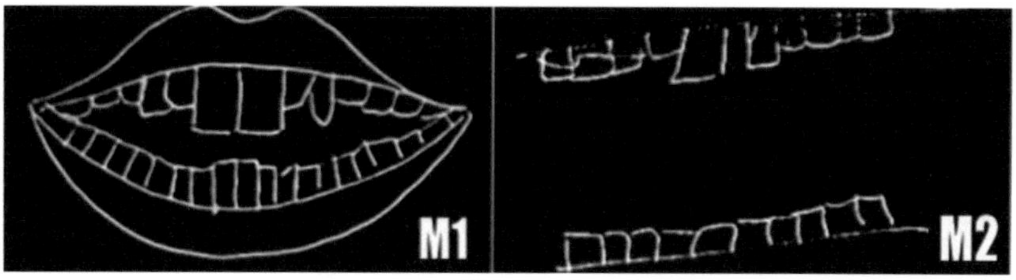

Figure 8.
The fractures of dental parts, mostly drawn in **M1**.

According to the literature study, the appearance of the smile and the subject's self-esteem are secondary determinants of psychosocial factors that support the use of an orthodontic device, with the degree of malocclusion serving as the primary predictor. According to some authors, malocclusions [2, 25–27, 30] appear to have a greater psychosocial impact than other parameters, indicating that subjects appear to be more concerned with the position of their teeth within the dental arch than with the esthetics of their smile [31].

According to the results of the current study, children and preteens typically seek orthodontic treatment to straighten their teeth. Additionally, missing teeth and a more pronounced anterior irregularity of the upper jaw are linked to children's and teenagers' unhappiness with their dental appearance. The survey also discovered that none of the participants in the adolescent age range cited the value of having attractive teeth as a justification for using an orthodontic device. These findings, however, appear to be at odds with those of Bica et al. [32], who found that many patients cited improving their facial appearance as a reason for seeking orthodontic treatment. Elias et al. reported similar findings, [33] adding that, depending on the characteristics of the age group examined, oral health and esthetics predominated in the self-image of the face.

Therefore, our findings support the notion that, in addition to structural and functional considerations, orthodontic treatment in emerging adults may also be motivated by oral health and esthetics.

Emerging adults listed orthodontic treatment as a factor for straightening teeth, having the perfect smile, correcting malocclusion, and being attractive, in that order of importance, while dispensing with health and breathing improvement as categorical determinants. These results are consistent with Delalibera et al., [34] which linked orthodontic treatment with emerging adults' interpersonal interactions and concluded that people's perceptions of dentofacial abnormalities could influence how well they fit in with society. Accordingly, teenagers (2.3%) and emerging adults (2.9%) cite the enhancement of self-esteem as the driving force for the usage of an orthodontic device, which may raise questions about the relationship between oral cavity esthetics and self-esteem, as suggested by Basha et al. [35]. The most common justification for receiving orthodontic treatment across all age groups was to straighten out crooked teeth.

4. Conclusion

The esthetic harmony that the smile offers is crucial for patients since having a beautiful face increases one's acceptability in their psychosocial environment.

It should be noted that when subjects underwent orthodontic treatment, they appeared to associate the smile category with the social display of a look with perfectly healthy teeth—the so-called "orthodontic smile"—in blatant contrast to the idea that the ideal smile from an esthetics point of view was less important for the use of an orthodontic appliance than the good functioning of the oral cavity. Therefore, the orthodontic smile—in the sense of a sound, functional smile—was more significant than the ideal esthetic smile.

These findings add to our understanding of the role that the mouth and smile play in a person's sense of self and psychological health, and they suggest that a perfect smile may be defined in new ways by enlarging our mental image of an orthodontic smile.

Therefore, we suggest the orthodontic smile as a new category for smiles used in orthodontic settings, considering the socially acceptable exhibition of excellent teeth.

Acknowledgments

We thank Aucéane Karramkan, a student in the 5th year of Dental Medicine at Egas Moniz, School of Health and Science for providing help in the formatting, the references and the revision of the article.

Conflict of interest

The authors declare no conflict of interest.

Author details

Maria do Rosário Dias*, Valter Pedroso Alves, Gunel Mammadova Kizi and Ana Sintra Delgado
Egas Moniz, School of Health and Science, Monte de Caparica, Portugal

*Address all correspondence to: mariadorosario.dias@gmail.com

References

[1] Dias MR, Ferreira A, Pires M, Alves V, Delgado A. Orthodontic smile: A new categorization of the perfect smile. European Journal of General Dentistry. 2020;**9**:129-133. DOI: 10.4103/ejgd.ejgd_157_20

[2] Dias MR, Ferreira A, Alves V, Delgado A. The mental representation of the smile before and after orthodontic treatment. In: Anand A, editor. Highlights on Medicine and Medical Research. 1st ed. 2021. pp. 1-9. DOI: 10.9734/bpi/hmmr/v5

[3] Dias MR, Naben L, Monteiro A, Ferreira A, Alves V, Delgado A. When the silence speaks: The smile. Journal of Biosciences and Medicines. 2018;**6**:13-20

[4] Dias MR, Duque AF. Mens Sana in Corpore Sano: The concept of health and illness in childhood. IFPE—21st Annual Interdisciplinary Conference. Psychoanalysis: Not the Same Old Song and Dance? Conference Program/Abstract Membership Directory; 29-31 October 2010, Nashville, Tennessee. 2010. p. 10

[5] Ekman P, Davidson RJ, Friesen WV. The Duchenne smile: Emotional expression and brain physiology: II. Journal of Personality and Social Psychology. 1990;**58**(2):342-353. DOI: 10.1037/0022-3514.58.2.342

[6] Mendes JJ, Dias MR, Neves AC, Ferreira A, Maia P, Silva N, et al. The language of silence in the therapeutic setting of dental medicine. In: 2 Congresso International do CiiEM-Research and Innovation in Human & Health Science. Monte da Caparica, Portugal: Egas Moniz, School of Health & Science. 2017

[7] Sarver DM, Ackerman MB. Dynamic smile visualization and quantification: Part 1. Evolution of the concept and dynamic records for smile capture. American Journal of Orthodontics and Dentofacial Orthopedics. 2003;**124**:4-12

[8] Singh P, Sharma J. Principles of smile analysis in orthodontics- A clinical overview. Health Renaissance. 2011;**9**:35-39

[9] Dávila F. Elarte em la medicina: las proporciones divinas. Revista de divulgación científica y tecnológica de la Universidade Autónoma de Nuevo León. 2004;**7**:150-156

[10] Bolívar L, Mariaca B. La Sonrisa y sus dimensiones. Revista Facultad de Odontología Universidade de Antioquia. 2012;**23**:353-365

[11] Freitas-Magalhães A. A Psicologia do Sorriso Humano. Portugal: Edições Universidade Fernando Pessoa; 2009

[12] Freitas-Magalhães A. Expressão facial: O efeito do sorriso na percepção psicológica da afectividade. Revista da Faculdade de Ciências Humanas e Socias da Universidade Fernando Pessoa. 2007;**4**:276-284

[13] Sabri R. Orthodontic objectives in orthognatic surgery. State of the art today. World Journal of Orthodontics. 2006;**7**:177-191

[14] Tjan AH, Miller GD, The JG. Some aesthetic factors in a smile. Journal of Prosthetic Dentistry. 1984;**51**:24-28

[15] Miller CJ. The smile line as a guide to anterior esthetics. Dental Clinics of North America. 1989;**33**:157-164

[16] Janson G, Branco NC, Morais JF, Freitas MR. Smile attractiveness in patients with Class II division 1

subdivision malocclusion treated with different tooth extraction protocols. European Journal of Orthodontics. 2014;**36**:1-8

[17] Machado AW. 10 commandments of smile esthetics. Dental Press Journal of Orthodontics. 2015;**19**:136-157

[18] Gonçalves S. O tratamento ortodôntico e a sua influência na auto-imagem do rosto do sujeito. [Ph.D. dissertation] Instituto Superior de Ciências da Saúde Egas Moniz, Monte da Caparica. 2011

[19] Torres J. A influência do género na adolescência em dentisteria operatória. [Ph.D. dissertation] Instituto Superior deCiências da Saúde Egas Moniz, Monte da Caparica. 2011

[20] Goldstein RE. Change your smile. In: Discover how a New Smile Can Transform your Life. Chicago: Quintessence Publishing; 2009

[21] Sawyer MD, Susan M, Azzopardi P, Wickremarathme MD, Patton MD. The age of adolescence. Lancet Criança Saúde Adolesc. 2018;2:223-228

[22] Pikunas J. Desenvolvimento Humano. Uma Ciência Emergente. São Paulo: Mcgraw-Hill; 1979

[23] Mohktar HA, Abuljadayel LW, Al-Ali RM, Yousef M. The perception of smile attractiveness among Saudi population. Clinical, Cosmetic and Investigational Dentistry. 2015;7:17-23

[24] Tuzgiray YB, Kaya B. Factors affecting smile esthetics. Turkish Journal of Orthodontics. 2013;**26**:58-64

[25] Pithon MM, Bastos GW, Miranda NS, Sampaio T, Ribeiro TP, Nacimento LE, et al. Esthetic perception of black spaces between maxillary central incisors by different age groups. American Journal of Orthodontics and Dentofacial Orthopedics. 2013;**143**:371-375

[26] Lacerda-Santos R, Pereira TB, Pithon MM. Esthetic perception of the buccal corridor in different facial types by laypersons of different ages. Bioscience Journal. 2015;**31**:1283-1290

[27] Gerritsen AE, Sarita P, Witter DJ, Kreulen CM, Mulder J, Creugers NH. Esthetic perception of missing teeth among a group of Tanzanian adults. The International Journal of Prosthodontics. 2008;**21**:169-173

[28] Rodrigues Cde D, Magnani R, Machado MS, Oliveira OB. The perception of smile attractivness. The Angle Orthodontist. 2009;**79**:634-639

[29] Almeida A, Leite I, Melaço C, Marques L. Dissatisfaction with dentofacial appearance and the normative need for orthodontic treatment: Determinants factors. Dental Press Journal of Orthodontics. 2014;**19**:120-126

[30] Lukez A, Pavlic A, Trinajstic Zrinski M, Spalj S. The unique contribution of elements of smile aesthetics to psychosocial well-being. Journal of Oral Rehabilitation. 2015;**42**:275-281

[31] Peres KG, Traebert ESA, Marcenes W. Differences between self-perception and normative criteria in the identification of malocclusions. Revista de Saúde Pública. 2002;**36**:230-236

[32] Bica I, Cunha M, Costa J, Rodrigues V, Neves D, Albuquerque I, et al. Body Perception and satisfaction in adolescentes and the relationship to their oral health. Millenium. 2011;**40**:115-131

[33] Elias MS, Cano MA, Junior W, Ferriani M. The importance of oral health for adolescents from different social strats in the city if Riberirão Preto. Revista Latino-Americana de Enfermagem. 2001;**9**:88-95

[34] Delalibera H, Da Silva M, Pascotto R, Terada H, Terrada R. Aesthetic evaluation of patients undergoing orthodontic treatment. Acta Scientiarum Health Sciences. 2010;**32**:93-100

[35] Basha S, Mohamed RN, Swamy HS, Parameshwarappa P. Untreated gross dental malocclusion in adolescents: Psychological impact and effect on academic performance in school. Oral Health & Preventive Dentistry. 2016;**14**:63-69

Chapter 18

Clinical and Phonetic Features of Dentognathic Deformations, Their Orthodontic Treatment

Alona Melnyk and Valerii Filonenko

Abstract

The substantiation of the current task of modern dentistry is presented, which consists of increasing the effectiveness of the treatment of dentognathic deformations accompanied by phonetic disorders through a multidisciplinary approach to the development and application of a complex of diagnostic and therapeutic measures based on the determined and clarified etiopathogenetic connections of dentognathic deformations with phonetic disorders, and causal mechanisms of the development of dentognathic deformations in cleft lip and palate and the importance of anatomic and morphometric characteristics of the tongue as a prognostic indicator of an effective orthodontic treatment as well are substantiated.

Keywords: dentognathic deformations, phonetic disorders, cleft lip and palate, multidisciplinary approach, orthodontic treatment

1. Introduction

Statistical data given in scientific sources testify to the high level of prevalence of dentognathic pathologies in Ukraine and the tendency of its growth [1–5]. According to the authors' research, it reaches more than 80% [6–8]. During the examination of 462 children aged 6–12 years in schools of Kyiv, anomalies and deformations of the dentognathic apparatus were found in 385 subjects, which is 83.3%. Distal bite is the most common pathology of the dentognathic apparatus, found in 58.2% (n = 224). Distal deep bite was 26.2% (n = 101), mesial bite—10.1% (n = 39), open bite—4.7% (n = 18), and transverse bite—0.8% (n = 3) [8].

The majority of dentognathic deformations are accompanied by phonetic disorders. They are caused by changes in the dentognathic relations of indicators of the functional activity of the masticatory muscles, circular muscle of the mouth, lips, tongue, and ear, nose, throat (ENT) organs [8–13]. Out of 385 people with dentognathic deformities, speech disorders were identified in 306 people, which was 66.2%. Out of 77 people with no orthodontic pathology, only 16.7% had speech disorders. Phonetic disorders combined with bite pathologies were found in 96.1% (n = 294), of which 52.0% (n = 153) had a distal bite, 31.3% (n = 92)—distal deep bite, mesial—10.9% (n = 32), and open—5.8% (n = 17) [8].

The outlined factors are in a close cause-and-effect relationship; therefore, they require a simultaneous complex approach in solving the problem of treatment of patients with dentognathic deformities accompanied by speech disorders.

Today, new methods of orthodontic treatment have been developed for this category of patients using various types of myogymnastics and taping, as well as a program-methodical complex of corrective speech therapy work using innovative and original technologies [9, 12, 14, 15].

The use of well-known methods of orthodontic treatment without directed logopedic correction reduces the effectiveness of rehabilitation of patients with dentognathic deformations accompanied by phonetic disorders. Attention should be paid to the study of the relationship between inflammatory diseases of the nasopharyngeal and palatine tonsils, narrowing of the upper respiratory tract with dentognathic deformations accompanied by phonetic disorders, the state of functional activity of the temporal muscles, circular muscle of the mouth depending on the types of deformations and phonetic changes. Attempts to normalize the bite without logopedic correction and functional reconstruction complicate orthodontic treatment and make it impossible to achieve a stable result [16, 17].

When carrying out a complex medical and rehabilitation measures in children with congenital cleft lip and palate, it is important to study the conditions of development of the dentognathic apparatus. According to statistical data, more than 500 children with various types of cleft lip and palate and syndromes are born in Ukraine per year, in which almost 94% of cases have combined dentognathic deformations of varying severity in the sagittal, transverse, and vertical directions (**Figure 1**) [18, 19].

In order to increase the effectiveness of the complex treatment of children with cleft lip and palate, it is advisable to justify pathogenetically directed personalized orthodontic therapy based on the definition of medical and social predictors, a set of diagnostic criteria, experimental modeling of the stressed-deformed state of the tissues of the upper jaw, and its clinical and morphometric characteristics.

The tongue is a powerful muscular factor influencing the development of the dentognathic apparatus. In maintaining myodynamic balance, an important role is played by the functional state of the tongue muscles, which depends on its anatomical features, namely its shape, size, position, and hyper- or hypotonus. Anatomical and topographic features of the tongue are important when choosing the tactics of orthodontic treatment and corrective speech therapy work in children with deformities of the dentognathic apparatus [18, 20].

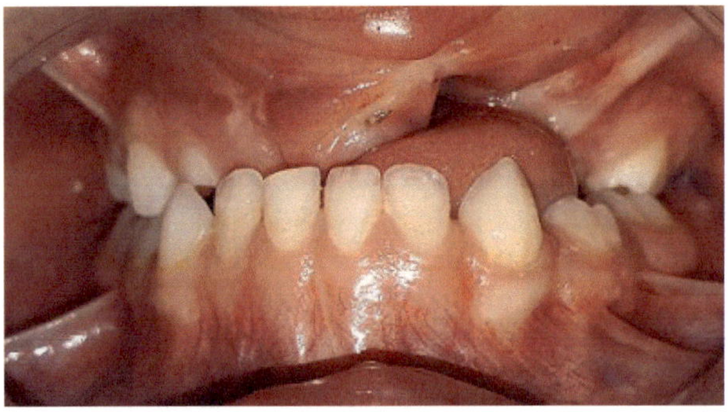

Figure 1.
Combined dentognathic deformity with unilateral cleft lip and palate.

Children with cleft lip and palate and violations of the anatomical and topographical parameters of the tongue have complex mechanisms of speech disorders, and the severity of violations of the phonetic component of speech varies from the damage of one group of sounds to their total violation.

In view of the aforesaid, there is a need for an in-depth study of the relationship between speech function disorders and deformations of the dentognathic apparatus, the development of modern preventive, diagnostic and therapeutic measures based on a multidisciplinary approach to overcome the identified problems, and in order to increase the effectiveness of orthodontic treatment—the development and justification of a complex diagnostic and therapeutic measures based on a multidisciplinary approach.

2. Clinical and laboratory research and principles of complex treatment of dentognathic deformations, and its results in accordance with the proposed complex of diagnostic and therapeutic measures

In order to study the influence of the state of the ENT organs on the formation of maxillofacial deformities and phonetic disorders, 155 children aged 6–12 years were subjected to rhinoscopy, pharyngoscopy, otoscopy, tonal and speech audiometry, tympanometry, and cone-beam computed tomography. Out of 155 subjects, 82 had dentognatic deformities accompanied by phonetic disorders, 73 subjects had phonetic disorders, orthodontic pathology was not observed [21].

Among 82 children with deformities of the dentognathic apparatus accompanied by phonetic disorders, hypertrophy of the nasopharyngeal tonsils (adenoid vegetations) of the I degree was diagnosed in 51.2% (n = 42), the II degree in 30.5% (n = 25), and the III degree—in 18.3% (n = 15). Among 73 children without orthodontic pathology with speech disorders, adenoid enlargement of the I degree was diagnosed in 43.8% (n = 32), the II degree (n = 15)—20.5%, and III degree (n = 2)—2.7%. An increase in the size of nasopharyngeal tonsils was not detected in 32.9% (n = 24) without orthodontic pathology.

Adenoid growths of the I degree in children with existing orthodontic pathology are observed 1.2 times more often than in children without orthodontic pathology, the II degree—1.5 times, and the III degree—6.8 times, respectively. Children without abnormalities in the size of the nasopharyngeal tonsils were not observed among people with deformities of the dentognathic apparatus.

During a pharyngoscopy to determine the size and condition of the palatine tonsils in 82 children with anomalies and deformations of the dentognathic apparatus, accompanied by phonetic disorders, hypertrophy of the I degree was diagnosed in 42.7% (n = 35), the II degree in 32.9% (n = 27), and the III degree—in 24.4% (n = 20). A different degree of hypertrophy of the palatine tonsils was detected in all subjects with deformities of the dentognathic apparatus. Among 73 children without orthodontic pathology with speech disorders, an increase in palatine tonsils of the I degree was determined in 38.4% (n = 28), the II degree—in 16.4% (n = 12), and the III degree—in 6.8% (n = 5). Disturbances in the size of palatine tonsils were not detected in 38.4% of children (n = 28) without orthodontic pathology.

Analysis of the state of the palatine tonsils indicates that hypertrophy of the I degree in children with existing orthodontic pathology is observed 1.1 times more often than in children without orthodontic pathology, the II degree—2 times, respectively, and the III degree—3.6 times.

The obtained data proved the presence of interrelationship of dentognathic deformations with inflammatory diseases of the ENT organs, which lead to changes in the volume of the upper respiratory tract, articulation zones, speech breathing and cause disorders of sound and speech.

Otoscopy of 155 children showed that their tympanic membranes are gray in color with a shiny shade, not retracted, mobile. The average thresholds of tonal hearing during tonal audiometry in the audiometer range of 125–8000 Hz ranged from 0 to 10dB. Suprathreshold tests corresponded to physiological values (Luscher test—1.5–2.0 dB, SiSi test—0%). In total, 50% legibility of air and bone conduction numerators was achieved at a sound pressure level of 15 dB and 100% legibility of the air conduction speech test at a sound test level of 30 dB. Therefore, according to the data of tonal and speech audiometry, no hearing disorders were detected, and speech tests proved normal hearing perception of speech in the subjects.

According to impedance measurements, 81.29% (n = 126) had a tympanogram of type "A" according to the Jerger classification, which indicates normal function of the auditory tube and auditory nerve. In 18.71% (n = 29), a tympanogram of type "C" was found, which indicates a violation of the function of the auditory tube with the occurrence of negative pressure in the middle ear cavity. Identified minor disorders of the condition of the middle ear did not significantly affect the hearing function. Compliance values ranged from 0.71 to 1.11 cm^3 (with an average statistical value of 0.77±0.12 cm^3), and intratympanic pressure ranged from −35 to +25 dRa. At a sound pressure level of 95 dB, ipsilateral and contralateral acoustic reflexes were recorded in all subjects [21, 22].

After an otolaryngological examination in order to study the influence of the state of the ENT organs on the formation of dentognathic deformations and, if necessary, after conservative and/or surgical treatment (adenoiditis, tonsillitis, etc.), 82 patients aged 6–12 with dentognatic deformations and phonetic disorders were given a detailed clinical dental orthodontic examination according to the generally accepted scheme with the usage of objective and additional research methods [22–24].

Anthropometric measurements were carried out using the 3Shape Viewer computer program on scanned models of the jaws of 82 patients with the determination of the length of the front part of the tooth rows before and after treatment by the method of M. Mirgazizov (n = 328), the transverse dimensions of the tooth rows using the Moorrees method before treatment in all 82 people (n = 164), after—in 71 (n = 142), because 11 lost temporary canines as a result of physiological changes in the teeth [25–27].

Out of 82 patients with dentognatic deformities, with the consent of the parents and on the condition that there were no general contraindications, cephalometric study was performed in 45 patients, and cone-beam computed tomography in 30 patients before and after orthodontic treatment.

A cephalometric study was carried out on 45 patients (n = 90) by the method of A. Schwarz using the computer program RadiocefStudio2, and superimposition of cephalometric images before and after orthodontic treatment was carried out according to the structural landmarks of the supraorbital plane Cl-RO and Se [26].

Cone-beam computed tomography of the skull of 30 patients was performed to determine the volume of the respiratory tract (paranasal sinuses) before and after treatment (n = 60) using a Gendex by iCAT CB-500 tomograph. A systematic analysis of changes in the respiratory tract was carried out by comparing tomograms before and after complex treatment, and the obtained data were processed in the graphic dental program SIMPlant (Materialise Software, Belgium) with the construction of multiplanar, panoramic, and 3D reconstructions.

The functional state of the group of muscles, the activity of which suffers the most due to certain orthodontic pathology, was determined by the method of total (surface) electromyography using the eight-channel electromyograph "BioEMG III" of the company "BioResearch Inc." (USA). Electromyograms of the surface part of the proper masticatory, anterior bundles of the temporal, sternoclavicular-mastoid, and anterior bellies of the biventricular muscles and the circular muscle of the mouth of 44 patients before and after treatment (n = 440) were studied and analyzed. The state of physiological rest, volitional contraction, and swallowing were subject to analysis [22, 28].

Diagnosis of the phonological aspect of speech was carried out in 155 patients, which were mentioned before, with phonetic disorders. The indicators of sound speech, the ratio of the most frequently detected distortions of sounds, and the average number of sound speech violations per child were determined [23, 24].

The 82 patients (38 patients with distal deep, 16—with distal, 18—with open, and 10—with mesial bites) were subjected to orthodontic treatment using removable and fixed orthodontic appliances for 10–12 months followed by a retention period [22].

Mechanically active elements in Schwarz', Andresen-Haupl's and Flis PS.–Filonenko VV. appliances (Patent of Ukraine for the utility model "Orthodontic appliance by Flis PS.–Filonenko VV. for the treatment of an open bite" No. 69548 A61C7/00 dated 04.25.12) (**Figure 2**) were used for narrowing of tooth rows, inclined or biting planes in Schwartz appliances on the upper jaw—in the treatment of distal and distal deep bites, occlusal linings in Bruckl's-Reeykhenbach appliances depending on the degree of reverse incisor overlap—mesial bite, Flis PS.–Filonenko VV. appliances—open bite. Marco Ross-fixed appliances (**Figure 3**) were used for narrowing of the upper tooth row, lack of space for upper incisors, correction of malocclusion associated with narrowing of the upper dental arch against the background of difficult nasal breathing [22, 29].

Removable orthodontic appliances were recommended to be used freely for the first one or two days, after the adaptation period—necessarily during sleeping and all free time, with the exception of periods of staying at school, eating, and practicing sports. The total time of usage during the day should have been at least 14–16 hours [29].

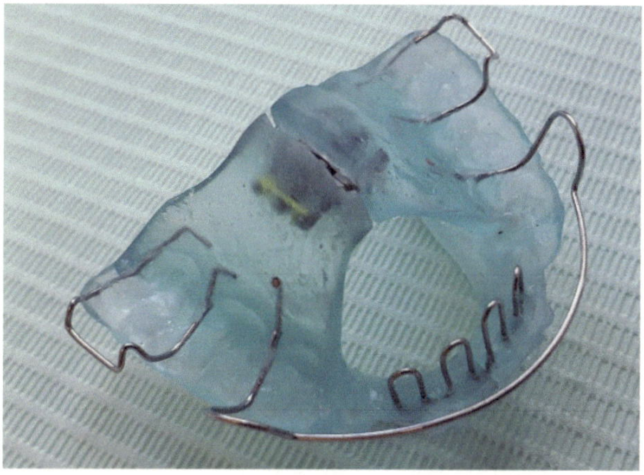

Figure 2.
Orthodontic appliance by Flis PS.–Filonenko VV. for the treatment of an open bite.

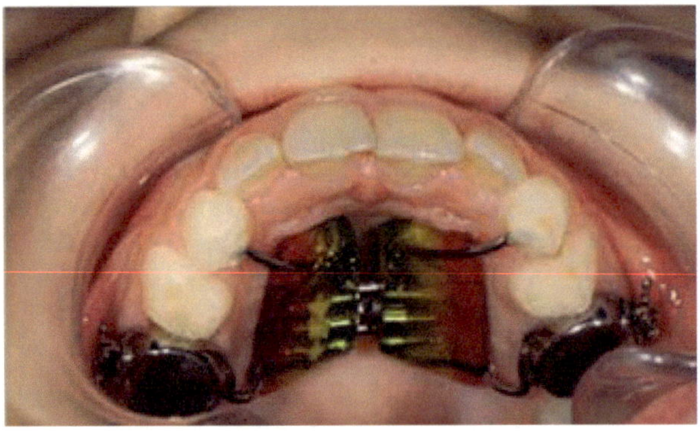

Figure 3.
Marco Ross appliance in the oral cavity.

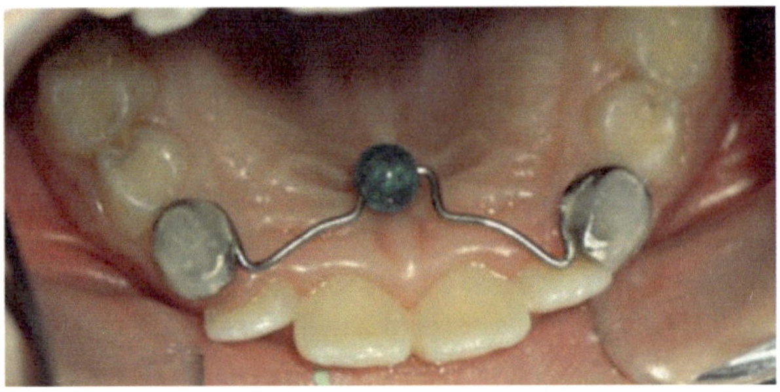

Figure 4.
Appliance for elimination and prevention of harmful tongue habits and training the muscle structures of the articulating apparatus.

In accordance with the proposed complex of diagnostic and therapeutic measures, the speech therapist carried out individual corrective work to overcome phonological disorders of speech, studied the semiotic component of speech, conducted speech therapy research methods (psychological-pedagogical—analysis of the child's speech therapy examination cards, observations, conversations with children and parents; neuropsychological—tests to determine the level of formation of oral kinetic and kinesthetic praxes; tests to determine the state of sound speech formation).

Children without orthodontic pathology were prescribed 10 classes of logopedic correction three times a week, with deformations of the dentognathic apparatus and speech disorders—2–3 courses of 10 classes with breaks of 1–2 months. In 12 cases, standard vestibular plates of Dr. Hinz—MUPPY-P with a bead were used in addition to speech therapy classes, in eight cases—removable appliances with a bead, in nine cases—fixed Bluegrass appliances, in six cases—appliances for eliminating and preventing harmful tongue habits and training the muscle structures of the articulating apparatus (Patent of Ukraine for the utility model "Appliance for elimination and prevention of harmful tongue habits" No. 126393 A61C7/00 dated 11.06.18) (**Figure 4**) [22, 30].

We offered a set of diagnostic and treatment measures for patients with dentognathic deformities accompanied by phonetic disorders, which consists of motivational, diagnostic, and treatment blocks [22, 25].

The motivational block envisages the formation of a positive result of orthodontic and speech therapy treatment; creating an atmosphere of emotional comfort among the orthodontist, the speech therapist, the child and his/her parents; formation of personally oriented treatment and corrective training.

The diagnostic block provides for the establishment of the type of dentognathic deformations based on clinical examination, anthropometry, electromyography, cephalometry, cone-beam computed tomography; determination of the state of formation of the phonetic side of speech using neuropsychological and speech therapy tests; examination of the state of the ENT organs by the methods of rhinoscopy, pharyngoscopy, otoscopy, tonal and speech audiometry, tympanometry, cone-beam computed tomography, and impedance measurement.

The treatment block included orthodontic treatment with the usage of removable and non-removable orthodontic appliances depending on the type of deformities, the age of the patient, the degree of formation of the dentognathic apparatus, etiology; phonetic correction with mandatory (orofacial gymnastics, formation of speech breathing) and correction-oriented (setting automation and differentiation of sounds) tasks; otolaryngological conservative and/or surgical treatment of adenoids and tonsillitis; control of the level of oral hygiene, therapeutic treatment of diseases of the hard tissues of the teeth, inflammatory processes of periodontal tissues, and the mucous membrane of the oral cavity. This made it possible to carry out complex multi-vector treatment of dentognathic deformations accompanied by phonetic disorders.

The criteria for evaluating the effectiveness of complex treatment of children with dentognathic deformities accompanied by phonetic disorders were based on the results of additional diagnostic methods, namely anthropometric, electromyographic, radiographic, and speech therapy.

After the orthodontic treatment, the results of the anthropometric measurements of the scanned models of the jaws according to the method of M. Mirgazizov and Moorrees showed a change in the dimensions of the tooth rows. A statistically significant decrease in the length of the front part of the upper tooth row was noted in the treatment of patients with a distal bite by 2.51±1.39 mm, with distal deep bite by 1.06±1.05 mm, in the treatment of a mesial bite, a decrease in the lower one by 1, 72±1.79 mm and an increase of the upper one by 3.43±1.36 mm; expansion in the area of canines on the upper jaw of patients with a distal occlusion by 3.32±1.03 mm; and a distal deep bite by 2.59±1.04 mm compared with the initial clinical picture.

Analysis of cephalograms using the A. Schwarz technique confirms positive changes after the treatment. The most informative improvements are related to the placement of the apical base of the lower jaw in relation to the base of the skull in the sagittal direction in the treatment of distal and distal deep bites (angle SeNB), the vertical position of the jaws in the treatment of open bite (angle B). The inclination of the axes of the teeth (angle 1SpP, angle 1MP) relative to the planes of the base of the jaws improved (**Figure 5**).

The narrowing of the upper jaw leads to forced retention of the tongue in the lower position, which interferes with its function and leads to speech defects.

The expansion of the upper jaw leads to the expansion of the bottom of the nasal cavity, which results in the expansion of the respiratory tract at all levels, which in turn causes an increase in the flow of air inhaled through the nose.

With the help of cone-beam computed tomography, it was objectively proven that the volume of the upper respiratory tract increased by 53.80±4.21% in patients from 11.82±2.06 ml to 18.01±3.84 ml after expansion of the upper jaw at the alveolar level, which are shown on figures (**Figures 6** and 7). These leads to a change in the position

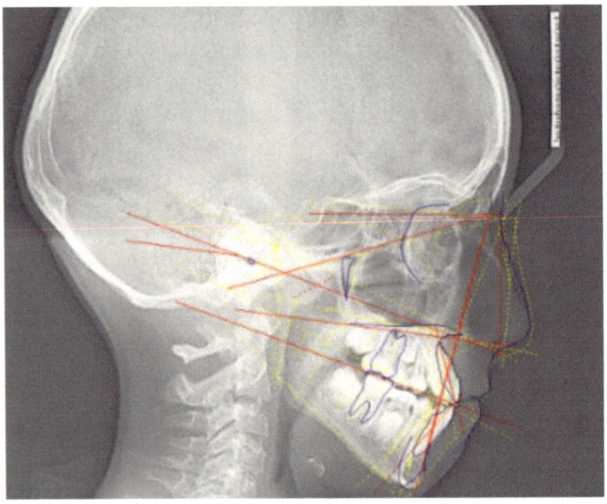

Figure 5.
Superimposition of cephalograms with an open bite before and after treatment.

of the tongue with its dislocation to the hard palate, which improves the results of orthodontic treatment and correction of speech. 3D reconstruction of the upper respiratory tract and superimposition of volumes before and after treatment were shown on figures (**Figures 8** and **9**).

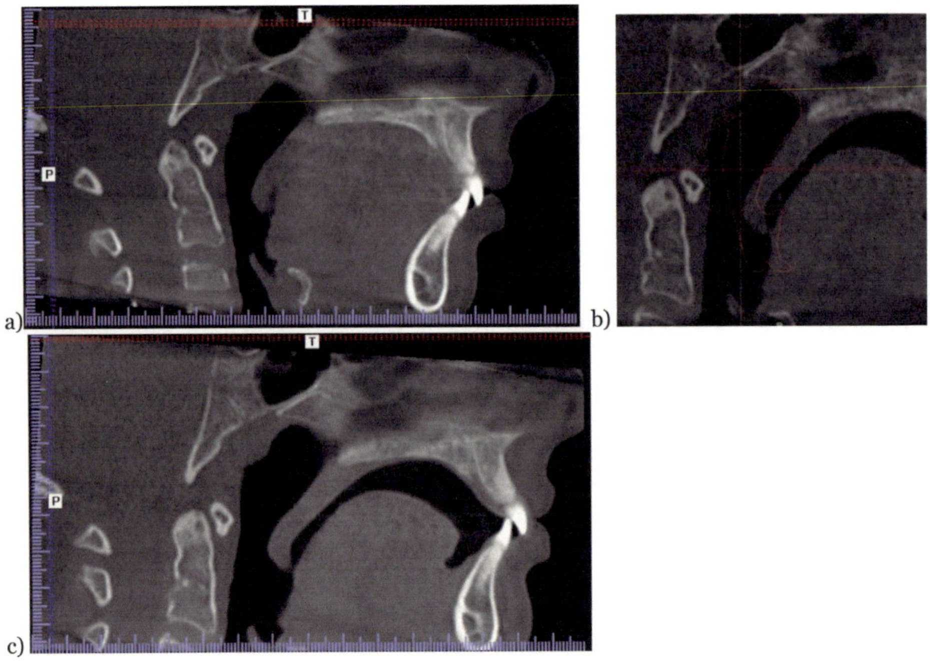

Figure 6.
Analysis respiratory tract, coronary section. Before treatment (a), after treatment (b), superimposition of pre-treatment airway profile on cone-beam computed tomography of post-treatment image (c) shows an increase in airway volume in the anteroposterior direction and in length.

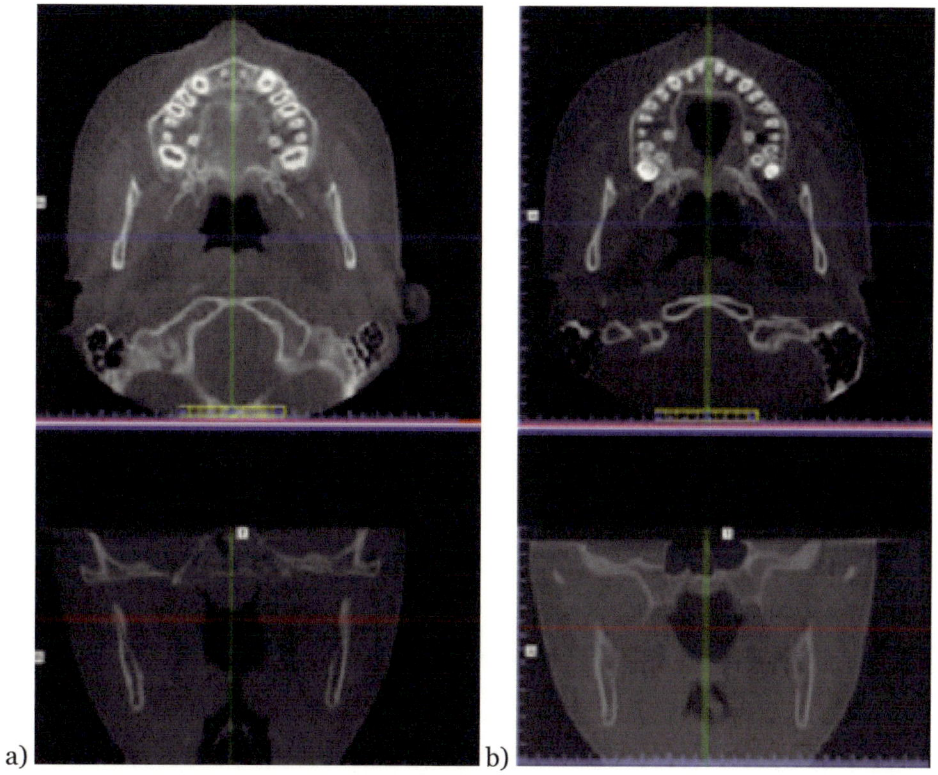

Figure 7.
Images of axial and frontal sections from the cone-beam computed tomography data before the start of treatment (a) and after treatment (b), demonstrating an increase in the area of the respiratory tract.

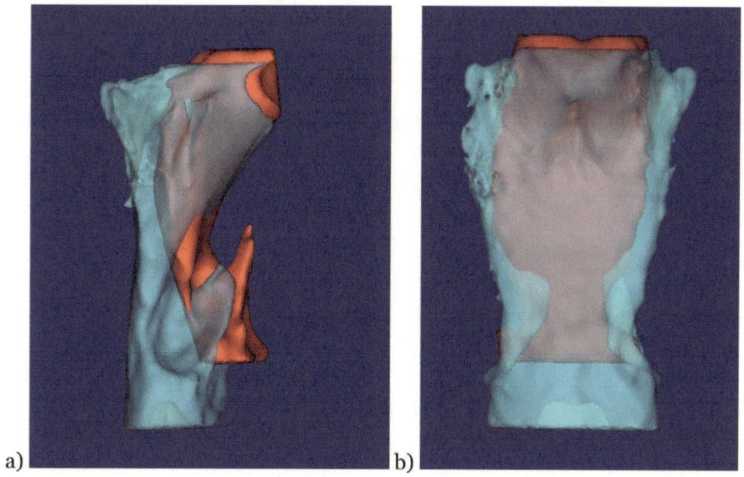

Figure 8.
3D reconstruction of the upper respiratory tract and superimposition of volumes (before the start of treatment—volume reconstruction in red, after treatment—in blue; side view (a), front view (b).

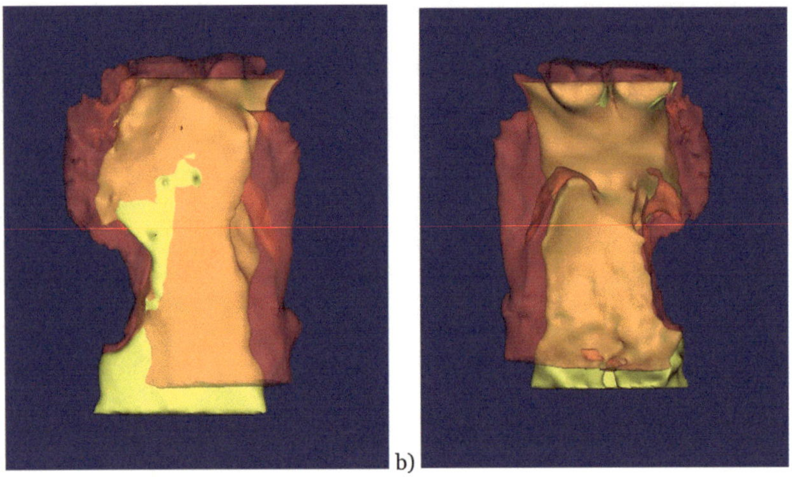

a) b)

Figure 9.
3D reconstruction of the upper respiratory tract and superimposition of volumes (before the start of treatment—volume reconstruction in red, after treatment—in yellow; side view (a), front view (b).

Changes in indicators of biopotentials of muscular structures of the articulating apparatus depending on the forms of dentognathic deformations, accompanied by phonetic disorders, testify to their correlation. The results of the electromyographic analysis of the state of the surface parts of the proper masticatory, anterior bundles of the temporal muscles, sternoclavicular-mammoid, anterior belly of the biventricular muscles, and the circular muscle of the mouth before orthodontic treatment indicate the group of muscles with the biggest functional impairment according to established orthodontic pathology and direct the work of the orthodontist and speech therapist to restore the neuromuscular balance of this particular group.

In patients with a mesial bite prior to orthodontic treatment, an increase in the bioelectric activity of the anterior temporal bundles (7.81±2.07 mV) and the proper masticatory muscles (2.29±1.03 mV) at rest and a significant decrease in the contractile activity of the latter during volitional compression (19.94±7.37 mV) were recorded; with an open bite—the amplitude of the biopotentials of the upper part of the circular muscle of the mouth is reduced at rest (3.21±1.07 mV) and during volitional compression as well (9.25±2.38 mV), an increase in the bioelectric activity of the anterior bundles of the temporal muscles at rest (4.62±1.13 mV); with distal and distal deep bites—the amplitude of biopotentials of the upper part of the circular muscle of the mouth at rest is the highest (5.24±1.17 mV and 5.04±2.01 mV, respectively).

After orthodontic treatment at the same time as logopedic correction, changes in electromyogram indicators were detected, which proved an improvement in the functional state of muscles and the effectiveness of the treatment by an average of 2.5 times. Patients with a mesial bite at rest had a decrease in the bioelectric activity of the anterior bundles of the temporalis (2.11±0.97 mV) and the masticatory muscles (1.32±0.78 mV), an increase in their contractile activity during volitional compression (44.48±6.33 mV) as well; with an open bite—increase in the amplitude of the biopotentials of the upper part of the circular muscle of the mouth during volitional compression (12.84±3.51 mV) and decrease in the bioelectric activity of the anterior bundles of the temporal muscles at rest (0.87±0.22 mV); with distal and distal deep bites—a decrease in the amplitude of biopotentials of the upper part of the circular muscle of the mouth at rest (3.22±1.37 mV and 2.76±1.02 mV, respectively).

Diagnostic screening of speech disorders was carried out and their dependence on dentognathic deformations was established. The average number of violations of the pronunciation of whistling sounds per child is the highest in mesial bite at the age of 6–8 years—2.4 and in open bite at the age of 9–12 years—2.0, and sonorous sounds in the distal and distal deep bites at the age of 6–8 years old—1.7, at the age of 9–12 years—1.1. The ratio of sonorous to whistling speech disorders in mesial bite at the age of 6–8 years is 6.0, in open bite—4.7, at the age of 9–12 years, 4.5 and 3.0, respectively. The lowest average number of sibilant sound-speech disorders was determined at the age of 6–8 years—0.4, at the age of 9–12 years—0.6 per child with all bite pathologies.

Staged work was carried out to overcome phonetic disorders of speech, which contributed to the normalization of the sound pronunciation of whistling, and sonorous and hissing sounds in various types of dentognathic deformations. The expediency and effectiveness of using the proposed non-removable appliance, which contributes to the training of the muscular structures of the articulating apparatus in combination with speech therapy correction, to improve kinetic and kinesthetic praxes in the temporary and first period of variable bite, was proven. The appliance includes a bracket soldered to two thin-walled cast perforated crowns designed to be fixed on temporary canines with a composite flowable photopolymer material (e.g., Filtek flow 3M ESPE, USA) or hybrid glass ionomer cement with a triple curing mechanism (e.g., Vitremer 3M ESPE, USA), in the middle part to which a functionally active element in the form of a bead is attached.

The advantage of the offered appliance, in our opinion, is the following: Fixation on temporary canines makes the appliance compact due to the reduction of the structure size, which in turn allows to facilitate hygienic care and improve the hygienic condition of the oral cavity, reduces the risk of caries in the non-mineralized cervical area at the stage of permanent teeth unformed root; compactness helps to increase the articulation zones of the tongue, trains the tip of the tongue, which speeds up the speech therapist's work at the stage of producing sounds, and eliminates their interdental pronunciation; the use of a functional element in the form of a bead allows to control the usual palatal position of the tongue and activate the work of the root of the tongue, because during a conversation the child involuntarily rolls it in the area of the palate, stimulating the muscles of the tongue.

The evaluation of the effectiveness of the diagnostic and treatment complex of measures offered on the basis of a multidisciplinary approach for patients with dentognathic deformities accompanied by phonetic disorders proved the need to determine the condition of the nasopharyngeal and palatine tonsils and the effectiveness of orthodontic treatment at the same time as logopedic correction as well. Improvements in electromyography, anthropometric measurements of scanned jaw models, and cephalometry were noted in 86.6% of patients; analysis of cone-beam computed tomography data showed a significant increase in the volume of the upper respiratory tract by 53.8±4.2% in patients after orthodontic treatment.

3. Predictors of the development of dentognathic deformations in cleft lip and palate, the basis of orthodontic correction

In most cases with cleft lip and palate, combined deformations of the dentognathic apparatus are observed. A congenital defect is the basis for the development of upper jaw deformity. Children with cleft lip and palate have at least one or more dentognathic deformities due to the development or position of the teeth themselves or the development and mutual arrangement of the jaws.

With cleft lip and palate, and unilateral cleft lip and palate, the following anomalies or deformations are most often observed: primary adentia, more often upper lateral incisors; presence of natal/neonatal teeth; presence of supplemental teeth; anomalies of the eruption of teeth; abnormalities in tooth morphology, e.g., fused teeth; tooth size abnormalities, e.g., microdontia; abnormalities of individual teeth, e.g., enamel hypoplasia; abnormalities in the position of the teeth; crowding of teeth or vice versa—tremas; mesial bite; cross bite; deep bite or open bite.

It should be noted that changes in the development of the upper jaw quite often affect the position and size of the lower jaw.

An integral component of deformations is the functional component.

A comprehensive approach to the orthodontic rehabilitation of children, taking into account the multifactorial determination of unilateral cleft lip and palate, allows to reduce the severity of dentognathic deformities caused by both congenital defects and surgical intervention. Orthodontic support during all periods of the formation of the dentognathic apparatus in children with congenital malformations of the lip and palate makes it possible to restore the size and shape of the dental arches, ensuring the stability of the complex treatment results.

The reason for the formation of deformations of the dentognathic apparatus in cleft lip and palate is pathogenetic predictors, which lay the foundation for the formation of deformations from the birth of a child. They can be divided into groups: primary congenital defects of soft (upper lip, soft palate) and hard tissues (defect of the alveolar process, hard palate); the influence of the tongue and cheek muscles, feeding method (pacifier, presence of an obturator); primary surgical interventions (cheilo-rhinoplasty, veloplasty, uranoplasty). The cascade of changes forms cause-and-effect mechanisms for the development of dentognathic deformations.

When studying the predictors, morphometric measurements of photograms and scanned models of the upper jaws of 72 children with unilateral cleft lip and palate were carried out.

Primary congenital defects of soft and hard tissues determine the presence of non-unions of the tissues of the lip and palate. It was showed that the largest soft tissue defect was located in the area of the nasolabial complex, non-union fragments of the soft palate. The largest defect of hard tissues was found at the border of the hard and soft palate. The distance between the ends of the alveolar process varied between 2.31 and 15.32 mm. The alveolar process of the large fragment deviated outward, while the small fragment shifted backward, which is due to the absence of both the muscular closure of the upper lip and the lack of its fusion with the vomer.

The influence of the pacifier on the mutual location of the fragments of the upper jaw and the position of the horizontal plate of the palatine bone was established. In most children with cleft lip and palate, who were artificially fed with the help of a pacifier, the shift of the alveolar process in the sagittal plane and the movement of the horizontal plates of the palatine bone to a vertical position were noted.

The development of the upper jaw in case of unilateral cleft lip and palate is also influenced by primary surgical interventions (cheilo-rhinoplasty, veloplasty, uranoplasty, uranostaphyloplasty), which restore the anatomical integrity of unfused fragments of the lip and palate, and create conditions for their functional capacity. The foundation for optimal orthodontic treatment is laid already at the stages of surgical interventions.

Cheilo-rhinoplasty is the primary surgical intervention performed on a child with cleft lip and palate. After primary cheilo-rhinoplasty, which is usually performed in 3–5 months for unilateral nonunions, the main muscle constrictor of the oral cavity is restored. It creates the conditions for regulating the mutual location of the unfused fragments of the upper jaw: It limits their further movements relative to each other, and optimizes and prepares them for the second stage of surgical interventions—veloplasty. Studying the dynamics of changes in the morphometric indicators of the upper jaw after veloplasty in children with cleft lip and palate showed changes in sagittal and transverse indicators of its dimensions. It was proven that cheilo-rhinoplasty, veloplasty, and uranoplasty affect the size of the defect by reducing it, but the risk zones for the growth of the upper jaw are the distances between the distal edges of the canines and between the points of the transition of the gingival mucosa to the hard palate in the area of the first molar, since in these areas there is a decrease in transverse sizes.

For orthodontic treatment of children with cleft lip and palate, orthodontic appliances are used, the scientific justification of which took place at the beginning of the twentieth century [29, 31]. However, there is still no universal orthodontic protocol for the treatment of dentognathic deformities in cleft lip and palate. To eliminate bite deformities caused by congenital defects of the palate, to obtain a satisfactory overjet (sagittal overlap) and overbite (vertical overlap), expansion of the upper jaw (dental arch) is usually performed. For this purpose, a number of appliances are used, which include the following:

- sectional removable appliances with one or more screws, occlusive side plates, spring pushers if necessary;

- appliancess for rapid expansion of the upper jaw depending on the type and volume of the required expansion (Hyrax appliance with screw, Mcnamara appliance with occlusive side plates, Hyrex-expanderr, Mini expander, Fan-type expander, Bobded Fan Expander with occlusive side plates, Expander with differential opening (EDO), etc.);

- stationary (non-removable) quad-helix/tri-helix appliances, etc. [26, 31–34].

The choice of orthodontic appliances in children with cleft lip and palate, especially in the period of variable bite, depends on the type of dentognathic deformities, existing conditions for fixation and socio-economic components.

Corrective speech therapy work with children with cleft lip and palate is carried out during the entire treatment and rehabilitation period and includes preparatory and main stages. The preparatory stage precedes the main one and is carried out in all age periods, and it involves the formation and development of basic psychomotor (kinetic and kinesthetic praxes), cognitive processes (visual, auditory gnosis, perception, auditory-speech memory, different types of thinking, spatial representation, attention), which are the basis for the child's speech development. The main stage is aimed at the formation and development of all constituent components of speech: speech breathing, phonetics, phonemic processes, vocabulary, grammar, coherent speech.

4. Interrelationship of the development of dentognathic deformations with anatomical and topographic indicators of the tongue, and prevention of dentognathic deformations in macroglacia

The action of the muscles of the maxillofacial area is decisive in the process of development of the dentognathic apparatus. The shape and size of the tooth rows are determined by the direction of growth and the influence of muscle forces, which depend on the anatomical features of the muscles and their functional state. The functional state of the tongue muscles plays an important role in maintaining myodynamic balance and depends on its anatomical features. They include the shape and size of the tongue, its position, and hyper- or hypotonus.

Deformations of the dentognathic apparatus in children of various ages develop as a result of congenital and acquired causes. One of them is macroglossia, which can be a manifestation of syndromic diseases, vascular malformations, congenital malformations, etc., which leads to a forced position of the tongue between the teeth rows, a change in the functional capacity of the tongue during articulation and eating, as well as the formation of conditions of constant excessive pressure on the lingual surface of the teeth and the alveolar process of the lower jaw.

Determining the state of formation of the tongue muscle tone, its anatomical and morphometric indicators, and its position are important for choosing the tactics of orthodontic treatment and corrective speech therapy work in children with dentognathic deformities [35].

Cone-beam computed tomography, electromyography, and anatomic and morphometric indicators of the tongue of 72 children aged 9–12, 44 with distal and distal deep bites and 28 without orthodontic pathology were performed. Adapted speech therapy functional tests were used to diagnose tongue muscle tone [20].

When examining children without orthodontic pathology, we determined the average thickness of the front, middle and back 1/3 of the tongue, which were 29.1±0.16 mm, 42.3±0.06 mm, and 44.4±0.08 mm, respectively. Average length m. genioglossus was 17.9±1.41 mm, and the total length of the tongue was 60.6±0.12 mm.

Anatomical and morphometric indicators in children with distal and distal deep bites showed that the thickness of the front, middle, and back 1/3 of the tongue were 22.9±0.11 mm, 42.1±0.09 mm, and 44.1±1.05 mm, respectively. Average length m. genioglossus—19.7±1.08 mm, total tongue length—54.1±2.16 mm.

The given data indicate a significant decrease in the thickness of the front 1/3 of the tongue, the total length of the tongue, and a significant increase in the length of m. genioglossus in children with distal and distal deep bites compared with children without orthodontic pathology.

Adapted speech therapy functional tests revealed an increase in tongue muscle tone by 65% among children with distal and distal deep bites. Hypertonus of the tongue was indicated by a change in its position and shape, increased mobility, and subtle differentiated movements, especially its tip. Displacement of the tongue backward to varying degrees was determined in all subjects with distal and distal deep bites. Among them, in 85% the tongue took the form of a "hill," in 15%—a "sting". In all patients, when trying to hold the tongue in a resting position, a muscle contraction was observed on the lower lip.

The results of electromyographic studies indicate an increased muscle tone of the tongue in children with distal and distal deep bites, which coincides with certain changes in mimic muscles.

Changes in the anatomical and morphometric indicators of the tongue are the main factors for moving its root to the back-lower position, which contributes to reducing the volume of the oral cavity, reducing the length of the lower dental arch, and increasing the tone of the tongue. The functional state of the muscles of the tongue has a direct correlation with the shape and size of the tooth rows and their type of closure.

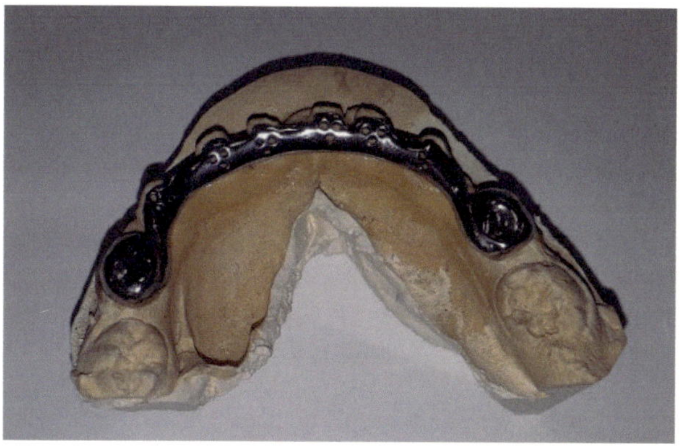

Figure 10.
Appliance for the prevention of bite deformations in children with macroglossia.

To prevent the development of bite deformations caused by a violation of the tone and position of the tongue, well-known orthodontic appliances are used, for example, standard vestibular plates of Dr. Hinz, removable orthodontic appliances with support for tongue (the tongue guard), glued to the inner surface of the upper or lower teeth medical alloy spikes that force the tongue to seek the correct position during swallowing, speaking, and at rest.

The action of preventive appliances in children with macroglossia of various origins is aimed at limiting the effect of excessive pressure of the tongue on the teeth and the alveolar process of the lower jaw from the lingual side, which contributes to the stabilization of the position of the teeth and prevents their movement at the alveolar level without injuring the tissues of a significantly enlarged tongue. A device comprising an element for limiting excessive tongue pressure and support elements is offered. The element for limiting excessive pressure of the tongue is made in the form of a cast-perforated buckle modeled after the shape of the lingual surfaces of the lower teeth, and the supporting elements are made in the form of cast crowns (Patent of Ukraine for the utility model "Appliance for the prevention of bite deformations in children with macroglossia" No. 146224 A61C7/00 dated 27.01.21) (**Figure 10**).

Clinical indications for the usage of the offered appliance are the prevention of bite deformities in children with macroglossia of various origins, by limiting excessive tongue pressure and stabilizing the conditions for movement at the dentoalveolar level of the lower jaw at the stages of surgical treatment in temporary and variable periods of bite.

5. Conclusions

In order to timely detect dentognathic deformities with phonetic disorders and carry out preventive and therapeutic measures, it is necessary to conduct preventive examinations of children in preschool and school, followed by their referral to orthodontists, speech therapists, children's dentists, otolaryngologists, and mandatory monitoring and summarization of statistical data.

In case of detection of hypertrophy of nasopharyngeal and palatine tonsils of II-III degrees, children with dentognathic deformities accompanied by phonetic disorders require otolaryngological treatment simultaneously with orthodontic correction.

For children with dentognathic deformations accompanied by phonetic disorders, it is advisable to determine the bioelectrical activity of the chewing, and temporal and circular muscles of the mouth in a state of relative physiological rest and during voluntary compression, which is necessary to determine the volume of muscle load of the articulating apparatus during corrective work of a speech therapist.

In the final period of temporary and the first period of variable bites, for training the muscular structures of the articulating apparatus in combination with speech therapy correction, it is advisable to use a fixed appliance containing a bracket soldered to two thin-walled cast perforated crowns intended for fixation on temporary canines, in the middle part of which a functionally active element in the form of a bead is attached.

Developed on the basis of a multidisciplinary approach, the diagnostic and therapeutic complex of measures for patients with dentognathic deformities accompanied by phonetic disorders proved the need for coordinated corrective and developmental speech therapy work.

Dentognathic deformations in children with cleft lip and palate are cause-and-effect and depend on the type of defect, method of feeding, methods and age of surgical interventions.

The study of anatomical and morphometric characteristics of the tongue is a prognostic indicator of effective orthodontic treatment and requires differentiated corrective speech therapy work to develop kinetic and kinesthetic praxes.

Author details

Alona Melnyk* and Valerii Filonenko
Bogomolets National Medical University, Kyiv, Ukraine

*Address all correspondence to: melnik.alona@gmail.com

References

[1] Bezvushko EV, Miskiv AP. Dynamics of prevalence of dental jaw abnormalities in children of Lviv region. Bulletin of Problems in Biology and Medicine. 2015;**2**(2):21-24

[2] Boytsanyuk SI, Falinskyi MM, Ostrovsky PY. Prevalence of dental anomalies among school-age children of Ternopil. Young Scientist. 2017;**5**:57-60

[3] Kuroedova VD, Makarova AN. Prevalence of dental anomalies and the proportion of asymmetric forms in adults. World of Medicine and Biology. 2012;**4**:31-35

[4] Zaiats OR, Ozhogan ZR. Advancement of dental anomalies among children of the Ivano-Frankivsk region. Actual Dentistry. 2020;**1**:68-72. DOI: 10.33295/1992-576X-2020-1-68

[5] Drok VO. Prevalence of dental anomalies and periodontal diseases among adolescents. Ukrainian Dental Almanac. 2018;**1**:72-74. DOI: 10.31718/2409-0255.1.2018.17

[6] Flis PS, Leonenko GP, Filonenko VV. The frequency of abnormalities in the position of separate teeth and bite during the period of interchangeable bite. Actual Orthodontics. 2008;**1**(11):6-7

[7] Flis PS, Filonenko VV, Doroshenko NM. Frequency and prevalence of anomalies and deformations of dentognathic apparatus in the transitional period of occlusion. Ukrainian Dental Almanac. 2016;**1**(1):75-78

[8] Flis PS, Rashchenko NV, Filonenko VV, Melnyk AO. Advancement of powerable anomalies and speech disorders among children with 6-12 years old. Actual Dentistry. 2018;**4**:54-57

[9] Trofimenko MV. Orthodontic Treatment of Patients Aged 6-9 Years with Impaired Swallowing and Speech [autoabstract]. Poltava: Ukrainian Medical Stomatological Academy; 2010

[10] Bida OV, Zabuga YI. Functional state of the masticatory muscles of patients with defects of dentition, which are complicated by dentoalveolar deformities. Ukrainian Dental Almanac. 2016;**3**(2):27-31

[11] Khokhlich OY. Rehabilitation of patients with dental diseases in the conditions of their professional phonatory performance. Ukrainian Dental Almanac. 2012;**2**(2):67-71

[12] Petrychenko OV. Justification of the Effectiveness of Early Treatment and Prevention of Dental Anomalies in Children with Speech Disorders [autoabstract]. Kyiv: Shupyk National Medical Academy of Postgraduate Education of the Ministry of Health of Ukraine; 2018

[13] Zabolotny DY, Mytin YV, Bezshapochny SB, Deeva YV. Otorhinolaryngology: Textbook for Students of Higher Medical Institutions. Kyiv: Kyiv Medicine; 2010. p. 494

[14] Sheremet MK, Bazima NV, Moroz OV. The problem of speech anxiety in children with severe speech disorders. Speech Therapy. 2015;**5**:93-99

[15] Yakovenko AO. Formation of Speech Readiness of Older Preschoolers with Speech Pathology for Integrated Learning. Kyiv: National Pedagogical Dragomanov University; 2018

[16] Kuroyedova VD, Dmytrenko MI, Makarova OM, Stasyuk OA.

Comprehensive Orthodontics. Poltava: Astraya; 2016. p. 88

[17] Kuroyedova VD, Vyzhenko EE, Doroshenko SI, Makarova OM. Physiotherapy in Orthodontics. Poltava: Astraya; 2019. p. 135

[18] Yakovenko LM, Cherkasov VG, Chekhova IL. Surgical Dentistry and Maxillofacial Surgery of Pediatric Age: Textbook. 2nd ed. Kyiv: Kyiv Medicine; 2022. p. 496

[19] Sharma G. Orthodontic management of cleft lip and palate patients. In: Ayşe G, editor. Current Treatment of Cleft Lip and Palate. 2020. pp. 19-33. DOI: 10.5772/intechopen.77803(Chapter 2) DOI: 10.5772/intechopen.90076

[20] Melnyk AO, Shafeta OB, Yakovenko AO. Correlation of bite pathologies with anatomical and topographic indicators of the tongue. In: Materials of the Scientific and Practical Conference "Modern Dentistry and Maxillofacial Surgery". Kyiv; 2020. pp. 105-106

[21] Flis PS, Filonenko VV, Melnyk AO. Study of the state of ent-organs in children with anomalies and deformations of dentoalveolar apparatus and speech disorders. Ukrainian Dental Almanac. 2018;**3**:26-32. DOI: 10.31718/2409-0255.3.2018.04

[22] Flis P, Yakovenko L, Filonenko V, Melnyk A. Validation of the diagnostic and treatment for patients with orthognathic deformities and phonetic disorders. Georgian Medical News. 2019;**9**(294):62-68

[23] Flis PS, Rashchenko NV, Filonenko VV, Melnyk AO, Lopoha AB. Algorithm of patient's treatment with the presence anomalies of bite and speech disorders. Ukrainian Dental Almanac. 2018;**4**:60-64. DOI: 10.31718/2409-0255.4.2018.11

[24] Flis PS, Filonenko VV, Nemyrovych YP, Melnyk AO, Lopoha AP. Algorithm for speech disorders correction using proprietary construction device. Bulletin of Scientific Research. 2018;**4**:145-151. DOI: 10.11603/2415-8798.2018.4.9780

[25] Melnyk AO. Evaluation of the effectiveness of the complex of diagnostic and therapeutic measures in patients with anomalies of the dentoalveolar apparatus accompanied by phonetic disorders. Actual Dentistry. 2018;**5**(94):72-75

[26] Flis PS. Orthodontics: Textbook. Kyiv: Kyiv Medicine; 2008. p. 336

[27] Moorrees CF, Gron AM, Lebret LM, Yen PK, Frohlich FJ. Growth studies of the dentition: A review. American Journal of Orthodontics. 1969;**55**(6):600-616

[28] Flis PS, Rashchenko NV, Filonenko VV, Melnyk AO, Etnis LO. Functional state of the muscles of the maxillofacial apparatus before and after child treatment with bite anomalies and speech disorders. Clinical Dentistry. 2018;**4**(25):76-83. DOI: 10.11603/2311-9624.2018.4.9630

[29] Flis PS, Leonenko GP, Filonenko VV, DoroshenkoNM.Orthodontics.Dentognathic anomalies and deformations: Textbook. Kyiv: AUS Medicine Publishing; 2015. p. 175

[30] Flis PS, Filonenko VV, Melnyk AA. Correction of speech disorders with the use of a device of own construction. Azərbaycan Tibbi Jornali. 2018;**4**:109-113

[31] Koskela A, Neittaanmäki A, Rönnberg K, Palotie A, Ripatti S, Palotie T. The relationship of severe malocclusion to patients' mental and behavioral disorders, growth, and speech problems. European Journal of Orthodontics. 2020;**43**(2):1-6. DOI: 10.1093/ejo/cjaa028

[32] Aura SC, Chen A. Orthodontic interventions in Cleft Lip and Palate Individuals [Internet]. 2019 . Available from: https://www.oralhealthgroup.com/features/orthodontic-interventions-in-cleft-lip-and-palate-individuals-an-overview-of-treatment-protocol/ [Accessed: September 10, 2019]

[33] Makeev VF. Orthodontic methods of treatment of patients with Clift Lip and Palate: Textbook. Lviv: Quart; 2013. p. 164

[34] Kovach IV, Khaletska VM. Features of the effectiveness of treatment of narrowing of the upper jaw in the transversal area in children with complete cleft palate in the state after uranoplasty in the early mixed dentition. Actual Dentistry. 2015;2:114-116

[35] Yakovenko LM, Chekhova IL, Yefimenko VP, Kiselyova NV, Egorov RI, Kovtun TO, et al. Examination of Children with Surgical Diseases of the Maxillofacial Area: Textbook. Kyiv: Book-plus; 2022. p. 164

Section 5

Advances and Future Perspectives

Chapter 19

Practical and Theoretical Considerations for Dental Restorative Materials

Tanvi Satpute

Abstract

Dental materials are essential for most of dental treatment modalities. Understanding the science and chemistry behind the materials and their properties can enable the operator to employ the dental material to its maximum advantage. Contemporary dental materials have evolved significantly from the conventional variety, but there is always room for refinement since the inadequacies of the current dental materials in function are recognized only with the advent of advanced dental materials testing methods. As a result, continuous improvement and modification of dental materials is essential. Caries is a process of continuous demineralization and re-mineralization. Recurrent caries is a common occurrence around the tooth-restoration margin. It most likely indicates that the current dental materials are inadequate in their applications. As a result, augmenting conventional dental materials with additional advantageous properties is critical. This chapter aims to reflect on the empirical status of direct restorative materials frequently used in the field of restorative dentistry.

Keywords: dental materials, material science, permanent dental restoration, dental restoration failures, operative dentistry

1. Introduction

Direct restorative materials are classified by the ADA Council on Scientific Affairs into four categories: amalgam, resin-based composites, glass ionomer, and resin-modified glass ionomer (**Figure 1**) [1].

Amalgam is particularly well suited for Class I and II restorations in teeth subjected to high chewing forces [2–4]. Amalgam restorations have several advantages over other direct-placement materials, which include wear resistance, tolerance to a wide range of clinical placement conditions and excellent load-bearing properties contributing to its high survival rate [2–4]. However, some primary issues limiting the longevity of amalgam restorations are secondary caries, increased incidence of bulk and tooth fracture, cervical overhang, and marginal degradation [2]. Until the late 1960s, when resin-based composites were introduced, amalgam was the material of choice for all but the most esthetically demanding restorations. Dental composites are the most esthetic direct filling material available, as they mimic the color and

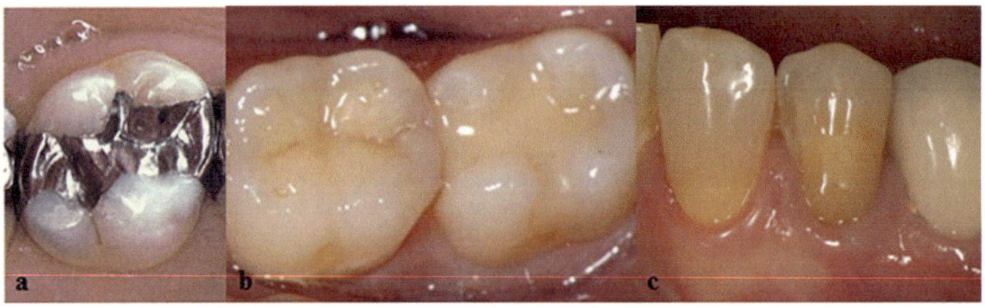

Figure 1.
Direct restorative materials a. Class II amalgam restoration b. Class II composite restoration c. Class V Glass Ionomer restoration.

translucency of natural teeth. Originally, these esthetic materials were only intended for anterior restorations. As their popularity grew and the materials improved, they were used in nearly all classes and types of dental restorations. Glass ionomers are tooth-colored filling materials that can be used to fill cavities with low load requirements. They are frequently used to repair non-carious erosion or abrasion defects in the tooth near the gingiva [5]. They are also used for pediatric restorations, which have low service longevity requirements [6]. Glass ionomers are also commonly used as cavity liners or bases, protecting the underlying tooth pulp in deep fillings. Resin-modified glass ionomers as compared to conventional glass ionomers have superior physical and mechanical properties and better handling characteristics [7]. Unlike traditional glass ionomers, which have short working and long setting times, the dentist has more control over the working and setting times of resin-modified glass ionomers. This removes some of the material's technique sensitivity, making it easier to achieve a successful restoration. They are used for Class I, Class II, and Class III restorations, primarily in the primary dentition, as well as Class V restorations, liners, and bases. Other applications include fissure sealants and bonding agents for orthodontic brackets.

2. Properties of dental materials

The selection of a dental material is based on the physical properties required and the unique functional demands placed upon it in a specific clinical application. For instance, in evaluating a fiber post, a clinician would not be that interested in characteristics such as abrasion resistance, solubility, or even compressive strength, but would be highly interested in these characteristics if evaluating a composite filling material. Similarly, when a clinician is faced with the selection of dental materials for use in the permanent restoration of severely broken down teeth, esthetics should be secondary to the mechanical and physical properties necessary for that particular application [8].

Since many complex forces occur and tend to deform the material (tensile, compressive, shear, bending forces), the knowledge and interpretation of how these materials behave under such forces are relevant to understand the performance of the material. When a specific force or stress is applied to a body, it causes a reaction of equal intensity but in the opposite direction, which can be quantified. Stress can be calculated using the force-unit-area relationship because the shape and dimensions of the specimens under test can be measured. The stress can alter the original dimensions structurally. Strain is defined as the rate at which the original dimension

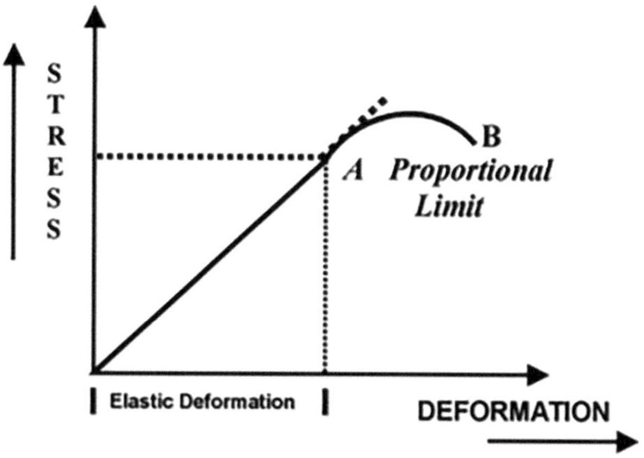

Figure 2.
Stress–Strain curve. Proportional limit (A), elastic deformation (point A), and plastic deformation (between points A and B). Point B represents the moment of rupture of the material under tensile condition.

gets altered due to deformation. The stress–strain ratio of a material is important in determining its mechanical behavior. Each material has a stress–strain proportional relationship, resulting in a stress–strain curve. If there is stress relief during loading and no permanent deformation occurs, it demonstrates elasticity. This proportion continues until a limit point, defined as a proportional limit, is reached, and deformation is defined as elastic deformation (**Figure 2**). This is the greatest amount of stress a material can withstand without permanently deforming. Because stress–strain is proportional until this point, there is a constant proportionality. It is the ratio of the stress–strain curve within the elastic limit that determines a material's elasticity. The modulus of elasticity, also known as Young's modulus, is a measure of this proportionality. Young's modulus represents the stiffness of the material [9]. When the applied load exceeds this point, however, irreversible deformation takes place, resulting in permanent or plastic deformation. Each material has a resistance to deformation, and after that point, it will rupture. The ultimate strength value is obtained at this point.

Other clinically relevant properties of direct restorative materials include tensile strength, diametral compressive strength, compressive strength, Poisson's ratio, flexural strength, resistance to fatigue, and hardness strength.

Tensile strength refers to the resistance of the material to a load when a body is subjected to axial forces in a straight line and opposite directions (**Figure 3**) [8, 10].

It is an important feature of metallic materials because they can deform under tensile forces until fracture occurs, indicating the workability of an alloy. Brittle rupture under low tension is a characteristic of fragile materials [11]. Tensile strength is not recommended in such cases to evaluate the material's reaction due to its low cohesive condition. An alternative method of tensile strength is calculated by compressive testing. It is also known as the diametral compression test for tension or the indirect tension test [7–9]. Materials must be investigated under this condition because the majority of mastication forces are compressive in nature [10]. As a result, this test is used to contrast dental amalgam, impression materials, investments, and cement [12]. When a material is subjected to axial loading, it also experiences lateral strain [10, 11]. Poisson's ratio indicates that during the elastic range, cross-sectional change is proportional to deformation. Brittle materials show little permanent reduction in cross-section during tensile test situations than more ductile materials (**Figure 4**) [7, 8].

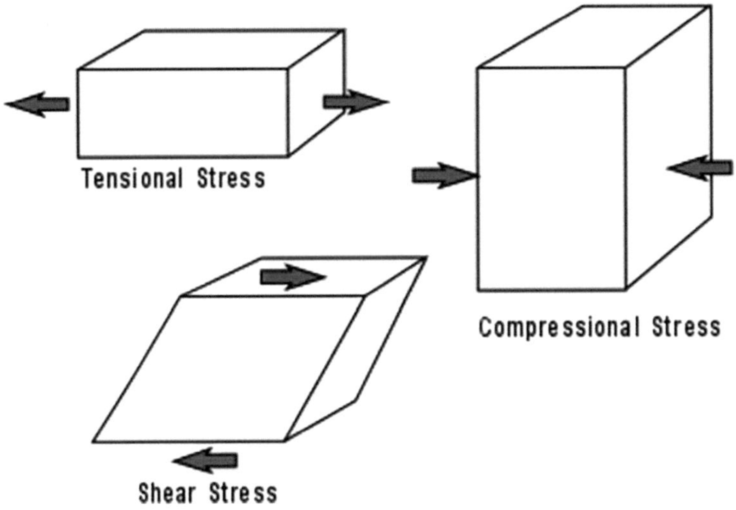

Figure 3.
The direction of forces in tensile, compressive, and shear stress.

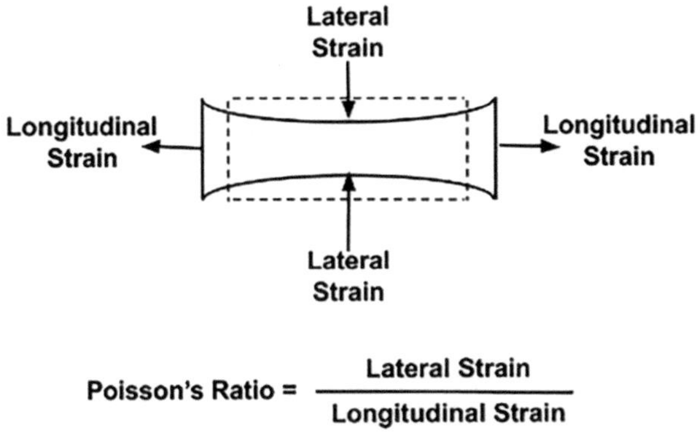

Figure 4.
Diagrammatic representation of Poisson's ratio.

A material's flexural strength is its ability to bend before breaking. It is obtained when a material's ultimate flexibility is reached before its proportional limit [13, 14]. Clinical situations generate flexural forces, and dental materials must withstand repeated flexing, bending, and twisting. Because dental materials are subjected to chewing stresses that can result in permanent deformation, high flexural strength is desirable. The behavior of these materials under the action of relatively low but intermittent stresses demonstrates their fatigue resistance [7]. Cracks form when defects in the microstructure of a restoration or specimen are subjected to high or low stresses, and these cracks can lead to material fracture. Hardness is not a material property that can be precisely defined in terms of fundamental mass, length, and time units. A specific measurement procedure is used to determine the value of a hardness property. The depth or area of an indentation left by a specific shape indent with a specific force applied for a specific time is the most common way to determine hardness.

The four most common standard test methods for expressing a material's hardness are Brinell, Rockwell, Vickers, and Knoop. Each of these methods is divided into scales based on the applied load and indenter geometry. Hardness tests are widely used in dentistry and have important applications. A hardness test, for example, can determine how mineralized a dental substrate is [15, 16]. This test can also be used to determine the polymerization level of resin composites and resin cement.

Another property relevant to composite resin is the thermal expansion coefficient, which may influence composite adaptation to cavity walls [17]. When there is a mismatch between the thermal expansion coefficients of composites and enamel and dentin, stresses can be generated at the interfacial bond during exposure to hot and cold food and drinks [18]. Color stability and viscosity are also important properties for composites because they affect the esthetics, handling, and placement of the material, respectively [19].

The complexity of the oral environment, as well as the geometric diversity of cavities filled with restorations, make it difficult to precisely define clinical failure processes and associate routinely measured mechanical properties with dental material performance. It is not an easy task to identify the relevant laboratory tests to predict the clinical performance of restorative material. The clinical difficulties are multifactorial, and there are likely to be significant interactions between the factors. As a result, one must have a thorough understanding of the clinical factors and the magnitude of their impact on long-term performance [13]. Even though mechanical tests have not yet reached the level of clinical simulation, they are an important parameter in analyses, so familiarity with the major laboratory tests for evaluating dental materials is essential [12].

3. Dental materials science: research, testing, and standards

Because they serve different purposes, the materials used in the mouth interact with different tissues and are exposed to different environments, resulting in a wide range of chemistries. Such materials must be manufactured, processed, and tested to be clinically safe. A 'test' is used in the context of a statistical examination of data to determine the probability of an outcome against a formal 'null hypothesis,' and thus potentially to falsify it. The term "test" can be used to characterize a dental material or product in direct comparison with others, but it can also be used to compare with some defined or threshold value of the tested property, and it is in this context that international standards are frequently and correctly used. The term "research" can refer to a very basic approach, such as developing new chemistry or discovering new mechanisms of action. A clear demarcation between the two domains is not always possible [20].

When designing the testing, careful consideration must be given. Test planning for all materials must be meticulous and specific to the material type and use. Blanket or routine testing with no understanding of why the testing is being performed is not cost-effective and does not provide useful information to readers or users of these materials. There are certain interactions of the material being tested with the substrate which can be difficult to simulate in vitro [21]. The context of the material used, location, and exposure are important during laboratory testing and experiment design, and this must be recognized and implemented through simulation for all tests, whether physical, chemical, mechanical, or biological.

The test suite must be comprehensive and integrated to cover all relevant aspects—single-factor work that ignores system complexity and interdependence is rarely helpful and may result in the "salami slicing" of the research [22]. Material characterization and reference to literature about the material's composition are essential in every experimental plan, as is covering theoretical expectations for the subsequent testing. Hence surface testing should also be included in addition to bulk (object) testing because changes and effects may occur only at the surface.

The longevity of the material should also be taken into account since the age at which failure occurs is also important, hence data from surveys of the ages of functional restorations should also be scrutinized. Ideally, longitudinal studies should be performed to study the longevity of a specified population of restorations. The longevity of restorations has been recorded in different ways previously e.g., by noting the percentage of restorations remaining after a specified number of years or by recording the mean or median age. All methods may be useful, but a common method needs to be selected to allow for comparison between and among data from different surveys. The median longevity has been concluded to be the most common way of recording longevity, and it is recommended that this value should always be recorded to allow for comparative estimates [23].

For biological testing, the choice of bacterial strains and cell lineages should be appropriate to the location of material placement, and, whenever possible, both microbiological and biological testing is required to ensure that, while the material may be antimicrobial, it is not toxic to the host [24]. Hence one has to keep in mind to modify or update the currently established protocols for testing newer materials.

4. Failure of restorations

A complex of bacterial cells adhering to one another in an enclosed polymeric extracellular matrix is known as a biofilm [25]. Replacement of the existing restoration and further tooth structure loss are two effects of oral infections brought on by biofilms involving previous or old restorations. Dental caries is caused by pathogenic biofilms that attach to the tooth surface or restorative materials using specific binding proteins [26]. Plaque buildup at the tooth/restoration interface enables microbial invasion and biofilm formation that results in another carious lesion. The recurrence of carious lesions around restorations is influenced by a variety of factors, including dysbiotic biofilm growth, difficult access for cleaning between teeth, and surface characteristics of the dental restorative material [27]. The production of acids by plaque biofilms near the restoration margins is a risk factor for restoration failure [28]. Failure-related restoration replacement accounts for 50–70% of all routine restorative procedures. Many approaches have been investigated to reduce biofilm formation over polymeric restorative materials and at tooth/material interfaces [27]. For example dental composite restorations are the first line of minimally invasive options for the treatment of dental caries in tooth structure However, polymeric materials are highly susceptible to bacterial attachment and colonization, leading to dental diseases. In previous reports, the prevalence of secondary caries associated with polymeric restorative materials has reached 60%, and it has been recognized as the most common reason for resin composite restoration failure and replacement [27].

One strategy to provide a durable, long-lasting restoration is the incorporation of antimicrobial agents to control and/or eliminate these secondary infections. Similarly, any strategy that could disrupt the formation of biofilms would be considered clinically valuable as a route to control infections related to biofilm accumulation [28, 29]. Different classes of agents in the development of antibacterial materials are- Contact based anti-bacterial materials, release-based anti-bacterial materials, dual contact, and release-based anti-bacterial materials, on-demand anti-bacterial materials, materials with bacterial resistant surfaces, and materials with bacterial release surfaces (**Figure 5**) [27].

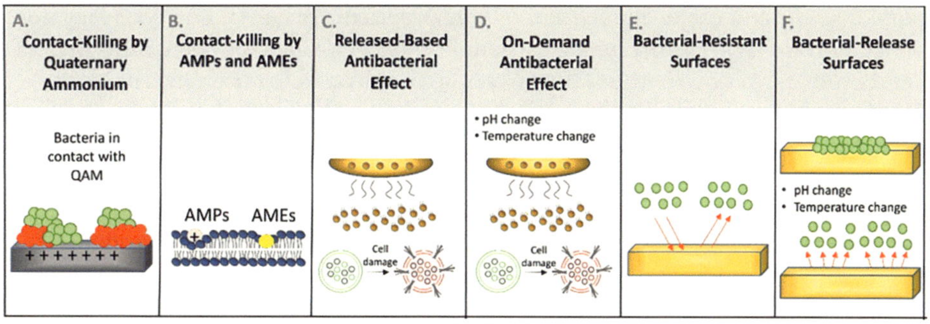

Figure 5.
Antibacterial killing strategies can be achieved via different approaches. The contact-killing mechanism can be provided by quaternary ammonium compounds posing highly positive charged surfaces to disrupt accumulated microorganisms (A). Antimicrobial peptides (AMPs) and antimicrobial enzymes (AMEs) can conduct contact-killing by invading the cellular membrane and targeting the main cellular components (B). Antimicrobial peptides can also conduct antibacterial action via their positively-charged surface. The antibacterial action via ion release can be provided by release-based (C) and on-demand (D) antibacterial materials. Materials interfering with bacterial adhesion can be designed using bacterial-resistance and bacterial-release surfaces (E, F) [30].

5. Incorporation of practice-based groups

Given that the majority of dental treatments in the world are performed in dental offices, there is therefore an imbalance between the treatment output and research output. It can be assumed that dental practice can have an increasing impact on clinical dental research. Since dental practice is "real world", a technique or material must be accepted under practical conditions for it to be successful and hence they should be evaluated in this context [31].

Research methods such as meta-analyses, systematic reviews, or randomized controlled clinical trials are less likely to be used by general dentists, as they require knowledge of statistics. and research methods which are often not within the reach of the physician. The use of practice-based networks to evaluate the effectiveness of materials and techniques in dental practice can be very productive. Several practice-based research groups are presently in operation in the UK and the USA, generally carrying out evaluations of the handling of materials, with increasing emphasis on the clinical evaluation of restorations. PREP (Product Research and Evaluation by Practitioners), BRIDGE (Birmingham Research in Dental General Practice), and GRID (Glasgow Research Initiative in Dental Practice, A West of Scotland-based practitioner research group) are a few of the practice-based groups. Perhaps the best-known group of practice-oriented researchers is the Clinical Research Association (CRA), founded by Gordon Christensen 30 years ago. To undertake research work in dental practice, trainees must be trained in the standardization of procedures, calibration in prosthetic evaluation, and the scientific method. All these groups include general practitioners and they hold annual meetings to propose and discuss ideas for new projects. They perform fact-based assessments of a variety of dental materials, as well as laboratory assessments [32].

Hence randomized clinical trials and retrospective and prospective clinical evaluations which can be readily carried out in dental practice, where the patient base is likely to be substantially greater than in dental schools or hospitals can prove to be very helpful. Moreover, the patient base in dental practice represents patients of various patterns of attendance, from different walks of life, and different levels of oral hygiene and caries experience. Whereas patients in dental hospitals may not be considered to represent a typical patient population, since they generally elect to attend such institutions because they have the time available for treatment by students (i.e., the retired or unemployed) and may often have attended many courses of treatment, in which their oral hygiene will be reinforced (i.e., their oral hygiene may be better than that in the general population).

Cross-sectional studies have also been used in general dental practice to produce useful and meaningful results, with Ivar Mjör being the main proponent of this research approach [33]. These cross-sectional studies have the advantage over randomized controlled clinical trials in terms of generating data that may include large numbers of recoveries in many different patients, performed by dentists with many different qualifications and experience. A recent example of this type of study examined the impact of different UK rehabilitative care funding methods on the age of recovery at replacement, with rehabilitations placed in the National Health Service being replaced at a lower age than the restoratives placed in other funding methods [34].

It is essential to follow an established protocol to standardize and eliminate any deviations. Board members must use the materials as directed. This way, the manufacturer can get feedback on the company's material handling from this particular group of practitioners. Members can place the restoration under typical routine conditions, perform a baseline assessment, and arrange to recall the patient for review by a trained and calibrated independent assessor who will review the restoration using the modified United States Public Health Service criteria. Patients and practitioners can be reimbursed for their costs, and the time the practitioner's office is inactive during a patient examination can be considered the primary expense incurred by the practitioner [32].

6. Conclusions

The way dentistry is practiced worldwide is changing as a result of the daily advances made in science. Without scientific research, clinical trials to evaluate the benefits and drawbacks of particular materials and methods would not be feasible, and we might end up using therapies that are ineffective and potentially harmful. Since dental well-being is crucial for general health, maintaining national and international standards for dental health depends on the development of new treatments and medications. The creation of a new generation of enhanced biomaterials with the potential to change how dental caries are currently managed has a bright future.

Conflict of interest

The author declares no conflict of interest.

DOI: http://dx.doi.org/10.5772/intechopen.109741

Author details

Tanvi Satpute
M.G.M. Dental College and Hospital, Navi Mumbai, India

*Address all correspondence to: tanvisatpute@gmail.com

References

[1] ADA Council on Scientific Affairs. Direct and indirect restorative materials. Journal of the American Dental Association 1939. 2003;**134**(4):463-472

[2] Manhart J, García-Godoy F, Hickel R. Direct posterior restorations: Clinical results and new developments. Dental Clinics of North America. 2002;**46**(2):303-339

[3] Yap AUJ, Teoh SH, Chew CL. Effects of cyclic loading on occlusal contact area wear of composite restoratives. Dental Materials: Official Publication of the Academy of Dental Materials. 2002;**18**(2):149-158

[4] Leinfelder KF. Do restorations made of amalgam outlast those made of resin-based composite? Journal of the American Dental Association (1939). 2000;**131**(8):1186-1187

[5] Perez Cdos R, Gonzalez MR, Prado NA, de Miranda MS, Macêdo Mde A, Fernandes BM. Restoration of noncarious cervical lesions: When, why, and how. International Journal of Dentistry. 2012;**2012**:687058

[6] Scott JM, Mahoney EK. Restoring proximal lesions in the primary dentition: is glass ionomer cement the material of choice? The New Zealand Dental Journal. 2003;**99**(3):65-71

[7] Mitra SB. Adhesion to dentin and physical properties of a light-cured glass-ionomer liner/base. Journal of Dental Research. 1991;**70**(1):72-74

[8] Boksman L, Pameijer CH, Broome CJ. The clinical significance of mechanical properties in retentive posts. Compendium of Continuing Education in Dentistry Supplement. Jun 2013;**34**(6):446-455

[9] Quinn JB, Sundar V, Lloyd IK. Influence of microstructure and chemistry on the fracture toughness of dental ceramics. Dental Materials. 2003;**19**(7):603-611

[10] Craig RG. Restorative Dental Materials. St Louis: Mosby: Elsevier; 1997. pp. 56-103

[11] Darvell BW. Materials Science for Dentistry. 10th ed. United Kingdom: Woodhead Publishing; 2018. pp. 1-34

[12] Wang L, D'Alpino PHP, Lopes LG, Pereira JC. Mechanical properties of dental restorative materials: Relative contribution of laboratory tests. Journal of Applied Oral Science. 2003;**11**(3):162-167

[13] Cattani-Lorente MA, Dupuis V, Moya F, Payan J, Meyer JM. Comparative study of the physical properties of a polyacid-modified composite resin and a resin-modified glass ionomer cement. Dental Materials: Official Publication of the Academy of Dental Materials. 1999;**15**(1):21-32

[14] Anusavice KJ. Recent developments in restorative dental ceramics. Journal of the American Dental Association (1939). 1993;**124**(2):72-74, 76-78, 80-4

[15] Pereira PN, Inokoshi S, Yamada T, Tagami J. Microhardness of in vitro caries inhibition zone adjacent to conventional and resin-modified glass ionomer cements. Dental Materials: Official Publication of the Academy of Dental Materials. 1998;**14**(3):179-185

[16] Shinkai RS, Cury AA, Cury JA. In vitro evaluation of secondary caries development in enamel and root dentin around luted metallic restoration. Operative Dentistry. 2001;**26**(1):52-59

[17] Fróes-Salgado NR, Silva LM, Kawano Y, Francci C, Reis A, Loguercio AD. Composite pre-heating: Effects on marginal adaptation, degree of conversion and mechanical properties. Dental Materials. 2010;**26**(9):908-914

[18] Çelik Köycü B, İmirzalıoğlu P. Heat transfer and thermal stress analysis of a mandibular molar tooth restored by different indirect restorations using a three-dimensional finite element method: Heat transfer and thermal stresses at indirect restorations. Journal of Prosthodontics. 2017;**26**(5):460-473

[19] Sarrett D. Clinical challenges and the relevance of materials testing for posterior composite restorations. Dental Materials. 2005;**21**(1):9-20

[20] Schmalz G, Watts DC, Darvell BW. Dental materials science: Research, testing and standards. Dental Materials. 2021;**37**(3):379-381

[21] Bohner M, Lemaitre J. Can bioactivity be tested in vitro with SBF solution? Biomaterials. 2009;**30**(12):2175-2179

[22] Camilleri J. Materials for dentistry—Raising the bar. Frontiers in Dental Medicine. 2020;**1**:7

[23] Mjör IA. Problems and benefits associated with restorative materials: Side-effects and long-term cost. Advances in Dental Research. 1992;**6**:7-16

[24] Camilleri J, Arias Moliz T, Bettencourt A, Costa J, Martins F, Rabadijeva D, et al. Standardization of antimicrobial testing of dental devices. Dental Materials: Official Publication of the Academy of Dental Materials. 2020;**36**(3):e59-e73

[25] Neelakantan P, Romero M, Vera J, Daood U, Khan A, Yan A, et al. Biofilms in endodontics—Current status and future directions. International Journal of Molecular Sciences. 2017;**18**(8):1748

[26] Balhaddad AA, Melo MAS, Gregory RL. Inhibition of nicotine-induced *Streptococcus mutans* biofilm formation by salts solutions intended for mouthrinses. Restorative Dentistry & Endodontics. 2019;**44**(1):e4

[27] Mitwalli H, Alsahafi R, Balhaddad AA, Weir MD, Xu HHK, Melo MAS. Emerging contact-killing antibacterial strategies for developing anti-biofilm dental polymeric restorative materials. Bioengineering. 2020;**7**(3):83

[28] Nedeljkovic I, Teughels W, De Munck J, Van Meerbeek B, Van Landuyt KL. Is secondary caries with composites a material-based problem? Dental Materials : Official Publication of the Academy of Dental Materials. 2015;**31**(11):e247-e277

[29] Melo MAS, Guedes SFF, Xu HHK, Rodrigues LKA. Nanotechnology-based restorative materials for dental caries management. Trends in Biotechnology. 2013;**31**(8):459-467

[30] Francois P, Fouquet V, Attal JP, Dursun E. Commercially available fluoride-releasing restorative materials: A review and a proposal for classification. Materials. 2020;**13**(10):2313

[31] Mjör IA, Wilson NH. General dental practice: The missing link in dental research. Journal of Dental Research. 1997;**76**(4):820-821

[32] Burke FJT. Evaluating restorative materials and procedures in dental practice. Advances in Dental Research. 2005;**18**(3):46-49

[33] Deligeorgi V, Mjör IA, Wilson NH. An overview of reasons for the placement and replacement of restorations. Primary

Dental Care : Journal of the Faculty of General Dental Practitioners (UK). 2001;**8**(1):5-11

[34] Burke FJT, Wilson NHF, Cheung SW, Mjör IA. Influence of the method of funding on the age of failed restorations in general dental practice in the UK. British Dental Journal. 2002;**192**(12):699-702

Chapter 20

Human Teeth Is Useful Even after Its SHED! So, Why Discard It?

Meghna Bhandary, Rachaita Chhabra, K. Ananya Rao and Mohammed Shahid

Abstract

A few decades ago, if one underwent a knee injury that makes walking painful or had an atrophied kidney, then, he/she was condemned to a life hooked on to machines, or on constant medications. However, in today's era, teeth can be grown in a Petri dish; heart and liver replacements are possible with no risk of rejection because the organs are made of the patient's own cells. This is the promise of regenerative medicine and tissue engineering. The entire idea of regenerative medicine is based on the presence of stem cells in the body or the ability to introduce stem cells into the body without causing harm. These can be obtained from a variety of body and dental tissues. Deciduous teeth often discarded as biological waste is proven to possess Stem cells (SHED) that have promising applications in tissue engineering and regenerative medicine. Hence, their contribution toward the field of regenerative medicine and dentistry is immense. This chapter summarizes SHED's regenerative potentials and therapeutic applications; and also focuses on its potential future scope in regenerative dentistry. Furthermore, procedures involved in SHED-induced therapy, from SHED collection to SHED banking, have also been explained.

Keywords: stem cells, stem cell from exfoliated deciduous teeth (SHED), tooth, regeneration, repair, dentin pulp regeneration, therapeutic, biodentine

1. Introduction

> *"The Regenerative Medicine revolution is upon us. Like iron and steel to the industrial revolution, like the microchip to the tech revolution, stem cells will be the driving force of this next revolution".*
>
> *-Cade Hildreth.*

Staggering progress in the field of regenerative medicine has sowed the seeds of cell-based therapies for various diseases which cannot be cured by conventional methods. Stem cell therapy deals with the functional revival of specific tissue and/or organ in patients who are suffering from severe injuries / chronic diseased conditions, in a state where the body's own regeneration feedback is not satisfactory.

The entire argument of regenerative medicine is built on the presence of stem cells in the body, or the ability to institute stem cells into the body without causing harm. Given that stem cells can be obtained from a variety of sources, the search for an ideal source that offers excellent therapeutic potential while requiring less invasiveness and immune rejection is unending. Even a tooth which is naturally discarded can be used as a great source for stem cells. Hence a better understanding of the nature and mechanism of stem cell is crucial for their application in cell-based therapy.

1.1 What are stem cells and why is stem cell therapy so much of interest?

Our bodies are the ultimate factory. Every cell has its own function to play, and the fate of each cell is determined at the embryo stage which then cannot be changed. However, the discovery of stem cells has paved the way for regenerative medicine. Stem cells are those immature cells which can differentiate into any type of cells as they are not specialized [1]. Therefore, they can be used in the repair and regeneration of dysfunctional tissues. For instance, they can help treat neurological diseases by making new brain cells to treat people with Parkinson's disease, or they could be used to repair the damaged immune system, or even reverse paralysis/regrow lost limbs.

Therefore, stem cell research can help to:

- Understand how diseases and ailments manifest by observing the maturation of stem cells into cells of bone, heart muscle, neurons, and other organs and tissues.

- Stem cells can be directed to differentiate into specific cells capable of regenerating and repairing damaged or diseased human tissues (regenerative medicine). Hence, Stem cell therapy may benefit people with spinal cord injuries, type 1 diabetes, Parkinson's disease, Battens disease, Amyotrophic lateral sclerosis, Alzheimer's disease, heart disease, stroke, burns, cancer, and osteoarthritis [2].

- Scientists can employ some types of stem cells to evaluate the drugs' quality and safety before administering investigational medications to humans. This form of testing will most likely immediately affect medication development for cardiac toxicity testing first. New fields of research examine the efficacy of employing human stem cells that have been engineered to differentiate into tissue-specific cells for testing new medications. For the testing of novel medications to be accurate, the cells must be programmed to acquire the characteristics of the type of cells the drug is designed to target. For example, nerve cells could be generated to test a novel medicine for a nerve disorder. Tests could determine if the new medicine had any effect on the cells or if they were damaged.

2. Classification of stem cells

Depending on the origin/source of the stem cells, stem cells are divided into various types;

2.1 Embryonic stem cells (ESCs)

They are pluripotent stem cells derived from the blastocyst's inner cell mass. The blastocyst stage, with 50–150 cells, occurs 4–5 days after conception. Embryonic stem cells are able to develop into any type of cell, except those of the placenta (**Figure 1**).

ESCs derived from mouse blastocysts have been studied for 2 decades and shown to differentiate into various cell types including fat cells, brain, nerve, insulin-producing cells of the pancreas, bone cells, endothelial cells, and heart muscle cells [3]. Human ESCs under appropriate culture conditions have demonstrated remarkable abilities to self-renew and produce multipotent cells. The cells are studied extensively in the treatment of diabetes, heart diseases, genetic disorders spinal cord injury, muscular dystrophy, heart illness, and vision/hearing loss. However, it possesses the risk of developing adverse effects such as tumors and unwanted immune responses. Hence the scope of ESCs is still under debate and research to understand how to prevent the rejection of transplanted cells is fundamental.

2.2 Adult stem cells/ somatic or tissue-specific stem cells (ASCs)

ASCs, also called somatic stem cells, are undifferentiated cells that can self-renew and generate all the cell types of the organ from which they originate. An adult stem cell is derived from adult tissue samples. Hence, their use in research and therapy is not considered to be controversial unlike embryonic stem cells derived from embryos.

ASCs are the gold standard for clinical applications and are being tested and accepted for a growing number of conditions. ASCs have been shown to have therapeutic benefits in clinical trials and progress towards fully tested and approved treatments. Phase I/II trials conducted suggest potential cardiovascular benefits from bone

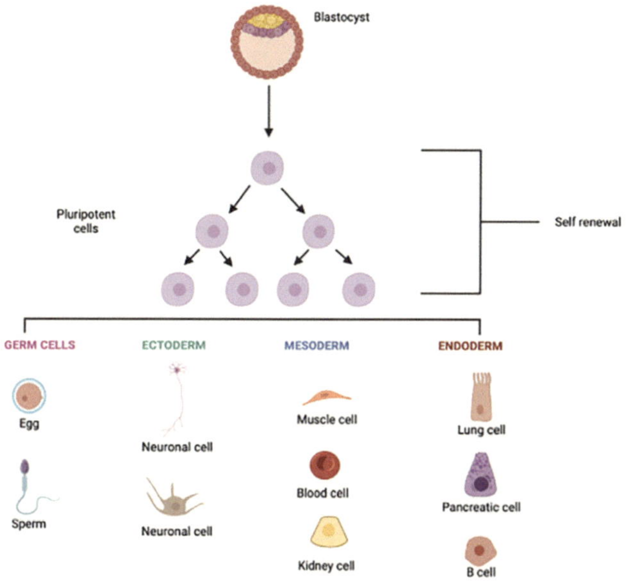

Figure 1.
Human embryonic stem cells differentiation. Image source: Biorender.com.

marrow–derived adult stem cells and umbilical cord blood–derived cells. Striking results have been reported using adult stem cells to treat neurological conditions, including chronic stroke. Positive long-term progression-free outcomes have been seen, including some remission for multiple sclerosis, as well as benefits in early trials for patients with type I diabetes mellitus and spinal cord injury. ASCs are also being used as vehicles for genetic therapies, such as for epidermolysis bullosa. One of the limitations of ASCs includes that they cannot be manipulated to produce all cell types, which limits their use in treating diseases (**Figure 2**).

2.3 Induced pluripotent stem cells (iPSCs)

The limitations in ASCs led to the creation of novel pluripotent cells termed induced pluripotent cells from the adult cells by the process of reprogramming the genes. ASCs can be fused with embryonic stem cells to generate induced pluripotent stem cells. Other somatic cells can also be altered to become pluripotent. iPSCs can differentiate from ESCs. Their gene expression and chromatin differ from ESCs. These cells are important because they may be utilized to create cells from almost all organs for each patient in therapeutic treatment. Besides, they also prevent the use of more ESCs which might cause ethical issues. It also helps to study new genetic diseases by generating iPSCs from their adult or somatic cells (**Figure 3**).

Hepatocyte–like cell derivatives, dendritic cells, macrophages, insulin–producing cell clusters similar to the duodenal islets of Langerhans, and hematopoietic and endothelial cells are currently produced from murine and human iPSCs, in addition to the already listed types of differentiated cells [4–7]. Reprogrammed iPSCs derived from peripheral blood cells could effectively develop into hematopoietic lineage cells [8]. Human β cell-derived iPSCs possess epigenetic memory and may develop into

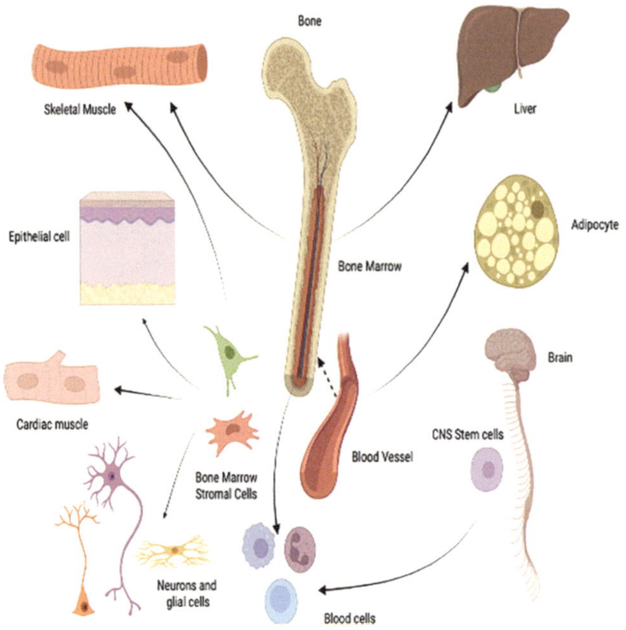

Figure 2.
Sources of adult stem cells. Image source: Biorender.com.

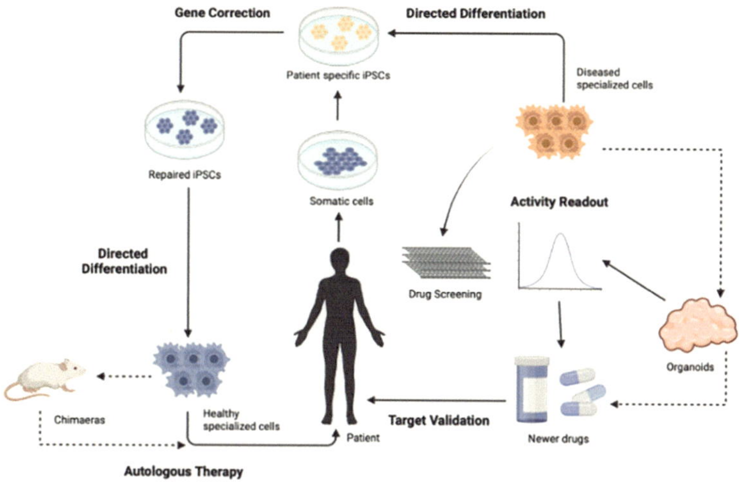

Figure 3.
Evolution of induced pluripotent stem cells. Image source: Biorender.com.

insulin-producing cells more readily [9]. Dopamine and motor neurons can also be produced from human iPSCs by directed differentiation *in vitro* [10, 11].

2.4 Mesenchymal stem cells (MSCs)

MSCs are a type of adult stem cell that can develop into mesodermal (osteocytes, adipocytes, and chondrocytes), ectodermal (neurocytes), and endodermal cell lines (hepatocytes). Some of the potent sources of MSCs include bone marrow, adipose tissue, synovial fluid, umbilical cord tissue, peripheral blood, placental tissue and dental pulp. MSCs can be extracted readily and yield more than other stem cells, making them beneficial for cell proliferation, differentiation, and tissue regeneration under severe immunological circumstances. These also have immunomodulatory features as they secrete cytokines and immune receptors, which regulate the microenvironment in the host tissue. MSCs can treat chronic diseases by producing cells of diverse cell lines, immunomodulating, and secreting anti-inflammatory chemicals. Thereby, showing promising results in preclinical studies for various medical conditions. Research continues to explore their potential in regenerative medicine (**Figure 4**).

MSCs have been studied for a wide range of therapeutic applications, including tissue repair, regenerative medicine, and cell-based therapies for various medical conditions. Some of the medical conditions that MSCs have been studied for include:

1. Myocardial infarct ion: MSCs have been studied for their potential to promote heart tissue repair following a heart attack [12].

2. Spinal cord injury: MSCs have been investigated for their potential to promote the repair of damaged nerve tissue and spinal cord injury [13].

3. Autoimmune diseases: MSCs have been demonstrated to have anti-inflammatory and immunomodulatory properties, which may make them helpful in treating autoimmune diseases, such as multiple sclerosis and lupus [14].

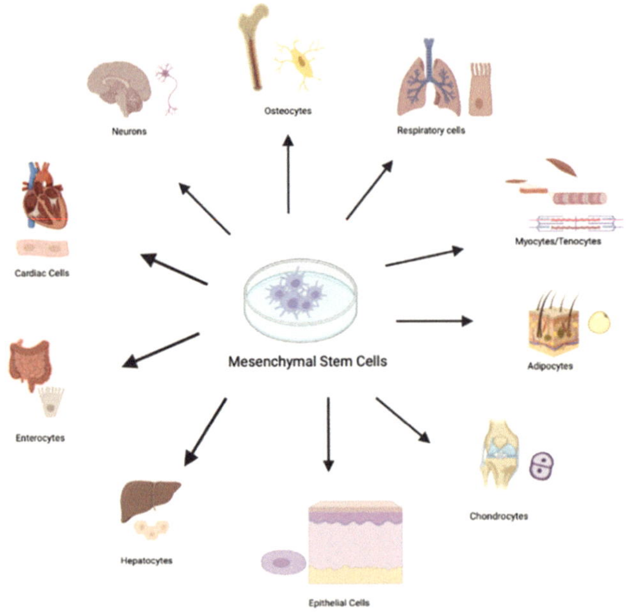

Figure 4.
Regenerating abilities of mesenchymal stem cells. Image source: Biorender.com.

4. Type 1 diabetes: MSCs have been examined for their potential to help preserve insulin-producing cells in type 1 diabetes patients

5. Lung diseases: MSCs have been reviewed for their potential to help repair lung tissue in conditions such as chronic obstructive pulmonary disease (COPD) and acute respiratory distress syndrome (ARDS).

6. Multiple Sclerosis (MS): MSCs have been shown to have anti-inflammatory and immunomodulatory properties, which may make them helpful in treating MS. This autoimmune disorder affects the central nervous system.

7. Lyme disease: MSCs have been analyzed for their potential to help repair tissue damage and reduce inflammation caused by Lyme disease, a bacterial infection transmitted by ticks.

8. Parkinson's disease: MSCs have been examined for their potential to help protect and repair nerve cells in the brain that are damaged in Parkinson's disease, a degenerative disorder that affects movement.

9. ALS (Amyotrophic lateral sclerosis): MSCs have been investigated for their potential to help protect and repair nerve cells in the spinal cord damaged in ALS, a progressive neurodegenerative disease that affects nerve cells in the brain and spinal cord [15].

10. Rheumatoid arthritis: MSC-based therapies via administration of exogenous MSCs or targeting of the endogenous MSCs in the joint are strategies that are being pursued to trigger/enhance repair of the damaged joint tissues, with the aim to restore joint homeostasis.

11. Osteoarthritis: Intra-articular injection of infrapatellar fat pad-derived mesenchymal stem cells is effective for reducing pain and improving knee function in patients being treated for knee osteoarthritis [16].

Various research on adult stem cells led to their discovery in different dental tissues. Stem cells extracted from dental tissue have been shown to possess similar properties to MSCs derived from other sources. Hence, considering dental stem cells are easily accessible. Currently, there is extensive research focusing on dental stem cells and their clinical applications.

2.5 Dental stem cells

Dental stem cells offer a very promising therapeutic approach to restoring structural defects. To date, eight unique populations of dental tissue-derived MSCs have been isolated and characterized. Postnatal dental pulp stem cells **(DPSCs)** were the first human dental MSCs to be identified from pulp tissue. Gradually, other dental MSC-like populations were also reported (**Figure 5**).

- **Dental Pulp Stem Cells (DPSCs):** The first dental MSCs from the dental pulp tissue of impacted third molars were isolated two decades ago [17]. These adhering cells are fibroblast-like and MSC-like [18]. They are valuable cells in regenerative medicine because of their strong proliferation capacity and multi-lineage differentiation potential [19].

- **Periodontal Ligament Stem Cells (PDLSCs):** PDL is a specialized issue located between the cementum and alveolar bone. Fibers of PDL, attach the tooth to the jaw and plays a vital role in maintaining and supporting the teeth. PDLSCs may be recovered from dental roots or the perivascular space of the periodontium. They possess characteristics of MSCs to self-renew and develop into cementum, PDL, alveolar bone, peripheral nerves and blood vessels. In vitro PDLSCs were able to differentiate into osteogenic, adipogenic and chondrogenic cells [20].

- **Dental Follicle Stem Cells (DFSCs):** The dental follicle surrounds the tooth germ before the eruption [21]. It has progenitor cells of the periodontal ligament,

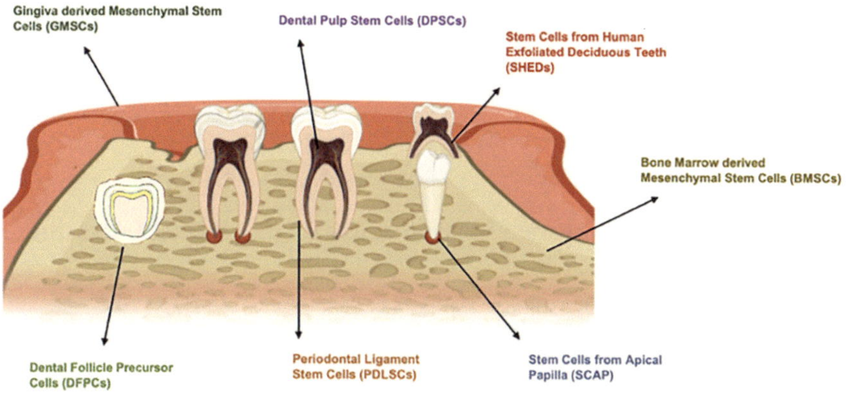

Figure 5.
Sources of dental stem cells. Image source: Biorender.com.

cementoblast, and osteoblast [20]. DFSCs are isolated from 3rd molars and expressed various biomarkers such as Notch 1, STRO-1 and Nestin. They demonstrated multilineage potential to undergo osteogenic, adipogenic, and neurogenic potential in vitro.

- **Alveolar Bone-derived MSCs (ABMSCs):** Alveolar bone is embryonically produced from the dental follicle and resembles a thickened ridge with tooth sockets [21]. ABMSCs show a similar osteogenic differentiation capacity to BMMSCs but lower chondrocyte and adipocyte differentiation [22]. They express surface markers CD73, CD90, CD105, and STRO-1 but do not express hematopoietic markers CD14, CD34 and CD45 [21]. They are used along with bioceramics scaffolds for bone tissue engineering applications.

- **Stem Cells from Apical Papilla (SCAP):** Stem cells from apical papilla (SCAP) from incompletely developed teeth were extracted in 2006 [23]. The apical papilla is loosely attached to the apices of immature permanent teeth and has fewer cells and vessels than pulp tissue [24]. These stem cells have higher proliferative potential than PDLSCs and DPSCs [25], self-renewal ability, and low immunogenicity. After implantation of SCAP into immunocompromised mice, in a carrier matrix, due to the presence of odontoblast-like cells typical dentin pulp-like structure was formed. In regenerative dentistry, SCAPs can generate osteogenic, odontogenic, neurogenic, adipogenic, and chondrogenic cells [26].

- **Tooth Germ Progenitor Cells (TGPCs):** TGSCs are found in the dental mesenchyme of the third molar tooth germ at the late bell stage [21]. TGSCs have similar multilineage differentiation capacity to other dental MSCs like differentiating into osteoblast/ odontoblast, chondrocytes, and neurons. TGPCs can differentiate into cells with morphological, phenotypic and functional characteristics of hepatocytes in vitro suggesting that TGPCs can be used to treat liver diseases [27, 28].

- **Gingival MSCs (GMSCs):** Gingival-derived mesenchymal stem cells (GMSCs) originated in the spinous layer of the human gingiva. They are multipotent, self-renewing, and immunomodulatory [29]. Regenerative dentistry uses gingiva stem cells because it is easily accessible during dental operations [20].

- **Stem Cells from Human Exfoliated Deciduous teeth (SHED):** SHED is a unique source of stem cells as deciduous teeth are usually discarded as biological waste, they can be readily accessible without any invasive procedures. Over the last decade, SHED have been identified to be highly proliferative, clonogenic cells capable of differentiating into a variety of cell types including neural cells, adipocytes, and odontoblasts [30].

3. Why there is such an interest in stem cells from exfoliated deciduous teeth?

Pulp from naturally exfoliated deciduous teeth may be like an umbilical cord providing a rich and distinctive source of stem cells showing a multipotent nature. SHEDs exhibit a much higher proliferation rate, faster population doublings and greater osteoinductive capacity than DPSCs, adult MSCs and PDLSCs. They can differentiate into a variety of cell types including odontoblasts, osteoblasts, adipocytes, chondrocytes, neural cells, hepatocytes, endothelial cells, β-cells, and dentin and pulp-like tissues. SHEDs express the same cell markers as ESCs such as OCT 4 and NONOG, which makes them have a significant impact on clinical applications. Evidence indicates that functional recovery and remodeling in lesions not only rely on their multipotency but also on their protective and anti-inflammatory action by the paracrine mechanism of grafted SHEDs. In this context, SHEDs have shown to function as an immunomodulator by suppressing T helper 17 cell functions. Transforming growth factors TGF-β1 and β2, fibroblast growth factor FGF-2, and Col I and III are highly expressed in SHED as compared to DPSCs.

The primary difference in the pulp of primary and permanent teeth is the occurrence of physiologic root resorption of the deciduous teeth. The transition from deciduous teeth to permanent teeth is a unique and dynamic process wherein the resorption of the deciduous teeth is coordinated with the development and eruption of permanent teeth. Deciduous teeth without any visible root resorption were unable to proliferate in vitro, whereas those in an advanced state of root resorption showed good proliferation and differentiation potential [31]. Due to its unique stemness of capability of multi-differentiation, self-renewal, developing into other cell lineages and easy accessibility, without major morbidity to host and minimal ethical concern, SHED has been widely investigated in the field of regenerative medicine and tissue engineering (**Figure 6**).

4. Applications of SHED in research

Owing to its multipotent, no/reduced ethical conflicts and minimally invasive to obtain has opened a wide area for research. Research spans across categories—Dental materials, wound healing, dental tissue engineering, treatment of chronic diseases like diabetes, and Wilson's disease, treatment of autoimmune diseases like SLE, and encephalomyelitis, adjunct to surgical treatment e.g., cleft lip and palate, pediatric surgeries like biliary atresia.

4.1 Cell culture studies

- The proliferative, osteogenic, and immunomodulatory potentials of MSCs isolated from the dental pulp of SHED and fragments of the orbicularis oris muscle (OOMDSCs) treated with an inflammatory IFN- γ stimulus were evaluated, and it was determined that SHED and OOMDSCs lack immunogenicity and have immunomodulatory properties that are enhanced by inflammatory stimulation with IFN- γ. This opens new perspectives for the therapeutic use of these cells [32].

- Response of stem cells from human exfoliated deciduous teeth (SHED) to three bioinductive materials conducted in vitro determined functional differentiation

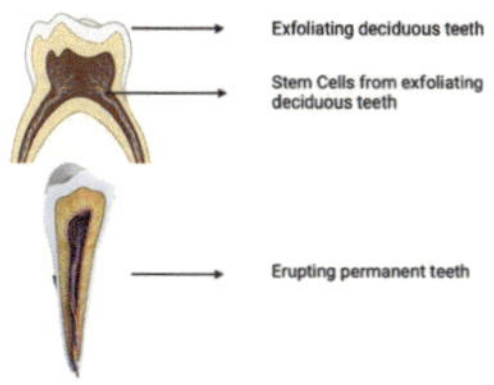

Figure 6.
Exfoliating deciduous teeth with root resorption possess SHED. Image source: Biorender.com.

potential (osteogenic/odontogenic) of various biomaterials on SHED and concluded that all the tested materials are bioinductive to SHED. Enamel Matrix Derivative (EMD) can be used in dentistry for various vital pulp therapies as that of Biodentine and Mineral Trioxide Aggregate (MTA) with predictable as well as enhanced success rates [33].

- Biological interactions of a calcium silicate-based cement (Biodentine™) with SHED were studied. SHED attached effectively to the crystalline surface of Biodentine specimens, exhibiting a spindle-shaped phenotype. Different concentration and time-dependent expression patterns of odontogenic genes were observed under non-inductive and inductive (osteogenic) conditions, with significant upregulation of DSPP and Runx2, BMP-2, BGLAP, and MSX. A gradual increase in the development of mineralized tissue was observed [34].

- Cytotoxicity and bioactivity of various pulpotomy materials on stem cells from human exfoliated primary teeth were investigated and SHEDs showed significant cell viability in the presence of Biodentine when compared to other materials. In addition, SHEDs maintained their mesenchymal phenotype in all conditions although their capacity to migrate was higher in the presence of Biodentine. Cytotoxicity and bioactivity of various pulpotomy materials on stem cells from human exfoliated primary teeth [35].

- The effects of MTA, Biodentine™, and calcium hydroxide on the viability, proliferation, migration, and differentiation of stem cells from human exfoliated deciduous teeth were investigated. The results demonstrated that the three capping materials are biocompatible, maintain viability, and stimulate proliferation, migration, and differentiation in a key dental stem cell population [36].

4.2 Animal studies

- SHED investigated for wound healing effect on mouse model revealed enhanced wound healing promotion [37].

- SHED conditioned medium Ameliorated Experimental Autoimmune Encephalomyelitis in a mouse model of multiple sclerosis [38].

- SHED teeth reduce tissue-infiltrating inflammatory cells, improving clinical signs in experimental autoimmune encephalomyelitis. SHED infusion improved EAE clinical score by reducing the number of tissue-infiltrating IFN-γ + CD8+, IL-4 + CD8+, IFN-γ + CD4+, and IL-4 + CD4+ T cells in the central nervous system (CNS). In addition, SHED can modify CD4+ T cell responses in the periphery, indicating that SHED may be investigated as a component of cellular treatment for autoimmune illnesses related to the CNS [39].

- SHED transplantation into the caudal vein of Diabetic Goto-Kakizaki (GK) rats revealed improved nerve function, thereby alleviating persistent neuropathic pain (mechanical hyperalgesia). SHED transplantation can prevent the development of DPN by participating in tissue regeneration, increasing local blood flow, and conferring neurotrophic protection [40].

- SHED and SHED-converted hepatocyte-like- cells (SHED-Heps) were transplanted into Wilson's disease model Atp7b-mutated Long-Evans Cinnamon (LEC) rats, reduced copper-induced oxidative stress via ATP7B- independent stanniocalcin 1 secretion, suggesting a possible role for paracrine effect. Therefore, SHED-Heps can be a novel effective source to rescue and prevent Wilson's Disease with fulminant hepatic failure by restoring the deficient ATB7B function and diminishing copper-induced oxidative stress [41].

- Intravenous infusions of SHED can effectively alleviate the autistic-like symptoms of impaired social novelty preference (SNP) and obvious social stress in SHANK3 mutant beagle dogs accompanied by an increase in the level of serum IL-10 and a decrease in the level of IFN-γ [42].

- SHED transplanted in carbon tetrachloride (CCl4)-induced liver fibrosis model mice directly transformed into hepatocytes without cell fusion and improved hepatic dysfunction [43].

- The application of implants pre-adhered with SHEDs improved and accelerated early osseointegration around the implant with an improved total bone-to-implant contact (BIC%) and interthread bone, resulting in thicker and denser trabecular bone in adult beagle dogs [44].

5. Therapeutical applications of SHED

It has been found that SHEDs showed alleviating effects on nervous system diseases, including Spinal cord injury, Parkinson's disease, Trigeminal neuralgia, Cerebral ischemia, Alzheimer's disease, and Encephalomyelitis. Owing to the capacity to interact with the local inflammatory microenvironment, SHEDs have also embraced remarkable modulatory effects in various autoimmune and inflammatory diseases such as rheumatoid arthritis, diabetes, acute kidney injury, liver fibrosis/ acute liver failure osteoarthritic, heatstroke, and acute respiratory distress syndrome (ARDS) could also benefit from SHEDs for the protective effects underlying immunomodulatory activities.

5.1 Evidence supporting SHEDs potential in autoimmune and nervous diseases via paracrine and immunomodulatory

Effects of SHED on Parkinson's disease: Transplantation of neural-like spheres derived from SHEDs into the striatum of parkinsonian rats significantly improved the behavioral disorders, the number of TH-positive (tyrosine hydroxylase) cells and the protective effect on endogenous dopaminergic neurons, indicating SHED spheres were of potential therapeutic value [45].

Effects of SHED on Acute liver failure: Intravenous administration of SHED-CM improved the condition of the injured liver and the animals' survival rate by induction of anti-inflammatory M2-like hepatic macrophages [46].

Effects of SHED on Heatstroke: Intravenous administration of SHED exhibited therapeutic benefits for heatstroke in mice, related to decreased inflammatory response, decreased oxidative stress, and increased hypothalamic pituitary adrenocortical axis activity.

Effects of SHEDs in the treatment of retinal degeneration: It has been confirmed through paracrine secreta that SHEDs exert neurotrophic, angiogenic, immunoregulatory, and antiapoptotic functions in injured tissues. SHEDs and SHED-CM showed therapeutic effects on Retinitis pigmentosa (RP) by improving retinal visual function and delaying the degeneration of photoreceptors by antiapoptotic activity. Therefore, SHEDs may be a promising stem cell source for treating retinal degeneration [47].

5.2 Other therapeutic effects of SHED

- Hair Regeneration: There is increasing evidence that mesenchymal-epithelial interactions in early morphogenesis stages of both tooth and hair follicles show many similarities. In vitro, SHED shortened the hair regeneration cycle and promoted the proliferation and aggregation of dermal cells. When epidermal and dermal cells were freshly extracted and co-cultured with SHED, several signaling molecules in hair follicle regeneration were detected and it was found that the expression of Sonic Hedgehog (Shh) and Glioma-associated oncogene 1 (Gli1) was up-regulated. It seems that SHED may boost the prosperity of hairs by increasing Shh/Gli1 pathway [48].

- Kidney Injury: SHED therapy is useful for ischemic kidney damage. *In vitro* studies showed that SHED significantly could reduce MCP-1 secretion in tubular epithelial cells caused by H_2O_2 and hence could be utilized for acute kidney injury [49].

- On wound healing and wound itching: SHED along with basic fibroblast growth factor (b-FGF) in a nude mouse full-thickness skin defect model significantly accelerated wound healing. SHED-derived exosomes were investigated for their contribution to immune response and wound itching during healing. The effects of SHED-derived exosomes on inflammatory wound healing were examined using lipopolysaccharide (LPS)-induced wounds in a mouse model. SHED-derived exosomes facilitated LPS-induced wound closure and relieved wound itching. SHED-derived exosomes containing miR-1246 also enhanced autophagy by regulating macrophage function through the AKT, ERK1/2, and STAT3 signaling pathways. Therefore, SHED-derived exosomes promote wound healing with less itching in an LPS-induced wound model by stimulating macrophage autophagy, which has implications for the treatment of inflammatory wound healing [50].

- On Photo aging: SHED were investigated for its effect on wrinkles caused by UV-B photodamage, SHED or SHED-conditioned medium injected subcutaneously reduced the wrinkles compared with the control group. In addition, SHED had effects on human dermal fibroblasts (HDFs) by increasing collagen synthesis and by activating the proliferation and migration activity of HDFs, suggesting that SHED or SHED conditioned medium can be used for the treatment of photoaging [37].

- On Type II Diabetes Mellitus: SHED administration reduced Glycosylated serum albumin and hemoglobin level significantly. Further research revealed there was a low reply to SHED administration in hypercholesterolemia and low C-peptide. As serum lipid level and baseline islet activity are major factors for treatment in Type 2 DM patients, it can be concluded that SHED administration is an assured and efficient treatment for islet activity and glucose metabolism recovery in T2DM patients [51].

- On Liver Impairment: Liver transplantation is an end treatment for incurable liver involvement. Stem cells are important as a suitable cell source for liver renewal. SHED administration significantly improved liver disorders and caused anti-fibrotic and anti-inflammatory influences. SHED could directly transform into hepatocytes and be suitable for liver renewal. SHED administration and bio three-dimensional printers that can produce scaffold-free three-dimensional images of the liver and diaphragm are innovative regenerative medicine treatments for uncontrolled pediatric surgery, such as biliary atresia and diaphragmatic hernia [43].

- On nerve impairment: Numerous common therapeutic applications of SHEDs have been identified, although their emphasis has been on neuroprotection rather than neuroregeneration. SHEDs and their medium may impact neural disorders by multiple mechanisms, including cell replacement, paracrine effects, angiogenesis, synaptogenesis, immunomodulation, and apoptosis inhibition. SHED-exos is a suitable regeneration agent for neuronal disorders [52]. SHED transplantation can inhibit peripheral c-Jun in the trigeminal ganglia (TG) and have analgesic effects in trigeminal neuralgia. The phosphorylation of c-Jun improved hyperalgesia and allodynia, suggesting that SHED transplantation could improve trigeminal neuralgia [53]. SHED-conditioned media promoted the regeneration of sciatic nerve defects in rats. SHED-CM promoted axon development, peripheral nerve tissue angiogenesis, SC migration, proliferation, and activation, and neuron survival. Therefore, SHED-CM promotes peripheral nerve regeneration through multiple processes, resulting in functional recovery, and maybe a promising strategy for the clinical treatment of Peripheral Nerve Injury [54].

- On Brain Injury: SHED-Exos could decrease neuroinflammation by replacing microglia M1/M2 polarization in animal models. SHEDs based therapies emerge as a potential therapy option for neurodegenerative disorders because of their homing, engraft, differentiate and generating factors for CNS improvement [55].

- On Bone formation: Transplanted SHED with hydroxyapatite/tricalcium phosphate into calvarial defect restored the parietal continuity dynamically contributing to bone formation [56].

- On Systemic Lupus Erythematosus: SHED administration could reverse SLE-associated defects in MRL/lpr mouse. SHED had significant effects on inhibiting T helper 17 (Th17) cells *in vitro.* At the cellular level, SHED transplantation elevated the ratio of regulatory T cells (Tregs) via Th17 cells. It can be concluded that SHED is an accessible and feasible mesenchymal stem cell source for treating immune disorders like SLE [57].

6. Scope of tissue engineering in dentistry using SHED

Recent advances in stem cell research especially in the field of dentistry have led to the onset of an entirely new era in which even an entire tooth can be regrown. This is just one of the several approaches that hold promise for tooth regeneration.

As far as pediatric dentistry is concerned, decay is one of the most common problems faced. Most often a pedodontist ends up performing a pulpectomy procedure which involves the complete removal of the pulpal tissue and filling it with an ideal obturating material. Now, with technological evolution, researchers are using stem cell therapy for regenerative pulpotomies which can restore the vital pulp, which bypasses the need of going through painful invasive dental procedures. Following are some scope and potential applications of SHED in regenerative dentistry.

6.1 Dentin Pulp complex regeneration (DPC)

The DPC consists of the outer hard tissue layer, which is composed of orientated cells (odontoblasts) that secrete a specific matrix to form new dentin, and the inner soft tissue layer, which is composed of vital pulp tissue that comprises a network of microvasculature, nerves, and fibrous elements. Therefore, regeneration of the Dentin Pulp Complex entails a cascade of events involving odontogenesis and angiogenesis. SHEDs have a greater capacity for the formation of Dentin Pulp Complex cells, including osteoblasts, chondroblasts, adipocytes, endothelial cells, nerve cells, and odontoblasts [58–60]. SHEDs have demonstrated the ability to develop into functional odontoblasts and endothelial-like cells [61]. SHED's capacity to develop into odontoblasts is defined by the expression of dentin matrix protein-1 (DMP-1) and Dentin Sialophosphoprotein (DSPP) [60, 62]. The goal of DMP-1 is to maintain dentin mineralization, DSPP stimulates odontoblast development in stem cells by phosphorylation of SMAD 1/5/8 and nuclear translocation via the P38 and ERK1/2 pathways [63, 64]. Regenerating the missing interface between two distinct tissues (dentin and pulp), as seen in the Dentin Pulp Complex, is one of the main challenges of regenerative dentistry. It is crucial to provide a perfect environment that encourages the aggregation, proliferation, and differentiation of these disparate tissues. The ultimate regeneration of the dentin-pulp complex requires successful innervation and revascularization. In this context, various scaffolds have been employed to support cell growth and functionality in the transplanted area. Tissue engineering methods involving SHED, growth factors and scaffolds have been researched to regenerate DPC. SHED has shown the ability to regenerate pulp- and dentin-like tissues utilizing scaffolds and stem cells in animal models. SHEDs are also able to increase the angiogenesis process by forming vascular connective tissue structures and expressing and synthesizing VEGF. This ability is crucial in maintaining pulp viability as it can supply oxygen and nutrients needed for cell metabolism for tissue regeneration. Also, Exosomes extracted from SHED aggregates (SA-Exo) showed to significantly improve pulp tissue regeneration and angiogenesis in vivo, it also promoted endothelial differentiation and enhanced the angiogenic ability of HUVECs [65, 66].

6.2 Dentin Pulp regeneration

Dentin Pulp Regeneration aims to revitalize necrotic, infected, or lost pulp teeth by restoring the morphology and function of the pulp. Ideal pulp regeneration should possess natural structures such as nerve, fibers and blood supply, allowing nutritional, defense, sensation, and immunological functions to be restored. SHED have been utilized for pulp revascularization in regenerative endodontic procedures over the years. The conventional endodontic treatment in an immature tooth with pulpal necrosis is often challenging owing to its open root apex and thin root canal dentin. In addition, there is a risk of obturating material overflowing into the periapex. Regenerative endodontics (pulp revascularization) in an immature tooth aims to promote continued root development by generating new tissues. SHED seeded onto the synthetic scaffolds formed well-vascularized pulp-like tissue in vivo on a tooth slice model [67].

To restore the vitality of a tooth, elements with regeneration properties in the pulp are required. SHED not only produced mesenchymal stem cell-specific markers, but it also caused odontoblastic differentiation and increased the formation of endothelium and fibroblasts. The regenerated pulp tissue built by SHED had similar cellularity and architecture of the physiological dental pulp [68]. The combination of SHED, Platelet Rich Fibrin, and Chitosan enhanced the migration, proliferation, and odontoblastic differentiation of dental pulp cells. Therefore, the combination of SHED, PRF, and Chitosan scaffold as a new method for pulp regeneration in a clinical environment appears promising as 3D tissue engineering. SHED were implanted in empty root canals of mini pigs to determine whether a full-length dental pulp is regenerated. After 3 months of implantation, the histological analysis showed that full-length dental pulp tissue was regenerated which contained the odontoblast layer and blood vessels. The regenerated pulp showed a similar tissue structure to the normal pulp. Furthermore, blood vessels and nerves were regenerated as confirmed by positive expressions of CD31 and neurofilament (NF) in regenerated pulp tissues. In addition, positive expressions of CGRP and TRPV1 cells indicated the regenerated pulp might have sensory nerves. These results indicated that implantation of SHED was capable of regenerating full-length dental pulp with blood vessels and nerves in a large animal model [69].

6.3 Bio root regeneration

Techniques based on cell-based tissue engineering have made significant advances in the field of tooth regeneration. However, regeneration of the complete tooth has proven to be laborious; consequently, tooth root regeneration is advocated as a more practical and promising alternative for tooth restoration than the regeneration of the entire tooth. The tooth root serves a vital role in chewing and maintaining the tooth's stability, which is the structural foundation of a functional tooth. Bio-engineered tooth root (bio root) mediated by stem cells has shown to be a promising treatment for tooth loss. Multiple studies have demonstrated that Dental Follicle cells are appropriate seeding cells for the development of bioroots. SHEDs can be regarded as a prospective seeding cell for use in bio root regeneration in the future [70]. The comparison of ultrastructure revealed that SHEDs participate in active cell metabolism and the autophagy process, which are essential for stem cell immunological defense, self-renewal, and apoptosis [71, 72]. In addition, they possessed protein synthesis and secretion capabilities that were superior to those of dental follicle cells, resulting in the establishment of a microenvironment that is favorable to tissue repair and regeneration, a breakthrough in the field of nerve regeneration [54, 73, 74].

6.4 Periodontal regeneration

SHEDs treated with dentin matrix can regenerate periodontal tissue composed of periodontal ligament fibers, blood vessels, and new alveolar bone [70]. Due to their high proliferative capacity, the strong immunosuppressive ability of multiple differentiation, and minimal carcinogenic potential, exfoliated human deciduous dental stem cells have been employed to restore periodontal tissue and repair alveolar bone abnormalities [75].

6.5 Reconstruction of Cleft lip/palate (CL/P)

Autogenous iliac bone grafting has been demonstrated to heal alveolar cleft defects; however, surgical intervention is required. Hence, the creation of a less invasive technique is anticipated. Consequently, Alveolar bone regeneration methods in patients with CL/P employing human bone marrow mesenchymal stem cells (hBMSCs) have been attempted, and the transplantation of hBMSC in a canine alveolar cleft model has demonstrated the ability to regenerate bone [76]. SHEDs have higher osteogenic potential compared to bone marrow stem cells [77]. SHEDs, human dental pulp stem cells (hDPSCs), and hBMSCs were utilized to induce bone regeneration in immunodeficient animals with calvarial bone abnormalities. However, animals treated with SHED scaffolds had the greatest amounts of osteoid and the widest distribution of collagen fibers. During cell culture, MSCs can secrete paracrine substances into a conditioned medium (CM). MSC-CM contain the growth factors insulin-like growth factor-1 (IGF-1), transforming growth factor 1 (TGF-1), and vascular endothelial growth factor (VEGF), which influence the features and behavior of regenerating bone cells [78–80]. It is possible that both transplanted MSCs and their paracrine actions contribute to tissue regeneration. SHED-CM demonstrated mature bone development and contained tissue-regenerating factors with functions in angiogenesis and osteogenesis. Deciduous dental pulp stem cells (DDPSC) associated with a hydroxyapatite-collagen sponge showed closure of alveolar defects during the secondary dental eruption in a clinical setting. Thus, SHED could be an ideal source of cells for alveolar cleft reconstruction due to its capacity to regenerate bone with minimal surgical invasion [81].

6.6 Temporomandibular Joint Osteoarthritis (TMJOA)

Exosomes secreted by SHEDs (SHED-Exos) demonstrated to suppress inflammation in TMJ chondrocytes. The anti-inflammatory effects of SHED-Exos were verified using western blotting and RT-qPCR. SHED-Exos down-regulated the expression of IL-6, IL-8, MMP1, MMP3, MMP9, MMP13, and dis-integrin and metalloproteinase with thrombospondin motifs 5 (ADAMTS5). Thus, they can be a novel therapeutic agent for TMJ inflammation [82].

7. Limitations / challenges of stem cell research

Even though stem cell offers a wide range of therapeutic potentials in regenerative medicine, one cannot deny the fact that it does possess limitations and challenges because of different ethical and other issues related to stem cell research. Some of them are listed below:

1. The most crucial challenge to stem cell research is the ethical issue related to the use of embryonic stem cells. Due to these issues, there are even political and religious obstructions to stem cell research. Stem cell research has come under some controversy due to the ethical problems associated with how stem cells are obtained. It is known that while obtaining the stem cells from an embryo, the embryo, in the end, is discarded, which raises ethical issues. However, due to the discovery and use of adult stem cells and induced pluripotent stem cells, the use of ESCs has decreased and so have the ethical issues.

2. The source of some stem cell lines might have mutations which increase the chances of mutations in the transplants.

3. It is also difficult to transplant the stem cells produced in the laboratories to the target cells.

4. ESCs also do not permanently renew themselves in vivo, but instead, differentiate soon into different lineage progenitor cells of the three embryonic germ layers.

5. Self-renewal of these cells can be achieved in vitro under artificial conditions, which inhibit their differentiation.

6. It is also challenging to obtain enough stem cells with the ability to differentiate into the desired cell type.

7. The differentiation of embryonic, as well as adult stem cells, even if guided by the addition of differentiation factors, inevitably involves a certain amount of spontaneous differentiation into various cell types

8. Additionally, the differentiation is not synchronizable yet, leading to a mixture of cells in various stages of development.

However, the advent of stem cells derived from less invasive tissues without immune rejection has widened the opportunities for cell-based therapies.

8. Then, why not discard primary teeth after they SHED?

According to the data presented above, SHED have regenerative abilities comparable to umbilical cord stem cells, implying that they have a high potential for treating some life-threatening diseases. Exfoliated teeth are also an unexpectedly unique resource for stem-cell therapies such as autologous stem-cell transplantation and tissue engineering due to their ease of access and lack of ethical concerns. Banking SHED cells is thus extremely beneficial (**Figure 7**).

9. SHED Banking and its advantages

Storing your child's own teeth stem cells could give them access to a huge range of treatment opportunities throughout their lifetime

- It provides a guaranteed matching donor (autologous transplant) for life. There are many advantages of autologous transplant including no immune reaction and

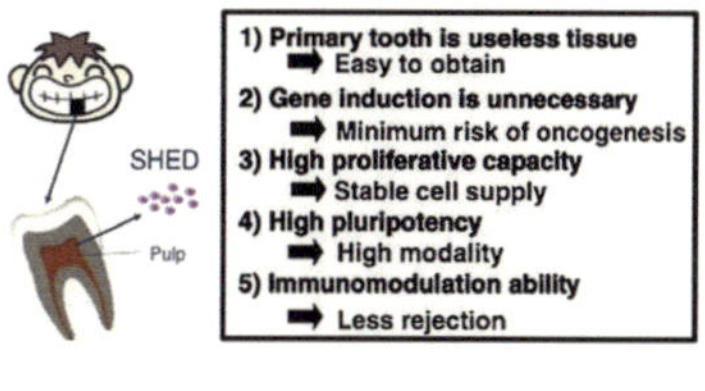

Figure 7.
SHED and its importance.

tissue rejection of the cells, no immunosuppressive therapy needed, and significantly reduced risk of communicable diseases [83, 84].

- Saves cells before natural damage occurs.
  - Simple and painless for both child and parent.
  - Less than one-third of the cost of cord blood storage.
  - SHED are adult stem cells and are not the subject of the same ethical concerns as embryonic stem cells [83, 84].
  - SHED cells complement stem cells derived from cord blood. Even though cord blood stem cells have proven useful in the regeneration of blood cell types, SHED is capable of regenerating solid tissue types that cord blood cannot, such as potentially rebuilding connective tissues, dental tissue, neural tissue, and bone [85–88].

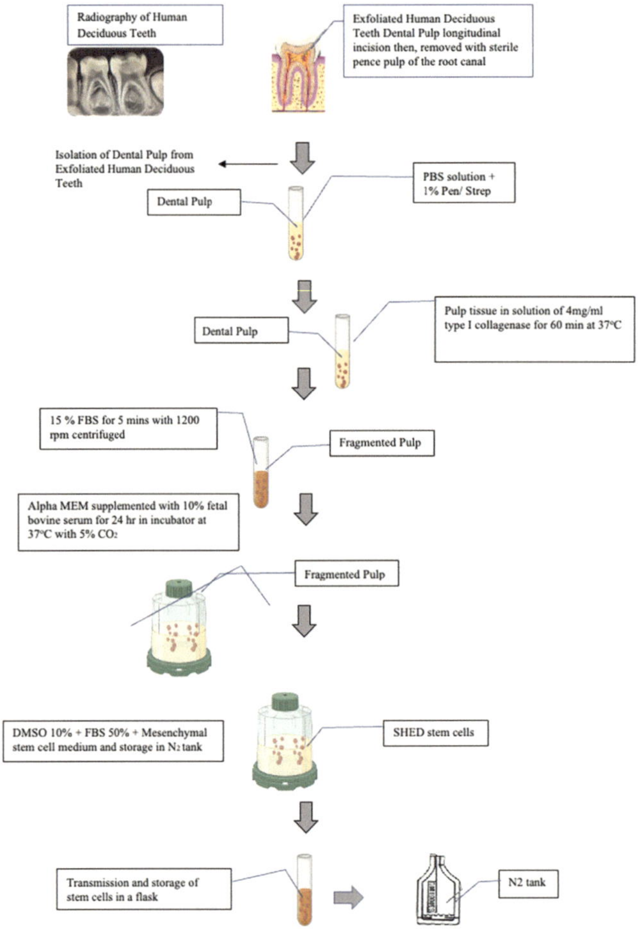

Figure 8.
Isolation protocol of SHED. Image source: Biorender.com.

- SHED may also be beneficial for the donor's immediate relatives, such as grandparents, parents, uncles, and siblings [84].

Certain tooth selection criteria must be followed before SHED banking, which is crucial for the successful isolation and characterization of cell lines to maintain viability.

1. A tooth that is extracted for orthodontic purposes. (Extraction to relieve crowding and allow permanent teeth eruption)

2. Tooth extracted/ exfoliated should be vital and devoid of any pulpal necrosis (indication of cell viability)

3. Tooth extracted/exfoliated should be free of any mobility either due to trauma/ pulpal pathology

4. The tooth extracted/exfoliated should be free of any periapical infection (**Figure 8**).

10. Procedure

We have outlined the detailed process involved in SHED banking for its application in stem cell therapy.

10.1 Collection, isolation, and preservation of SHED

1. Tooth collection

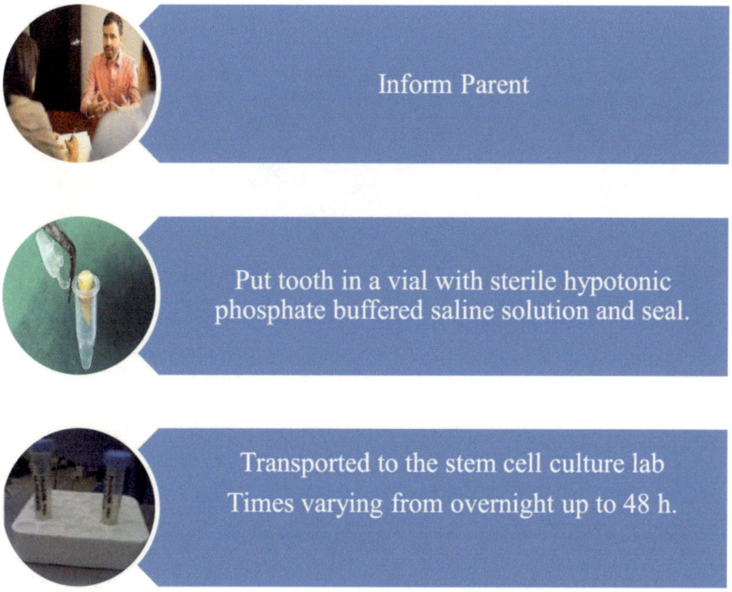

1. Stem cell isolation: Step by step of Stem cell Isolation is explained below.

2. Stem cell verification and viability testing

Figure 9.
Stem cell verification and validation.

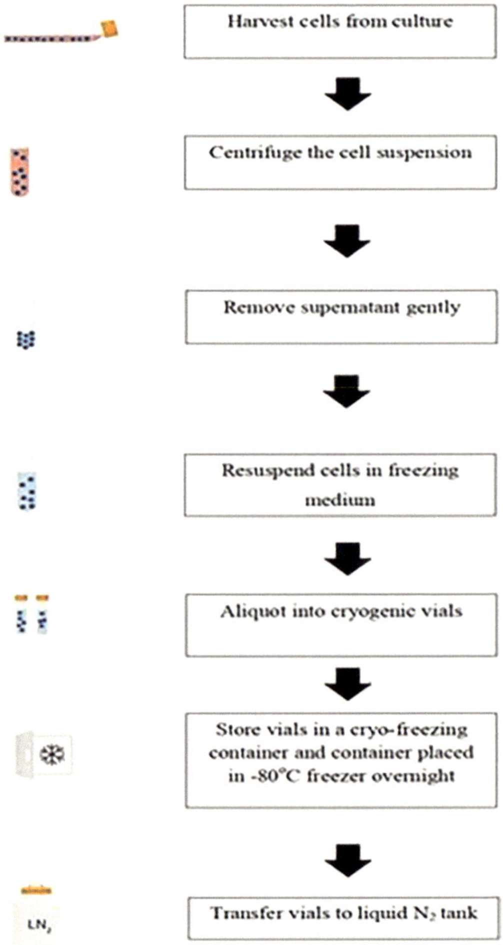

Figure 10.
Typical step-by-step cryopreservation protocol. Image source: Cryopreservation Basics: Protocols and Best Practices for Freezing Cells. Stem cell technologies.

The viability of the isolated cells is then tested (**Figure 9**).

3. Stem Cell Storage

Stem cell storage refers to the collection and cryopreservation of stem cells from source tissue for use in stem cell treatments or clinical trials in the future. Methods used for Stem cell storage are cryopreservation or magnetic freezing [89].

10.2 Cryopreservation

This routine procedure generally involves slow cooling in the presence of a cryoprotectant to avoid the damaging effects of intracellular ice formation (**Figure 10**).

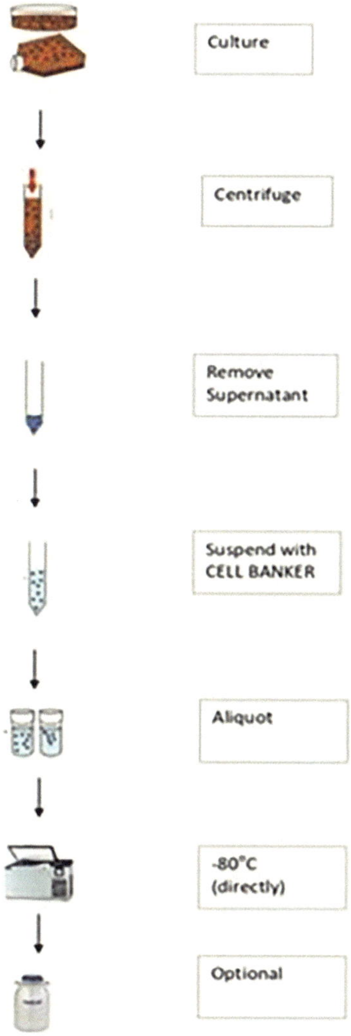

Figure 11.
Magnetic freezing step by step protocol.

10.3 Magnetic freezing

These above steps are followed by Stem cell differentiation, characterization, and validation of required cell types prior to their application in cell-based therapies (**Figure 11**).

11. Conclusion

Stem cell therapy has made significant advances in regenerative medicine and dentistry. It has simplified the treatment of many diseases that were previously difficult to treat. Cord blood stem cells are used to treat over 85 different blood and immune diseases, including Leukemia and Neuroblastoma. Considering the scope of stem cells in regenerative medicine, it is imperative to opt a source of stem cell which is less invasive and has promising future in regenerative therapies without immune rejection. SHED has been shown to be as distinct as cord blood stem cells. Multiple studies demonstrate that SHED can differentiate into odontoblasts, neurons, hepatocytes, endothelial cells, β-cells, and other cell types. This vast array of cell types generates an abundance of options for the application of SHED in tissue regeneration processes. SHED has demonstrated a wide range of therapeutic applications due to its ease of harvesting, lack of bioethical concerns, and excellent expansibility. Also, using one's own stem cells (SHED) reduces, if not eliminates, the risk of developing immune reactions or rejection after transplantation, as well as the risk of contracting disease from donor cells. However, Prior to SHED-based therapies becoming a clinical reality, it is necessary to have a deeper understanding of the mechanisms underpinning differentiation processes. Considering the tremendous use of SHED in stem cell therapy, will banking of SHED or the cost of stem cell banking justified and reasonable? The answer is—Why not, if money is not an issue? If a naturally discarded tooth can have such broad therapeutic applications in the future, why not save it? Regardless, the ultimate fate of SHED cell banking will be decided by the patient or parent.

Acknowledgements

We are extremely grateful for the generous contributions of our colleagues and friends to this work. Specially, Dr. B. Mohana Kumar, Associate Professor, Nitte University Centre for Stem Cell Research and Regenerative Medicine (NUCSReM), K. S. Hegde Medical Academy, Nitte (Deemed to be University), and his team for their expertise in stem cell isolation and characterization. Dr. Veena Shetty, PhD, Additional Professor, Department of Microbiology, K. S. Hegde Medical Academy, for providing emotional and scientific support for this work and also, Nitte (Deemed to be University) for providing access to all of the educational material required for this work.

Author details

Meghna Bhandary[1*], Rachaita Chhabra[2], K. Ananya Rao[1] and Mohammed Shahid[3]

1 Department of Pediatric and Preventive Dentistry, AB Shetty Memorial Institute of Dental Sciences, India

2 Pediatric Dental Practitioner, Mumbai, India

3 Department of Oral Pathology and Microbiology, A.J. Institute of Dental Sciences, India

*Address all correspondence to: meghnabhandary92@gmail.com

References

[1] Zakrzewski W, Dobrzyński M, Szymonowicz M, Rybak Z. Stem cells: Past, present, and future. Stem Cell Research & Therapy. 2019;**10**(1):68

[2] Ul Hassan A, Hassan G, Rasool Z. Role of stem cells in treatment of neurological disorder. International Journal of Health Science (Qassim). 2009;**3**(2):227-233

[3] Odorico JS, Kaufman DS, Thomson JA. Multilineage differentiation from human embryonic stem cell lines. Stem Cells. 2001;**19**(3):193-204

[4] Song Z, Cai J, Liu Y, Zhao D, Yong J, Duo S, et al. Efficient generation of hepatocyte-like cells from human induced pluripotent stem cells. Cell Research. 2009;**19**(11):1233-1242

[5] Senju S, Haruta M, Matsunaga Y, Fukushima S, Ikeda T, Takahashi K, et al. Characterization of dendritic cells and macrophages generated by directed differentiation from mouse induced pluripotent stem cells. Stem Cells. 2009;**27**(5):1021-1031

[6] Tateishi K, He J, Taranova O, Liang G, D'Alessio AC, Zhang Y. Generation of insulin-secreting islet-like clusters from human skin fibroblasts. Journal of Biological Chemistry. 2008;**283**(46):31601-31607

[7] Choi KD, Yu J, Smuga-Otto K, Salvagiotto G, Rehrauer W, Vodyanik M, et al. Hematopoietic and endothelial differentiation of human induced pluripotent stem cells. Stem Cells. 2009;**27**(3):559-567

[8] Kim K, Doi A, Wen B, Ng K, Zhao R, Cahan P, et al. Epigenetic memory in induced pluripotent stem cells. Nature. 2010;**467**(7313):285-290

[9] Bar-Nur O, Russ HA, Efrat S, Benvenisty N. Epigenetic memory and preferential lineage-specific differentiation in induced pluripotent stem cells derived from human pancreatic islet Beta cells. Cell Stem Cell. 2011;**9**(1):17-23

[10] Karumbayaram S, Novitch BG, Patterson M, Umbach JA, Richter L, Lindgren A, et al. Directed differentiation of human-induced pluripotent stem cells generates active motor neurons. Stem Cells. 2009;**27**(4):806-811

[11] Chambers SM, Fasano CA, Papapetrou EP, Tomishima M, Sadelain M, Studer L. Highly efficient neural conversion of human ES and iPS cells by dual inhibition of SMAD signaling. Nature Biotechnology. 2009;**27**(3):275-280

[12] White SJ, Chong JJH. Mesenchymal stem cells in cardiac repair: Effects on myocytes, vasculature, and fibroblasts. Clinical Therapeutics. 2020;**42**(10):1880-1891

[13] Qu J, Zhang H. Roles of mesenchymal stem cells in spinal cord injury. Stem Cells International. 2017;**2017**:1-12

[14] Figueroa FE, Carrión F, Villanueva S, Khoury M. Mesenchymal stem cell treatment for autoimmune diseases: A critical review. Biological Research. 2012;**45**(3):269-277

[15] Sykova E, Cizkova D, Kubinova S. Mesenchymal stem cells in treatment of spinal cord injury and amyotrophic lateral sclerosis. Frontiers in Cell and Development Biology. 2021;**6**:9

[16] Koh YG, Choi YJ. Infrapatellar fat pad-derived mesenchymal stem cell

therapy for knee osteoarthritis. The Knee. 2012;**19**(6):902-907

[17] Gronthos S, Mankani M, Brahim J, Robey PG, Shi S. Postnatal human dental pulp stem cells (DPSCs) *in vitro* and *in vivo*. Proceedings of the National Academy of Sciences. 2000;**97**(25):13625-13630

[18] Aydin S, Şahin F. Stem cells derived from dental tissues. 2019;123-132

[19] Takebe Y, Tatehara S, Fukushima T, Tokuyama-Toda R, Yasuhara R, Mishima K, et al. Cryopreservation method for the effective collection of dental pulp stem cells. Tissue Engineering. Part C, Methods. 2017;**23**(5):251-261

[20] Gan L, Liu Y, Cui D, Pan Y, Zheng L, Wan M. Dental tissue-derived human mesenchymal stem cells and their potential in therapeutic application. Stem Cells International. 2020;**2020**:1-17

[21] Liu J, Yu F, Sun Y, Jiang B, Zhang W, Yang J, et al. Concise reviews: Characteristics and potential applications of human dental tissue-derived mesenchymal stem cells. Stem Cells. 2015;**33**(3):627-638

[22] Matsubara T, Suardita K, Ishii M, Sugiyama M, Igarashi A, Oda R, et al. Alveolar bone marrow as a cell source for regenerative medicine: Differences between alveolar and iliac bone marrow stromal cells. Journal of Bone and Mineral Research. 2004;**20**(3):399-409

[23] Sonoyama W, Liu Y, Fang D, Yamaza T, Seo BM, Zhang C, et al. Mesenchymal stem cell-mediated functional tooth regeneration in swine. PLoS One. 2006;**1**(1):e79

[24] Sonoyama W, Liu Y, Yamaza T, Tuan RS, Wang S, Shi S, et al. Characterization of the apical papilla and its residing stem cells from human immature permanent teeth: A pilot study. Journal of Endodontia. 2008;**34**(2):166-171

[25] Patil R, Kumar BM, Lee WJ, Jeon RH, Jang SJ, Lee YM, et al. Multilineage potential and proteomic profiling of human dental stem cells derived from a single donor. Experimental Cell Research. 2014;**320**(1):92-107

[26] Kang J, Fan W, Deng Q, He H, Huang F. Stem cells from the apical papilla: A promising source for stem cell-based therapy. BioMed Research International. 2019;**2019**:1-8

[27] Krismariono A, Rubianto M, Maduratna E, Sari DR. A revolution of stem cell in periodontal regeneration. 2020;060013

[28] Ikeda E, Yagi K, Kojima M, Yagyuu T, Ohshima A, Sobajima S, et al. Multipotent cells from the human third molar: Feasibility of cell-based therapy for liver disease. Differentiation. 2008;**76**(5):495-505

[29] Zhang Q, Shi S, Liu Y, Uyanne J, Shi Y, Shi S, et al. Mesenchymal stem cells derived from human gingiva are capable of immunomodulatory functions and ameliorate inflammation-related tissue destruction in experimental colitis. The Journal of Immunology. 2009;**183**(12):7787-7798

[30] Miura M, Gronthos S, Zhao M, Lu B, Fisher LW, Robey PG, et al. SHED: Stem cells from human exfoliated deciduous teeth. Proceedings of the National Academy of Sciences. 2003;**100**(10):5807-5812

[31] Bernardi L, Luisi SB, Fernandes R, Dalberto TP, Valentim L, Bogo Chies JA, et al. The isolation of stem cells from human deciduous teeth pulp is related

to the physiological process of resorption. Journal of Endodontia. 2011;**37**(7):973-979

[32] Junior AL, Pinheiro CCG, Tanikawa DYS, Ferreira JRM, Amano MT, Bueno DF. Mesenchymal stem cells from human exfoliated deciduous teeth and the orbicularis Oris muscle: How do they behave when exposed to a Proinflammatory stimulus? Stem Cells International. 2020;**2020**:1-15

[33] Dahake PT, Panpaliya NP, Kale YJ, Dadpe M, Kendre SB, Bogar C. Response of stem cells from human exfoliated deciduous teeth (SHED) to three bioinductive materials – An in vitro experimental study. Saudi. Dental Journal. 2020;**32**(1):43-51

[34] Athanasiadou E, Paschalidou M, Theocharidou A, Kontoudakis N, Arapostathis K, Bakopoulou A. Biological interactions of a calcium silicate based cement (biodentine™) with stem cells from human exfoliated deciduous teeth. Dental Materials. 2018;**34**(12):1797-1813

[35] Collado-González M, García-Bernal D, Oñate-Sánchez RE, Ortolani-Seltenerich PS, Álvarez-Muro T, Lozano A, et al. Cytotoxicity and bioactivity of various pulpotomy materials on stem cells from human exfoliated primary teeth. International Endodontic Journal. 2017;**50**:e19-e30

[36] Araújo LB, Cosme-Silva L, FernandesAP,deOliveiraTM,CavalcantiB, Gomes Filho JE, et al. Effects of mineral trioxide aggregate, BiodentineTM and calcium hydroxide on viability, proliferation, migration and differentiation of stem cells from human exfoliated deciduous teeth. Journal of Applied Oral Science. 2018;**26**:6

[37] Nishino Y, Yamada Y, Ebisawa K, Nakamura S, Okabe K, Umemura E, et al. Stem cells from human exfoliated deciduous teeth (SHED) enhance wound healing and the possibility of novel cell therapy. Cytotherapy. 2011;**13**(5):598-605

[38] Shimojima C, Takeuchi H, Jin S, Parajuli B, Hattori H, Suzumura A, et al. Conditioned medium from the stem cells of human exfoliated deciduous teeth ameliorates experimental autoimmune encephalomyelitis. The Journal of Immunology. 2016;**196**(10):4164-4171

[39] Rossato C, Brandão WN, Castro SBR, de Almeida DC, Maranduba CMC, Camara NOS, et al. Stem cells from human-exfoliated deciduous teeth reduce tissue-infiltrating inflammatory cells improving clinical signs in experimental autoimmune encephalomyelitis. Biologicals. 2017;**49**:62-68

[40] Xie J, Rao N, Zhai Y, Li J, Zhao Y, Ge L, et al. Therapeutic effects of stem cells from human exfoliated deciduous teeth on diabetic peripheral neuropathy. Diabetology and Metabolic Syndrome. 2019;**11**(1):38

[41] Fujiyoshi J, Yamaza H, Sonoda S, Yuniartha R, Ihara K, Nonaka K, et al. Therapeutic potential of hepatocyte-like-cells converted from stem cells from human exfoliated deciduous teeth in fulminant Wilson's disease. Scientific Reports. 2019;**9**(1):1535

[42] Zhao L, Li Y, Kou X, Chen B, Cao J, Li J, et al. Stem cells from human exfoliated deciduous teeth ameliorate autistic-like Behaviors of *SHANK3* mutant beagle dogs. Stem Cells Translational Medicine. 2022;**11**(7):778-789

[43] Yamaza T, Alatas FS, Yuniartha R, Yamaza H, Fujiyoshi JK, Yanagi Y, et al. In vivo hepatogenic capacity and therapeutic potential of stem cells from human exfoliated deciduous teeth in liver

fibrosis in mice. Stem Cell Research & Therapy. 2015;**6**(1):171

[44] Cao X, Wang C, Yuan D, Chen S, Wang X. The effect of implants loaded with stem cells from human exfoliated deciduous teeth on early osseointegration in a canine model. BMC Oral Health. 2022;**22**(1):238

[45] Wang J, Wang X, Sun Z, Wang X, Yang H, Shi S, et al. Stem cells from human-exfoliated deciduous teeth can differentiate into dopaminergic neuron-like cells. Stem Cells and Development. 2010;**19**(9):1375-1383

[46] Matsushita Y, Ishigami M, Matsubara K, Kondo M, Wakayama H, Goto H, et al. Multifaceted therapeutic benefits of factors derived from stem cells from human exfoliated deciduous teeth for acute liver failure in rats. Journal of Tissue Engineering and Regenerative Medicine. 2017;**11**(6):1888-1896

[47] Li XX, Yuan XJ, Zhai Y, Yu S, Jia RX, Yang LP, et al. Treatment with stem cells from human exfoliated deciduous teeth and their derived conditioned medium improves retinal visual function and delays the degeneration of photoreceptors. Stem Cells and Development. 2019;**28**(22):1514-1526

[48] Zhang X, Lei T, Chen P, Wang L, Wang J, Wang D, et al. Stem cells from human exfoliated deciduous teeth promote hair regeneration in mouse. Cell Transplantation. 2021;**30**:096368972110429

[49] Hattori Y, Kim H, Tsuboi N, Yamamoto A, Akiyama S, Shi Y, et al. Correction: Therapeutic potential of stem cells from human exfoliated deciduous teeth in models of acute kidney injury. PLoS One. 2015;**10**(11):e0143561

[50] Xie Y, Yu L, Cheng Z, Peng Y, Cao Z, Chen B, et al. SHED-derived exosomes promote LPS-induced wound healing with less itching by stimulating macrophage autophagy. Journal of Nanobiotechnology. 2022;**20**(1):239

[51] Li W, Jiao X, Song J, Sui B, Guo Z, Zhao Y, et al. Therapeutic potential of stem cells from human exfoliated deciduous teeth infusion into patients with type 2 diabetes depends on basal lipid levels and islet function. Stem Cells Translational Medicine. 2021;**10**(7):956-967

[52] Anoop M, Datta I. Stem cells derived from human exfoliated deciduous teeth (SHED) in neuronal disorders: A review. Current Stem Cell Research & Therapy. 2021;**16**(5):535-550

[53] Bai X, Xiao K, Yang Z, Zhang Z, Li J, Yan Z, et al. Stem cells from human exfoliated deciduous teeth relieve pain via downregulation of c-Jun in a rat model of trigeminal neuralgia. Journal of Oral Rehabilitation. 2022;**49**(2):219-227

[54] Sugimura-Wakayama Y, Katagiri W, Osugi M, Kawai T, Ogata K, Sakaguchi K, et al. Peripheral nerve regeneration by Secretomes of stem cells from human exfoliated deciduous teeth. Stem Cells and Development. 2015;**24**(22):2687-2699

[55] Li Y, Yang YY, Ren JL, Xu F, Chen FM, Li A. Exosomes secreted by stem cells from human exfoliated deciduous teeth contribute to functional recovery after traumatic brain injury by shifting microglia M1/M2 polarization in rats. Stem Cell Research & Therapy. 2017;**8**(1):198

[56] Seo B, Sonoyama W, Yamaza T, Coppe C, Kikuiri T, Akiyama K, et al. SHED repair critical-size calvarial defects in mice. Oral Diseases. 2008;**14**(5): 428-434

[57] Yamaza T, Kentaro A, Chen C, Liu Y, Shi Y, Gronthos S, et al. Immunomodulatory properties of stem cells from human exfoliated deciduous teeth. Stem Cell Research & Therapy. 2010;**1**(1):5

[58] Smojver I, Katalinić I, Bjelica R, Gabrić D, Matišić V, Molnar V, et al. Mesenchymal stem cells based treatment in dental medicine: A narrative review. International Journal of Molecular Sciences. 2022;**23**(3):1662

[59] Han Y, Zhang L, Zhang C, Dissanayaka WL. Guiding lineage specific differentiation of SHED for target tissue/organ regeneration. Current Stem Cell Research & Therapy. 2021;**16**(5):518-534

[60] Anggrarista KA, Cecilia P, Nagoro AA, Saskianti T, Surboyo MD. SHED, PRF, and chitosan as three-dimensional of tissue engineering for dental pulp regeneration. Dental Hypotheses. 2021;**12**(1):43

[61] Kerkis I, Caplan AI. Stem cells in dental pulp of deciduous teeth. Tissue Engineering. Part B, Reviews. 2012;**18**(2):129-138

[62] Ganesh KishoreS O, Don KR, PriyaA J. Therapeutic potential of stem cells from human exfoliated deciduous teeth(Shed)-A review. Indian Journal of Forensic Medicine & Toxicology. **14**

[63] Khazaei S, Khademi A, Torabinejad M, Nasr Esfahani MH, Khazaei M, Razavi SM. Improving pulp revascularization outcomes with buccal fat autotransplantation. Journal of Tissue Engineering and Regenerative Medicine. 2020;**2020**:3094

[64] Vu HT, Han MR, Lee JH, Kim JS, Shin JS, Yoon JY, et al. Investigating the effects of conditioned media from stem cells of human exfoliated deciduous teeth on dental pulp stem cells. Biomedicine. 2022;**10**(4):906

[65] Wu M, Liu X, Li Z, Huang X, Guo H, Guo X, et al. SHED aggregate exosomes shuttled miR-26a promote angiogenesis in pulp regeneration via TGF-β/SMAD2/3 signalling. Cell Proliferation. 2021;**54**:7

[66] Rosa V, Zhang Z, Grande RHM, Nör JE. Dental pulp tissue engineering in full-length human root canals. Journal of Dental Research. 2013;**92**(11):970-975

[67] Demarco FF, Conde MCM, Cavalcanti BN, Casagrande L, Sakai VT, Nör JE. Dental pulp tissue engineering. Brazilian Dental Journal. 2011;**22**(1):3-13

[68] Sukarawan W, Osathanon T. Stem cells from human exfoliated deciduous teeth: Biology and therapeutic potential. In: Mesenchymal Stem Cells—Isolation, Characterization and Applications. London, UK: InTech; 2017

[69] Guo H, Zhao W, Liu A, Wu M, Shuai Y, Li B, et al. SHED promote angiogenesis in stem cell-mediated dental pulp regeneration. Biochemical and Biophysical Research Communications. 2020;**529**(4):1158-1164

[70] Yang X, Ma Y, Guo W, Yang B, Tian W. Stem cells from human exfoliated deciduous teeth as an alternative cell source in bio-root regeneration. Theranostics. 2019;**9**(9):2694-2711

[71] Guan JL, Simon AK, Prescott M, Menendez JA, Liu F, Wang F, et al. Autophagy in stem cells. Autophagy. 2013;**9**(6):830-849

[72] Boya P, Codogno P, Rodriguez-Muela N. Autophagy in stem cells: Repair, remodelling and metabolic

reprogramming. Development. 2018;**145**(4)

[73] Taghipour Z, Karbalaie K, Kiani A, Niapour A, Bahramian H, Nasr-Esfahani MH, et al. Transplantation of undifferentiated and induced human exfoliated deciduous teeth-derived stem cells promote functional recovery of rat spinal cord contusion injury model. Stem Cells and Development. 2012;**21**(10):1794-1802

[74] Inoue T, Sugiyama M, Hattori H, Wakita H, Wakabayashi T, Ueda M. Stem cells from human exfoliated deciduous tooth-derived conditioned medium enhance recovery of focal cerebral ischemia in rats. Tissue Engineering. Part A. 2013;**19**(1-2):24-29

[75] Gao X, Shen Z, Guan M, Huang Q, Chen L, Qin W, et al. Immunomodulatory role of stem cells from human exfoliated deciduous teeth on periodontal regeneration. Tissue Engineering. Part A. 2018;**24**(17-18):1341-1353

[76] Abe T, Sumi K, Kunimatsu R, Oki N, Tsuka Y, Awada T, et al. Bone regeneration in a canine model of artificial jaw cleft using bone marrow–derived mesenchymal stem cells and carbonate hydroxyapatite carrier. The Cleft Palate-Craniofacial Journal. 2020;**57**(2):208-217

[77] Lee JM, Kim HY, Park JS, Lee DJ, Zhang S, Green DW, et al. Developing palatal bone using human mesenchymal stem cell and stem cells from exfoliated deciduous teeth cell sheets. Journal of Tissue Engineering and Regenerative Medicine. 2019;**13**(2):319-327

[78] Ando Y, Matsubara K, Ishikawa J, Fujio M, Shohara R, Hibi H, et al. Stem cell-conditioned medium accelerates distraction osteogenesis through multiple regenerative mechanisms. Bone. 2014;**61**:82-90

[79] Inukai T, Katagiri W, Yoshimi R, Osugi M, Kawai T, Hibi H, et al. Novel application of stem cell-derived factors for periodontal regeneration. Biochemical and Biophysical Research Communications. 2013;**430**(2):763-768

[80] Ogata Y, Mabuchi Y, Yoshida M, Suto EG, Suzuki N, Muneta T, et al. Purified human synovium mesenchymal stem cells as a good resource for cartilage regeneration. PLoS One. 2015;**10**(6):e0129096

[81] Hiraki T, Kunimatsu R, Nakajima K, Abe T, Yamada S, Rikitake K, et al. Stem cell-derived conditioned media from human exfoliated deciduous teeth promote bone regeneration. Oral Diseases. 2020;**26**(2):381-390

[82] Luo P, Jiang C, Ji P, Wang M, Xu J. Exosomes of stem cells from human exfoliated deciduous teeth as an anti-inflammatory agent in temporomandibular joint chondrocytes via miR-100-5p/mTOR. Stem Cell Research & Therapy. 2019;**10**(1):216

[83] Mao JJ. Stem cells and the future of dental care. New York State Dental Journal. 2008;**74**(2):20

[84] Reznick JB. Continuing education: Stem cells: Emerging medical and dental therapies for the dental professional. Dentaltown Magazine. 2008;**2008**:42-53

[85] Arthur A, Rychkov G, Shi S, Koblar SA, Gronthos S. Adult human dental pulp stem cells differentiate toward functionally active neurons under appropriate environmental cues. Stem Cells. 2008;**26**(7):1787-1795

[86] Shi S, Bartold P, Miura M, Seo B, Robey P, Gronthos S. The efficacy of

mesenchymal stem cells to regenerate and repair dental structures. Orthodontics & Craniofacial Research. 2005;**8**(3):191-199

[87] Cordeiro MM, Dong Z, Kaneko T, Zhang Z, Miyazawa M, Shi S, et al. Dental pulp tissue engineering with stem cells from exfoliated deciduous teeth. Journal of Endodontia. 2008;**34**(8):962-969

[88] Mao JJ, Giannobile WV, Helms JA, Hollister SJ, Krebsbach PH, Longaker MT, et al. Craniofacial tissue engineering by stem cells. Journal of Dental Research. 2006;**85**(11):966-979

[89] Arora V, Arora P, Munshi A. Banking stem cells from human exfoliated deciduous teeth (SHED): Saving for the future. Journal of Clinical Pediatric Dentistry. 2009;**33**(4):289-294

Chapter 21

Artificial Intelligence in Dentistry

Seyide Tugce Gokdeniz, Arda Buyuksungur and Mehmet Eray Kolsuz

Abstract

Artificial intelligence (AI) is the branch of computer science dedicated to building systems to perform tasks that normally require human intelligence. AI tries to solve problems and learn similar to humans. The field of AI has experienced phenomenal development and growth over the past two decades; with the latest developments in digitized data collection, machine learning, and computing infrastructure, AI applications are expanding rapidly, especially in areas that are thought to be reserved for experts in their fields. Artificial intelligence has started to take place rapidly in dental clinical applications. The use of artificial intelligence in dentistry has enormous potential to improve patient care and drive major advances in healthcare. AI in dentistry is being researched for various purposes, such as identifying anatomical and pathological structures, diagnosing diseases and predicting treatment results, and selecting materials to be used. Advances in AI offer healthcare benefits, such as reducing postoperative complications, improving quality of life, and reducing the number of unnecessary procedures. It can also play a great helping role for dentists in increasing the accuracy of diagnosis. This chapter aims to explain the current applications and future predictions of artificial intelligence in dentistry, which is one of the most current topics of recent times.

Keywords: artificial intelligence, machine learning, deep learning, dental research, future dentistry

1. Introduction

The term artificial intelligence (AI), defined as "the ability of a device to perform functions normally associated with human intelligence, such as reasoning, learning, and self-development," continues to occupy a large place in our lives at an ever-increasing pace [1].

The first question was asked by Alan Turing in 1950 as "Can machine think?" and then the term was first used by John McCarthy as "Artificial Intelligence" [2].

The complexity of neural networks in a functional human brain has always been a topic of interest to many researchers in different fields [3]. With the development of science over time, many advanced technology products have emerged that imitate the functions of the human brain, but it is still not possible to fully simulate the human brain today [4]. Despite the many difficulties experienced, "artificial intelligence" (AI) has gained great importance in all areas of life with its irrepressible progress [5, 6].

Artificial intelligence (AI) studies, which continue to develop at an ever-increasing pace, are candidates to change and advance dentistry, as in many disciplines. Some of the situations that accelerate the developments such as the increase in computer power, the ease of access to universal information, and the availability of big data ready to be processed with AI applications in the field of health; it is the use of terms such as "digital transformations," "digitalized workflows," "technical developments" both in social life and in dentistry [7]. AI studies basically aim to imitate the cognitive processes of human intelligence by using machines and software-type algorithms to manage possible problems and complex tasks solved by human mind and skill [8, 9].

AI is actually a science under Computer Science or Computer Engineering. The sub-branches of AI are symbolic or machine learning. Machine learning is an algorithm that refers to the study of computer models that improve their performance by learning from experience without using explicit instructions, so it needs sample data to make predictions or decisions [10]. Preparing datasets for AI requires large amounts of data to be processed to improve accuracy. The experts must invest time and effort in making data suitable for effective learning. Deep learning and statistical learning are sub-branches of machine learning.

Machine learning (ML) is a branch of artificial intelligence in which systems learn to perform intelligent tasks without prior knowledge or hand-made rules, identifying patterns in samples from a large dataset without human assistance. This system can optimize its adjustable functions by defining a goal and achieving it. In this process, known as training, a machine learning algorithm is exposed to random samples. It then defines patterns that it can apply to new images as it gains experience by gradually adjusting the "tunables" toward the correct answer. Machine learning is basically like an adult showing a child several photos of cats, and eventually, the child learns the patterns of recognizing a cat in new images [11].

Deep learning (DL) is a sub-branch of ML. In this system, it is aimed to learn not only a model, but also a hierarchy of malleable models that are built on top of each other. The combination of dies creates a "deep" system that is much stronger than a flat, "shallow" system. For example, a child does not identify a cat in a single and indivisible pattern-matching step. The child sees the edges of the object first and defines a sketch with simple shapes such as eyes, nose, and ears, making a specific grouping. Among these components, larger groups emerge, such as the head, trunk, and legs. A particular grouping of these large groups also defines the whole cat [12].

Developments in computer technologies have increased the studies on AI and the interest in the subject. In this way, scientific studies have been facilitated, and neuropsychological, psychological, and similar behaviors can be perceived in three dimensions, especially for a better understanding of the brain. Thanks to the recording of sudden metabolic activities in the brain tissue with magnetic resonance imaging, some predetermined mental functions can be investigated and the results obtained make great contributions to a better understanding of the working principle of the brain. In the future, it will be possible to contribute more to the development of AI techniques with the results obtained from the studies carried out to illuminate the working principle of the brain.

Since dentistry is one of the areas where technology is most widely used, it is a very open area for developments in technology and the adaptation of AI applications. In recent years, AI applications have been widely used in dentistry, ranging from the diagnosis of caries, and the detection of various pathologies, to robotic surgery and dental implant construction [13–18].

The compatibility of dental radiology with image processing methods has made AI studies even more prominent in the field of radiology. There are many studies in AI that is applied on 2D and 3D (2D/3D) radiographic images. Detection of gingival diseases and evaluation of risk groups; automatic marking of anatomical structures and performing cephalometric analysis; diagnosis of certain diseases such as osteoporosis, which can be detected in X-rays of the jaw, etc., applications are just some of the studies on this subject [19].

With the rapid developments in the field, AI studies, which have started to radically change the sectors, will inevitably transform dentistry. The purpose of this section is to review the current and potential uses of AI applications in dentistry, to discuss the opportunities it will offer to improve oral and dental health of the community, and to examine its possible contributions to the economy and education.

2. Artificial intelligence applications in dentistry

As described in the previous section, AI is currently used in dentistry and in is explained in detail.

2.1 Dentomaxillofacial radiology

Dentistry examines the conditions and pathologies seen in the mouth, teeth, and jaws by establishing a relationship between clinical, systemic, and radiological findings. The patient's anamnesis and complaint history, digital radiography images, and, when necessary, intraoral/extraoral photographs are recorded. Oral diseases such as dental caries, gingival diseases, inflammatory conditions, cysts, and tumors are commonly evaluated with these instruments. The schematization of the detection of dental caries on radiographs by AI algorithms is shown in **Figure 1**.

Digital radiographs, radiographic images produced by X-ray irradiation are digitally encoded and easily translated into computational language [20]. Thanks to this digitalization in radiographs, the development of artificial intelligence in the field of radiology has accelerated and increased. Dental radiography, that is, intraoral radiographs, panoramic, cephalometric and advanced imaging technique, and cone beam computed tomography (CBCT), are collected for diagnosis, treatment planning, and treatment evaluation purposes during routine dental practice. As such, these large datasets offer a rich resource for scientific and medical research, especially for use in AI development work. In radiology practice, findings are visually evaluated and interpreted by radiologists according to the characteristics of the images; however, this assessment can often be subjective and time consuming. In contrast, AI methods provide automatic recognition and quantitative analysis of complex patterns in imaging data [21]. Therefore, AI can be used as an effective tool to help clinicians make more objective and reproducible assessments of radiological images.

While radiographs are interpreted by dentists, first of all, certain radiological features of normal tissues and pathologies are recognized and distinguished from each other, and then, preliminary diagnoses and differential diagnoses are created by overlapping the obtained radiographic information with clinical findings. The first recognition and discrimination process is basically a pattern recognition function. Some radiographic analyses can now be done automatically, thanks to the fact that machine learning techniques allow computers to recognize patterns [22]. On the other hand, the function of justifying the outputs obtained in the radiography

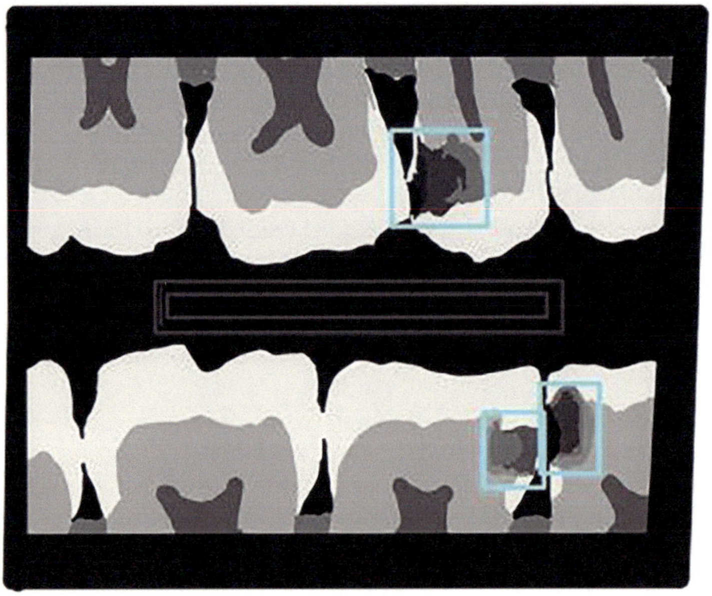

Figure 1.
Illustration of dentinal caries marked (blue boxes) on Bitewing radiographs.

interpretation process cannot yet be fulfilled due to the nature of existing machine-learning approaches [23].

Artificial intelligence applications are particularly promising in the field of dentomaxillofacial radiology. Recent research on artificial intelligence in the field of dentomaxillofacial radiology has mostly used algorithms capable of image classification, detection, segmentation, recording creation, and enhancement [24, 25]. AI-based algorithms developed in this field are generally aimed at diagnosis, image analysis, and image quality improvement. A large amount of data is needed to obtain accurate results. In addition, the participation of experienced radiologists is very important to create accurate and consistent datasets.

With artificial learning models, it is possible to detect the structures to be examined in radiography, to separate (segmentation) or classify the desired data in the image from other data [26]. These uses of artificial learning models are schematized in **Figure 2**. There are application examples of these usage areas for specific tasks in dentomaxillofacial radiology. For example, in the field of dentomaxillofacial radiology, various studies have been carried out on artificial intelligence in many different subjects such as dental caries [13, 27], periodontal diseases [28, 29], vertical root fractures [30], periapical pathologies [24], osteosclerosis, odontogenic cysts and tumors [31], maxillary sinus pathologies [32, 33], or diagnosis of temporomandibular joint diseases [25].

In the imaging of maxilla and mandible jaw bones, all existing teeth and surrounding support tissues, panoramic radiographs are the most widely used radiological diagnostic tools. Due to the complex anatomical structure of the region, examined by panoramic radiographs, displaying in 2D may cause superpositions and may cause incorrect and incomplete interpretations. In recent years, these radiographs are among the research data that have an important place in artificial intelligence studies. With a convolutional neural network (CNN) developed to detect benign tumors

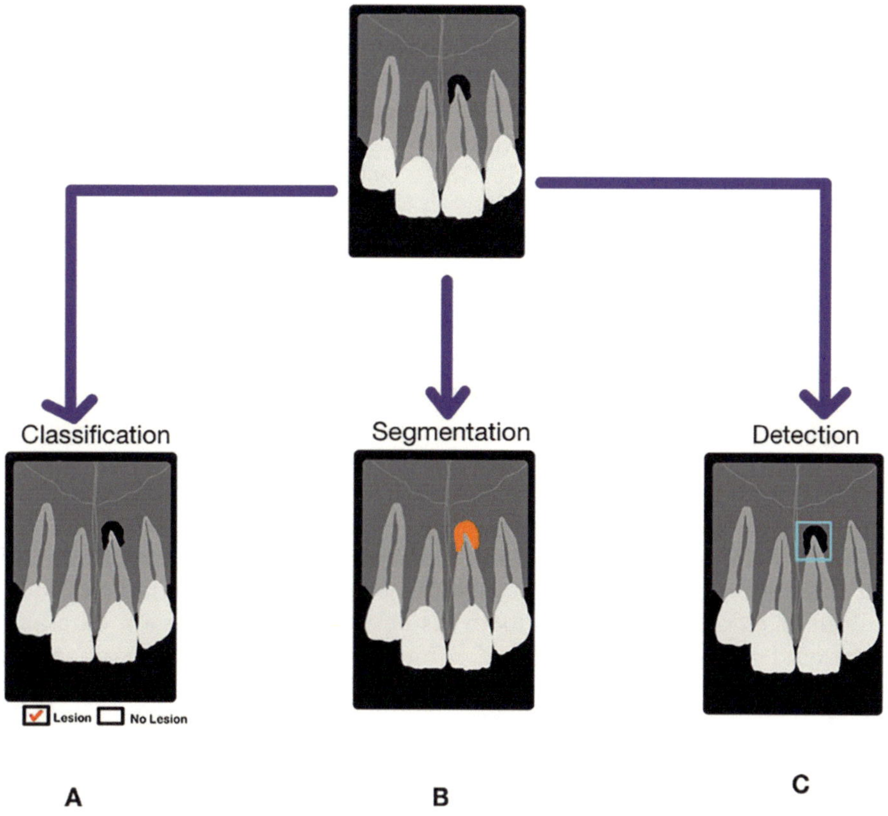

Figure 2.
(A) The classification (mark as yes or no) of the lesion on the periapical radiograph is schematized. (B) The segmentation (orange staining) of the lesion on the periapical radiograph is schematized. (C) The detection (blue box) of the lesion on the periapical radiograph is schematized.

in the jaws on panoramic radiographs, researchers have developed an algorithm that can diagnose with similar accuracy with specialist dentists [14]. In another study, an algorithm was developed for the diagnosis of osteoporosis by using the same 2D panoramic radiographs and it was stated that the radiographs determined by the algorithm distinguished osteoporosis with excellent accuracy when compared with specialist dentists [34].

In another study using panoramic radiographs, an AI system was created to detect idiopathic osteosclerosis (IO). As a result of the study results, the sensitivity, precision, and F-measure values were 0.88, 0.83, and 0.86, respectively; the developed AI algorithm has the potential to accurately detect IOs in panoramic radiographs [35].

In cases clinical examination and 2D radiographs are insufficient, the 3D imaging method commonly used by dentists is cone beam computed tomography (CBCT). In a study in which the detection of periapical lesions with AI algorithms was performed with an accuracy of 92.8%, with an algorithm that detects periapical lesions using CBCT images, the detection and numbering of teeth in volumetric data are also performed [24].

Radiographic imaging has a very important place in the field of dentistry, both for diagnostic purposes and to provide appropriate treatment. In cases such as implant planning, orthognathic surgery, and surgical applications planned for various pathologies, radiographic images are analyzed as part of treatment planning. The development of artificial intelligence in these areas is highly valuable. Dentomaxillofacial radiologists continue to play an important role in artificial intelligence research, as specialists, who know the basic principles and characteristics of radiographic imaging, can interpret radiographs for various diseases. The data detected by artificial learning will reduce both the rate of inaccurate and incomplete diagnosis and the daily workload of dentists, considering the increasing medical data in radiology practice.

2.2 Dentomaxillofacial surgery

It is the branch of dentistry that treats and rehabilitates pathological conditions, trauma, and developmental anomalies in the mouth, teeth, and jaws. Impacted third molar tooth extraction is one of the most common surgical procedures. One of the most common postoperative complications in third molar surgery is facial swelling. In a study by a group of researchers, an artificial neural network model was developed to predict the probability of facial swelling after extraction. In 400 extraction procedures performed by a surgeon, surgical factors (individual characteristics of the patients such as age, gender, medical and dental status, the relationship of the third molar with the second molar and ramus, the degree of impaction in the bone, the type of incision, the removal of the tooth in single or multiple pieces, the duration of the procedure, etc.) were determined. While 300 patient data were used in model training, 100 patient data were used in the testing phase. In this study, the estimation accuracy of the probability of swelling was reported to be 98% [36].

Investigating the ability of CNNs to automatically detect mandible fractures on panoramic radiographs is a current issue in oral and maxillofacial surgery [37]. In a study investigating the detection and classification of mandible fractures (e.g., condyle, coronoid, ramus) on panoramic radiographs, with F1 scores ranging from 0.6 to 0.87, the lowest fixation sensitivity was achieved in coronoid fractures, while the highest sensitivity was obtained in condyle fractures [38].

Diagnosis and treatment of temporomandibular joint (TMJ) diseases are one of the most difficult issues for dentists. In a study aiming to create an artificial neural network model that is predicted to distinguish between TMJ internal irregularities and normal joint, using TMJ clinical examination findings, the model was trained and tested to detect unilateral/bilateral anterior disc displacements with and without reduction. The sensitivity and specificity of this developed model range from 37% to 100% [39]. In a new diagnostic tool to accurately detect temporomandibular joint disc displacement with artificial intelligence, proposed by Kao et al., cited the InceptionV3 and DenseNet169 models as the best performing models. In this study, a two-step procedure was proposed involving the inference of probable location and binary classification of the disc, emphasizing that larger studies using data from different hospitals or countries are needed to fully validate their system. It has been stated that the use of deep learning neural networks in the automatic detection of temporomandibular joint disc displacement from sagittal MRI images are a promising technique to assist radiologists [25].

In a recent study using magnetic resonance imaging (MRI), in condyle, articular eminence and disc segmentation, the AI model has been reported to have a performance close to that of specialist doctors. In addition, this study also mentioned that CNN-based segmentation models can be a reliable tool to help clinicians define the basic anatomy in TMJ-MRIs [40]. Automatically locating the articular disc and surrounding structures with an AI algorithm has the potential to improve the reliability of TMJ-MRI interpretation and save time.

One of the most commonly used methods to eliminate tooth deficiencies is dental implant application. A dental implant is a treatment method developed to replace missing teeth, with long-term success and high survival rates. Although dental implants have high success, mechanical and biological complications can occur. A study was conducted in 1996 using the ML method to classify eight different dental implant systems based on various parameters such as implant diameter, length, and cross-sectional area [41]. In recent years, CNN has been used to classify different implant systems [42]. There are several studies showing that CNN is highly effective in classifying similar shapes of different types of implant systems in dental radiographic images [42–44]. The quality of the bone area to be implanted is one of the most important factors that directly affect the success of the implant. The sparseness or density of trabecular bone can directly affect the success of implant surgery. Because of the great importance of preoperative determination of trabecular bone morphology, a study group aimed to classify trabecular bone quality. In this study, bone morphology was classified using CBCT images and it was reported that the algorithm made predictions with twice the accuracy of expert dentists in the results obtained [45]. The lack of a gold standard in the classification of the groups and the heterogeneous distribution of the groups in the dataset were stated as the limitations of the study. Despite these limitations in the study, implant planning has been evaluated as a study that contributes to the computer-aided design and production of implant-based prostheses [45].

Robotic systems have become a part of daily life; although it has been on the agenda for a long time in the field of medicine, minimally invasive surgical procedures such as heart valve repair and gynecological interventions are applied worldwide with the developments in autonomous robot systems. Autonomous robots plan a sequence of actions with AI methods to perform a specific task on their own. Implant surgery is moving toward fully automating implantology upon the implementation of possible robotic applications. On the other hand, due to the inadequacy in the number of studies on this subject, these methods have not been widely used because accuracy and reliability results are not sufficient yet [17, 46].

Orthognathic surgery is one of the areas where AI applications find their place in oral and maxillofacial surgery. With the widespread use of intraoral scanners, digital imaging in orthognathia has changed radically. Use of AI applications in orthognathic surgery workflow; it can also provide improvements in areas such as obtaining and interpreting maxillofacial imaging, treatment planning, personalized orthodontic and surgical appliances, and treatment follow-up [47].

In light of these developments, we think that the use of AI in surgical applications will gradually increase and its current uses will gradually improve.

2.3 Prosthetic dentistry

In functional and esthetic losses caused by deficiencies in teeth and surrounding supporting tissues, it is the branch of dentistry in which applications are made to restore the lost esthetics and function of the patient by being treated with appropriate prosthetic materials. The preparation and construction stages of prosthetic restorations consist of a series of processes that require high precision, which can be carried out with conventional or digital methods, CAD/CAM (computer-aided design/computer-aided manufacturing), which has been widely used in the production of prosthetic restorations, especially in recent years. It refers to a manufacturing technique that uses computer skills in the design and production of materials required for prosthetic treatment.

Intelligent software is often one of the issues needed to optimize the digital steps of CAD/CAM systems. It is concluded that the novel soft computing optimization process described in a study will increase the efficiency of firms in optimizing machine parameters for industrial processes and will significantly reduce the costs of preparing and adjusting machine processes [48]. AI applications to the dentist in prosthesis production; it helps in designing the best and most esthetic prosthesis by evaluating many variables such as anthropological calculations, face measurements, and patient expectation [49].

When standard CAD/CAM technologies are applied in implant prosthesis cementation, various problems can arise, such as positional errors, cementation errors, and errors that may occur during occlusal correction with an abutment [50]. In a study aiming to eliminate these errors and time losses, 91% survival and 93% success rate using an AI model in the manufacture of zirconia implants for posterior teeth [51].

In a study in which the segmentation network structure for tooth preparations was designed with the CNN model, which can automatically extract the margin line by learning the characteristics of the margin line region of the tooth preparation, an accuracy rate of 97.43% was reached [52].

It has been reported that results with acceptable accuracy were obtained in a study aiming to design a single molar dental prosthesis that mimics the morphology of a natural healthy tooth by learning the tooth characteristics from the remaining teeth, thanks to a developed AI algorithm. According to the results of this study, the applicability of artificial intelligence in the design of single-molar dental prostheses has been demonstrated, and it has been emphasized that the accuracy of biomimetic artificial intelligence-designed dental prostheses can be further increased with further training and optimization of the algorithms [53].

Advances in AI applications promise to autonomously create innovative dental restorations that meet the highest standards in terms of fit, function, and esthetics. Advances in this area will also have a significant impact on orofacial and craniofacial prostheses.

2.4 Periodontology

It is a branch of dentistry that examines the health, diseases, diagnosis, and treatment of the soft and hard tissues that support the teeth. Periodontitis is one of the most common oral diseases that can cause alveolar bone loss, tooth mobility, and tooth loss [54]. The diagnosis of periodontitis can be made by clinical examination and radiographic examination of periodontal tissues [55]. Due to their complex structure and low resolution, detecting and analyzing periodontal bone loss (PBL) on radiographs is difficult even for experienced dentists [56]. Therefore, the application of AI for automated assist systems in dental radiographic image data may allow for more reliable and accurate assessments of PBL. Lin et al. developed a computer-aided diagnostic model that can measure the degree of PBL using a hybrid-featured engineering process [55, 57]. Promising results have been shown in these studies, as artificial intelligence models used in various subjects such as classification of periodontal conditions, detection of PBL, and classification of peri-implant diseases and conditions show comparable and even better results than manual analysis of PBL [29, 58, 59].

Microbial dental plaque accumulating on tooth surfaces is one of the main causes of gingival diseases. Therefore, it has an important place in the scoring of periodontal diseases. In one study, a CNN algorithm was developed to classify the amount of microbial dental plaque in intraoral photographs obtained with a Quantitative Light

Effect Fluorescence (QLF) camera. It has been reported that the model that performs high in the training set has lower success in the test set. On the other hand, although the intra-observer and interobserver agreement of dentists was high in manual QLF analyzes, it was emphasized that this CNN model was superior in terms of cost and time when the number of images was high [60].

We predict that artificial intelligence in dentistry will continue to maintain its popularity in the field of periodontology and will show greater developments in the coming years.

2.5 Orthodontics

It is the branch of dentistry that diagnoses and treats the placement irregularities of the teeth and the development and position disorders of the jaws. In a study, the extraction requirement before orthodontic treatment was modeled using an artificial neural network, and it was concluded that the accuracy increased significantly, especially as a result of the use of the decision-making model with pretraining [61].

In a study to estimate the mesiodistal widths of unerupted teeth, the deep learning model (DL) used and the Moyers table (MT) were compared. According to the results obtained in the study, the prediction performance of the DL system for unerupted mandibular canine and premolar teeth was at an acceptable level with an accuracy of 49.5%. It is thought that AI models may offer a potential alternative to existing methods in estimating erupted tooth size and provide diagnostic support for mixed dentition analysis in tooth patterns [18].

Cephalometric image analysis is widely used to evaluate the skeletal anatomy of the human skull for treatment planning and evaluation of treatment outcomes [62]. Many anatomical landmarks usually need to be defined manually for cephalometric analysis. AI technology has been developed in automatic cephalometric anatomical points and skeletal relationship classification to reduce the burden on the clinician and save time.

The developed AI models allowed not only to identify anatomical points, but also to measure or analyze anatomical points in cephalograms. In a study conducted in 2015 using knowledge-based algorithms, an AI model capable of automatic cephalometric measurement has been developed. According to the results of this study, there was no significant difference between automatic and manual measurements [63]. The highly successful results demonstrated by fully automated cephalometric analysis systems have demonstrated the potential of AI application as a cephalometric orthodontic diagnostic tool.

In a recent CNN-based AI study, Sella Turcica segmentation and classification using CBCT images yielded results with a high percentage of accuracy [64]. Using artificial intelligence algorithms, it can be predicted that the detection of anatomical signs with orthodontic importance will save orthodontists time and facilitate diagnosis.

2.6 Endodontics

It is a branch of dentistry that consumes all components of pulpal and periapical pathologies from their etiology to their treatment. Areas such as anatomical variations of root canals, canal shaping techniques, and materials used in treatments are some of the topics that are researched and continue to be developed in endodontics. The role of artificial intelligence in the diagnosis and treatment planning of diseases in endodontics is increasing.

Due to the clinical examination of the root canal morphology, radiological imaging is closely related to endodontic treatment. Panoramic radiographs, which are 2D imaging methods, may be insufficient to understand root canal morphology. In a study aimed at estimating the number of distal roots of mandibular first molars, which are single in the majority of the population and sometimes may have accessory canals, on panoramic radiographs, two separate CNN algorithms were created and their estimation performances were compared. There was no significant difference between the performances of the two models, and it was reported that both had high accessory root discrimination performance on panoramic radiographs [65].

Determining the location of the root canal openings is one of the important stages of root canal treatment. Clinicians use magnifying glasses or operating microscopes to detect these small canal entrances. In a study for the location and classification of canal orifices, researchers reported that the model detected canal orifices with 94% accuracy and differentiated upper and lower molars with 90% accuracy. The study indicates that root canals can be detected and classified in real-time *via* software [66].

One of the main stages that determine the success of root canal treatment is the determination of the working length of the canal. In another endodontic study, researchers tried a new approach to reduce the error in the step of determining canal working length on radiographs. In the experimental setup as *in vitro*, the extracted teeth with access cavities were placed in the extraction sockets of the cadavers, and intraoral films were taken by reaching the root tip with the canal tool. Then, two endodontists examined the apex of the teeth with a microscope and the apical foramen with radiographs, and the working length was divided into three classes as "short, *in situ* and long" according to the position of the canal file. The reliability of the measurement of the working length was questioned by training the model with tooth length and approximate and detailed images. It was concluded that the position of the file can be reliably determined using AI [67].

Correct determination of working length is very important for successful root canal treatment results [68]. A method used to evaluate working length, radiography, digital tactile sense, electronic apex locators, patient response to a paper point inserted into the root canal system, and CBCT imaging [67, 69–71]. Clinical dentists most commonly prefer radiography and electronic apex locators. The clarity of the image in digital radiography and several other factors that affect how radiographic interpretations are made are essential for the correct interpretation of the anatomy of the root canal system [72]. Otherwise, it may lead to misdiagnosis [73]. Saghiri et al. used a human cadaver model to replicate a clinical setting and examined the accuracy of run-length assessment by an artificial neural network. Compared to the actual working length measurements, where the stereomicroscope was used as the gold standard after tooth extraction, the model was more accurate than the determinations of the endodontists, and it was concluded that neural networks are an accurate method for determining the working length [74].

2.7 Restorative dentistry

It is a branch of dentistry that treats structure and deformities caused by dental hard tissue (enamel, dentin) diseases with protective, restorative, and esthetic methods. The fact that dental caries is still one of the most common diseases in the world causes studies in this field to be among the popular research topics. The most commonly used method to support the clinical examination in the diagnosis of dental caries is radiological examination.

AI can provide assistance in recognizing some pathologies such as proximal caries and periapical pathologies that cannot be detected on radiographs due to noise, artifact, and low contrast in images [75]. There are many studies that have achieved high-performance results (86–97% accuracy) in the classification of dental caries on radiographs [75–79]. Different DL-based CNN methods have also been developed to detect dental caries on periapical radiographs [13, 80].

Detection of caries that arise on the chewing surfaces (occlusal) of the teeth and have not yet progressed to the interfaces can be successfully made by clinical examination. The high enamel density due to the anatomy of the region may cause the demineralization area formed by caries to remain hidden in the radiography. In a study, a deep learning model was designed to diagnose caries in photographs of occlusal caries recorded with an intraoral camera. Since the samples of intraoral photographs obtained *in vivo* are low in number, the model has been strengthened with data augmentation and transfer learning methods [81]. In another pilot study, the transillumination method used for caries diagnosis was examined. Although the diagnostic performance of the designed CNN algorithm is not bad, the sensitivity and specificity tests indicate that it is not yet reliable enough to be used in clinical practice [82].

2.8 Pediatric dentistry

It is a branch of dentistry that deals with the examination and treatment of first and permanent teeth in infants, children, and young adults.

In a study on the assessment of oral health in children using machine learning, researchers developed an algorithm for assessing children's oral health status and treatment needs. Oral health status and treatment needs were estimated by processing the data obtained through a questionnaire prepared for children and their parents with machine learning algorithms. It has been proposed as a method that can be used for screening purposes in schools [83].

In the literature, there is a study stating that software that can predict individual pain level and analgesia responses for postoperative pain management has been developed by using artificial intelligence models in the field of anesthesia [84].

Assistive tools have been developed to facilitate communication in age groups who have difficulty in describing their pain, in children with it is difficult to communicate, and in individuals with disabilities [85].

One of the areas studied extensively by the European Academy of Pediatric Dentistry (EAPD) is molar-incisor-hypomineralization (MIH) [86].

In the study published in 2022, aiming to update the current European Academy of Pediatric Dentistry (EAPD) 2010 policy document on "Best Clinical Practice guideline for clinicians dealing with children presenting with molar-incisors-hypomineralization (MIH); " MIH etiology has a multifactorial etiology including systemic medical and genetic factors, and more focused laboratory research and prospective clinical studies are needed to elucidate additional factors. Successful prevention and treatment options have been researched and established. Regarding treatment options, there are many alternative options such as composite restorations, metal crowns, and laboratory indirect restorations. There are factors depending on the severity of the defects and the age of the patient that play a role in the selection of the appropriate treatment [87]. A deep learning-based convolutional neural network (CNN) was developed for the automatic detection and classification of teeth affected by MIH in intraoral photographs. It could accurately categorize teeth with MIH with an overall diagnostic accuracy of 95.2% [88].

The dental plaque that causes many oral diseases (e.g., caries, gingivitis, and periodontitis) is one of the important issues in protecting the oral and dental health of children [89, 90]. Dental plaque consists of bacterial masses on tooth surfaces; dental plaque is often difficult to distinguish, as these masses usually occur at the gingival margin and in the interproximal areas [91].

A deep learning-based artificial intelligence model has been used as a current approach to detect plaque in primary teeth. As a result of plaque detection examinations on intraoral photographs; it was stated that there was no significant difference between the developed AI model and the pediatric dentist in diagnosing dental plaque in deciduous teeth, and the AI model showed clinically acceptable performance in detecting dental plaque in primary teeth compared to an experienced pediatric dentist [92].

Determining the most appropriate treatment procedure for primary teeth in pediatric dentistry is directly related to the presence or absence of permanent tooth germ [93]. Developmentally missing or extra permanent teeth are an important issue in the selection of endodontic or restorative rehabilitation for primary teeth. Evaluation of the performance of the deep learning system for permanent tooth germ detection in panoramic radiographs is among the subjects studied. A total of 4518 panoramic radiographs of children aged 5–13 were used in a study addressing this issue. Permanent tooth germs were detected with a deep learning-based approach in panoramic X-rays, using the YOLOv4 model, with an extraction time of 90 ms, with an average precision of 94.16% and an F1 value of 0.90, with a high level of significance. In this way, it is thought by researchers that it can facilitate the early diagnosis of missing teeth or supernumerary teeth and help dentists find more accurate treatment options while saving time and effort [94].

2.9 Forensic dentistry

Determination of gender and age from skeletal remains is an important issue in forensic studies. Age estimation is an important aid, from identifying individuals and injured persons in natural disasters to resolving forensic cases [95, 96]. Teeth are hard and not easily destroyed tissues in the human body, and their shape remains unchanged and intact after the individual dies. For this reason, forensic dentistry has an important place in identification. Comparison of tooth numbers and shapes in a dental radiograph taken from a cadaver with previous records is one of the methods used effectively [97].

In forensic medicine literature, teeth have proven to be a meaningful tool in identifying people [98]. Studies suggesting that neural networks can serve as a reliable method for determining age and gender also indicate that more research is needed to determine their accuracy [99–101]. In one of the studies on this subject, it was found that neural networks showed 95% accuracy in distinguishing gender in the anthropological skull [99]. Successful results have also been obtained in studies using trained neural networks to calculate skeletal age from hand-wrist X-rays [100, 101]. AI algorithms can also be used to analyze bite marks and predict mandible morphology [102].

In a study for age and gender estimation, 1142 digital X-ray images, 80% of it were used for educational purposes and the remaining 20% of the dental images were used for test cases, achieving 96% accuracy [103].

In a recent study, Patil et al. discussed the use of root length in age determination, which has not been researched until that time. In the study, it was stated that the Deep Learning model outperformed the Machine Learning model and the right third molar tooth mesial root length is a good indicator of age. It is emphasized that the training dataset should be expanded to include more radiographs from multiple sources in order to further refine, diversify, and bring the algorithms into clinical practice [104].

3. Artificial Intelligence in dentistry education

Artificial intelligence applications, which aim to change the traditional structure of dental education in faculties, are still an area open to research. Intelligent teaching systems provide new opportunities for education in universities, improving the quality of learning in dentistry education. It is obvious that learning, teaching, measurement-evaluation, and feedback methods have changed and improved. Today, many technological developments supported by artificial intelligence are rapidly taking place in the challenging education process of dentistry. Clinical decision support systems are effective information technology that reduces clinical errors, helps dentists make better medical decisions, and thus increases the quality of treatment. This technology holds the strongest position among promising fields of study, as it can help inexperienced dentists and students to form stronger convictions about the treatment process.

The field of intelligent teaching systems has made significant progress today. AI applications are often used to create scenarios that mimic clinical work on patients and minimize any educational hazards. It is observed that the preclinical virtual patient feedback given to the students has improved significantly. The main goal is to create high-quality learning environments, allowing students to evaluate their work and compare it to the ideal one. Numerous studies have been conducted on the effectiveness of these systems, and it has been shown that students develop a competency-based skill level more quickly with these systems than with traditional simulator units [49].

Dentistry students receive preclinical training to develop their manipulative skills before they come into contact with the patient. In the skill development studies on the traditional phantom models used during this training, the student learns by receiving feedback from the trainer. Instead, when the same practices are experienced with a virtual patient under the supervision of a trainer, the feedback that the virtual patient provides to students can improve the quality of learning. Studies examining the effectiveness of these systems have shown that students reach the expected skill level more quickly [49].

Today, along with the development of robotic models that can verbalize pain, move the head with pain, make jaw and tongue movements, and simulate functions such as bleeding and saliva flow. In dentistry education, the use of robotic models has increased in preclinical laboratories to develop students' basic motor skills [46]. Increasing studies in this field and decreasing software and hardware costs of robotic instruments will enable more widespread use of robotic applications in dental faculties. In AI applications in the field of education, as well as improving education, the inclusion of AI-based subjects in the curriculum is also a very important issue. Considering the ongoing research in almost every branch and the applications that are starting to come into use, it is clearly revealed that dental candidates will be both users and developers of these applications in the future. Introducing dentistry students to basic AI terminology and working principles will contribute to the advancement of this field developed with human intelligence.

4. Conclusion and evaluations

4.1 Future prospects

Hybrid intelligence is a term that refers to the combination of human and machine intelligence with the aim of advancing human intelligence. It is based on the principle that humans can work in harmony with artificial intelligence to solve difficult problems. The use of a newly developed hybrid intelligent image fusion method to combine multimodal images for better diagnosis and treatment planning will become widespread [21].

The diagnostic performance of AI models varies between the different algorithms used. Although the AI models used have high success rates, it is still necessary to verify the generalizability and reliability of these models using adequate, representative images from multiple institutions before using these models in clinical applications.

Artificial intelligence digital systems have indisputably changed the future predictions and development direction of dentistry [46]. It is obvious that computer-based neural networks play a supporting role in the decision-making and minimizing of errors during the execution of dental treatment planning for dentists, but it should be noted that more research and application are needed.

In short, although there are both technical and ethical difficulties in artificial intelligence, the fact that it is a very open field for development and progress makes artificial intelligence worth researching. The risks are especially high in the health field. There are great concerns about data protection, data security, and the transfer of critical medical decisions to computers. However, artificial intelligence has the potential to revolutionize both healthcare and dentistry. In the upcoming period, it seems inevitable that artificial intelligence and dentists must work together in cooperation. It is becoming a necessity of the age for dentists to follow current developments and integrate them into their clinical lives. On the other hand, the support of dentists is absolutely needed in order to carry the achieved developments further. Thanks to this link between dentists and technology, current developments will progress rapidly and become more useful.

Author details

Seyide Tugce Gokdeniz[1]*, Arda Buyuksungur[2] and Mehmet Eray Kolsuz[1]

1 Faculty of Dentistry, Dentomaxillofacial Radiology Department, Ankara University, Ankara, Turkey

2 Faculty of Dentistry, Basic Medical Sciences Department, Ankara University, Ankara, Turkey

*Address all correspondence to: seyidetugce@gmail.com

References

[1] ANSDIT (American National Standard Dictionary of Information Technology. New York: Springer; 1996

[2] Grudin J. Turing maturation: Separation of artificial intelligence and human-computer interaction. Interactions. 2006;**13**(5):54-57. DOI: 10.1145/1151314.1151346

[3] Richardson K. Genes, Brains, and Human Potential: The Science and Ideology of Intelligence. Columbia University Press; 2017

[4] Lebedev MA, Nicolelis MA. Brain-machine interfaces: From basic science to neuroprostheses and neurorehabilitation. Physiological Reviews. 2017;**97**(2):767-837. DOI: 10.1152/physrev.00027.2016

[5] Stuart R, Peter N. Artificial Intelligence: A Modern Approach. 3rd ed. Berkeley; 2016

[6] Ishwarya TA et al. A modern approach to design and integrate conceptual methods in video games with artificial intelligence. Materials Today: Proceedings. 2017;**4**(8):9100-9106

[7] Joda T, Bornstein MM, Jung RE, Ferrari M, Waltimo T, Zitzmann NU. Recent trends and future direction of dental research in the digital era. International Journal of Environmental Research and Public Health. 2020;**17**:1987. DOI: 10.3390/ijerph17061987

[8] Shan T, Tay F, Gu L. Application of artificial intelligence in dentistry. Journal of Dental Research. 2021;**100**:232-244. DOI: 10.1177/0022034520969115

[9] Joda T, Yeung A, Hung K, Zitzmann N, Bornstein M. Disruptive innovation in dentistry: What it is and what could be next. Journal of Dental Research. 2020;**2020**:0022

[10] Bishop CM. Pattern Recognition and Machine Learning. New York: Springer; 2006

[11] James G, Witten D, Hastie T, Tibshirani R. An Introduction to Statistical Learning with Applications in R. New York: Springer; 2013

[12] Goodfellow I, Bengio Y, Courville A. Deep Learning. 1st ed. Cambridge, MA: MIT Press; 2016

[13] Lee JH, Kim DH, Jeong SN, Choi SH. Detection and diagnosis of dental caries using a deep learning-based convolutional neural network algorithm. Journal of Dental. 2018;**77**:106-111

[14] Poedjiastoeti W, Suebnukarn S. Application of convolutional neural network in the diagnosis of jaw tumors. Healthcare Information Research. 2018;**24**(3):236-241. DOI: 10.4258/hir.2018.24.3.236

[15] Faber J, Faber C, Faber P. Artificial intelligence in orthodontics. APOS Trends Orthod. 2019;**9**(4):201-205

[16] Woo SY, Lee SJ, Yoo JY, Han JJ, Hwang J. Autonomous bone reposition around anatomical landmark for robotassisted orthognathic surgery. Journal of Cranio-Maxillo-Facial Surgery. 2017;**45**(12):1980-1988. DOI: 10.1016/j.jcms.2017.09.001

[17] Revilla-León M, Gómez-Polo M, Vyas S, Barmak BA, Galluci GO, Att W, et al. Artificial intelligence applications in implant dentistry: A systematic review. The Journal of Prosthetic Dentistry. 2023;**129**(2):293-300. DOI: 10.1016/j.prosdent.2021.05.008

[18] Camcı H, Salmanpour F. Estimating the size of unerupted teeth: Moyers vs deep learning. American Journal of Orthodontics and Dentofacial Orthopedics. 2022;**161**(3):451-456

[19] Hwang JJ, Jung YH, Cho BH, Heo MS. An overview of deep learning in the field of dentistry. Imaging Science Dental. 2019;**49**(1):1-7. DOI: 10.5624/isd.2019.49.1.1

[20] Thrall JH, Li X, Li Q, Cruz C, Do S, Dreyer K, et al. Artificial intelligence and machine learning in radiology: Opportunities, challenges, pitfalls, and criteria for success. Journal of American College Radiology;**15**(3):504-508. DOI: 10.1016/j.jacr.2017.12.026

[21] Hosny A, Parmar C, Quackenbush J, Schwartz LH, Aerts HJWL. Artificial intelligence in radiology. Nature Reviews. Cancer. 2018;**18**:500-510. DOI: 10.1038/s41568-018-0016-5

[22] Jha S, Topol EJ. Adapting to artificial intelligence: Radiologists and pathologists as information specialists. JAMA. 2016;**316**(22):2353-2354. DOI: 10.1001/jama.2016.17438

[23] Chan S, Siegel EL. Will machine learning end the viability of radiology as a thriving medical specialty? The British Journal of Radiology. 2019;**91**:20180416

[24] Orhan K, Bayrakdar IS, Ezhov M, Kravtsov A, Özyürek T. Evaluation of artificial intelligence for detecting periapical pathosis on cone-beam computed tomography scans. International Endodontic Journal. 2020;**53**:680-689. DOI: 10.1111/iej.13265

[25] Kao ZK, Chiu NT, Wu HH, Chang WC, Wang DH, Kung YY, et al. Classifying temporomandibular disorder with artificial intelligent architecture using magnetic resonance imaging. Annals of Biomedical Engineering. 2023;**51**(3):517-526. DOI: 10.1007/s10439-022-03056-2

[26] Yaji A. Artificial intelligence in dento-maxillofacial radiology. Acta Scientific Dental Science. 2019;**3**:116-121

[27] Casalegno F, Newton T, Daher R, Abdelaziz M, Lodi-Rizzini A, Schürmann F, et al. Caries detection with near-infrared transillumination using deep learning. Journal of Dental Research. 2019;**98**:1227-1233. DOI: 10.1177/0022034519871884

[28] Thanathornwong B, Suebnukarn S. Automatic detection of periodontal compromised teeth in digital panoramic radiographs using faster regional convolutional neural networks. Imaging Science Dental. 2020;**50**:169-174. DOI: 10.5624/isd.2020.50.2.169

[29] Kim J, Lee H-S, Song I-S, Jung K-H. DeNTNet: Deep neural transfer network for the detection of periodontal bone loss using panoramic dental radiographs. Scientific Reports. 2019;**9**:17615. DOI: 10.1038/s41598-019-53758-2

[30] Fukuda M, Inamoto K, Shibata N, Ariji Y, Yanashita Y, Kutsuna S, et al. Evaluation of an artificial intelligence system for detecting vertical root fracture on panoramic radiography. Oral Radiology. 2020;**36**(4):337-343. DOI: 10.1007/s11282-019-00409-x

[31] Kwon O, Yong TH, Kang SR, Kim JE, Huh KH, Heo MS, et al. Automatic diagnosis for cysts and tumors of both jaws on panoramic radiographs using a deep convolution neural network. Dentomaxillofacial Radiology. 2020;**49**(8):20200185

[32] Murata M, Ariji Y, Ohashi Y, Kawai T, Fukuda M, Funakoshi T, et al. Deep learning classification using

convolutional neural network for evaluation of maxillary sinusitis on panoramic radiography. Oral Radiology. 2019;**35**:301-307

[33] Kim Y, Lee KJ, Sunwoo L, Choi D, Nam CM, Cho J, et al. Deep learning in diagnosis of maxillary sinusitis using conventional radiography. Investigative Radiology. 2019;**54**:7-15

[34] Jae-Seo L, Adhikari S, Liu L, Jeong HG, Kim H, Yoon SJ. Osteoporosis detection in panoramic radiographs using a deep convolutional neural network-based computer-assisted diagnosis system: A preliminary study. Dento Maxillo Facial Radiology. 2019;**48**:120170344

[35] Yesiltepe S, Bayrakdar IS, Orhan K, Çelik Ö, Bilgir E, Aslan AF, et al. A deep learning model for idiopathic osteosclerosis detection on panoramic radiographs. Medical Principles and Practice. 2022;**31**(6):555-561. DOI: 10.1159/000527145

[36] Zhang W, Li J, Li ZB, Li Z. Predicting postoperative facial swelling following impacted mandibular third molars extraction by using artificial neural networks evaluation. Scientific Reports. 2018;**8**(1):1-9

[37] Warin K et al. Assessment of deep convolutional neural network models for mandibular fracture detection in panoramic radiographs. International Journal of Oral and Maxillofacial Surgery. 2022;**51**(11):1488-1494. DOI: 10.1016/j.ijom.2022.03.056

[38] Vinayahalingam S, van Nistelrooij N, van Ginneken B, Bressem K, Tröltzsch D, Heiland M, et al. Detection of mandibular fractures on panoramic radiographs using deep learning. Scientific Reports. 2022;**12**(1):19596. DOI: 10.1038/s41598-022-23445-w

[39] Bas B, Ozgonenel O, Ozden B, Bekcioglu B, Bulut E. Use of artificial neural network in differentiation of subgroups of temporomandibular internal derangements: A preliminary study. Journal of Oral and Maxillofacial Surgery. 2012;**70**(1):51-59

[40] Li M, Punithakumar K, Major PW, Le LH, Nguyen KT, Pacheco-Pereira C, et al. Temporomandibular joint segmentation in MRI images using deep learning. Journal of Dentistry. 2022;**127**:104345. DOI: 10.1016/j.jdent.2022.104345

[41] Lehmann T, Schmitt W, Horn H, Hillen W. Idefix: Identification of dental fixtures in intraoral X rays. Proc. SPIE: Medical Imaging 1996: Image Processing. 1996;**2710**:584-595

[42] Kim J-E, Nam N-E, Shim J-S, Jung Y-H, Cho B-H, Hwang JJ. Transfer learning via deep neuralnetworks for implant fixture system classification using periapical radiographs. Journal of Clinical Medicine. 2020;**9**:1117. DOI: 10.3390/jcm9041117

[43] Sukegawa S, Yoshii K, Hara T, Yamashita K, Nakano K, Yamamoto N, et al. Deep neural networks for dental implant system classification. Biomolecules. 2020;**10**:984-913. DOI: 10.3390/biom10070984

[44] Lee J-H, Jeong S-N. Efficacy of deep convolutional neural network algorithm for the identification and classification of dental implant systems, using panoramic and periapical radiographs: A pilot study. Medicine. 2020;**99**:e20787. DOI: 10.1097/MD.0000000000020787

[45] Nicolielo LFP, Van Dessel J, van Lenthe GH, Lambrichts I, Jacobs R. Computer-based automatic classification of trabecular bone pattern can assist radiographic bone quality assessment

at dental implant site. British Journal of Radiology. 2018;**91**:20180437. DOI: 10.1259/bjr.20180437

[46] Grischke J, Johannsmeier L, Eich L, Griga L, Haddadin S. Dentronics: Towards robotics and artificial intelligence in dentistry. Dental Materials. 2020;**36**(6):765-778

[47] Bouletreau P, Makaremi M, Ibrahim B, Louvrier A, Sigaux N. Artificial intelligence: Applications in orthognathic surgery. Journal of Stomatological Oral Maxi. 2019;**120**(4):347-354

[48] Vera V, Corchado E, Redondo R, Sedano J, Garcia AE. Applying soft computing techniques to optimise a dental milling process. Neurocomputing. 2013;**109**:94-104

[49] Khanna SS, Dhaimade PA. Artificial intelligence: Transforming dentistry today. Indian Journal of Basic Applied Medical Research. 2017;**6**(4):161-167

[50] Pareek M, Kaushik B. Artificial intelligence in prosthodontics: A scoping review on current applications and future possibilities. IJAM. 2022;**9**:367

[51] Lerner H, Mouhyi J, Admakin O, Mangano F. Artificial intelligence in fixed implant prosthodontics: A retrospective study of 106 implant-supported monolithic zirconia crowns inserted in the posterior jaws of 90 patients. BMC Oral Health. 2020;**20**(1):80. DOI: 10.1186/s12903-020-1062-4

[52] Zhang B, Dai N, Tian S, Yuan F, Yu Q. The extraction method of tooth preparation margin line based on S-Octree CNN. International Journal of Numerical Methods in Biomedical Engineering. 2019;**35**(10):e3241. DOI: 10.1002/cnm.3241

[53] Chau RCW, Chong M, Thu KM, Chu NSP, Koohi-Moghadam M, Hsung RT, et al. Artificial intelligence-designed single molar dental prostheses: A protocol of prospective experimental study. PLoS One. 2022;**17**(6):e0268535. DOI: 10.1371/journal.pone.0268535

[54] Tonetti MS, Jepsen S, Jin L, Otomo-Corgel J. Impact of the global burden of periodontal diseases on health, nutrition and wellbeing of mankind: A call for global action. Journal of Clinical Periodontology. 2017;**44**:456-462. DOI: 10.1111/jcpe.12732

[55] Lin PL, Huang PY, Huang PW. Automatic methods for alveolar bone loss degree measurement in periodontitis periapical radiographs. Computer Methods and Programs in Biomedicine. 2017;**148**:1-11. DOI: 10.1016/j.cmpb.2017.06.012

[56] Åkesson L, Håkansson J, Rohlin M. Comparison of panoramic and intraoral radiography and pocket probing for the measurement of the marginal bone level. Journal of Clinical Periodontology. 1992;**19**:326-332. DOI: 10.1111/j.1600-051X.1992.tb00654.x

[57] Lin PL, Huang PW, Huang PY, Hsu HC. Alveolar bone-loss area localization in periodontitis radiographs based on threshold segmentation with a hybrid feature fused of intensity and the H-value of fractional Brownian motion model. Computer Methods and Programs in Biomedicine. 2015;**121**:117-126. DOI: 10.1016/j.cmpb.2015.05.004

[58] Lee J-H, Kim D-H, Jeong S-N, Choi S-H. Diagnosis and prediction of periodontally compromised teeth using a deep learning-based convolutional neural network algorithm. J Periodontal Implant Sci. 2018;**48**:114-123. DOI: 10.5051/jpis.2018.48.2.114

[59] Chang H-J, Lee S-J, Yong T-H, Shin N-Y, Jang B-G, Kim J-E, et al.

Deep learning hybrid method to automatically diagnose periodontal bone loss and stage periodontitis. Scientific Reports. 2020;**10**:7531. DOI: 10.1038/s41598-020-64509-z

[60] Imangaliyev S, van der Veen MH, Volgenant CMC, Keijser BJF, Crielaard W. Deep learning for classification of dental plaque images. In: International Workshop on Machine Learning, Optimization, and Big Data. Cham: Springer; 2016. pp. 407-410

[61] Xie X, Wang L, Wang A. Artificial neural network modeling for deciding if extractions are necessary prior to orthodontic treatment. The Angle Orthodontist. 2010;**80**:262-266

[62] Devereux L, Moles D, Cunningham SJ, McKnight M. How important are lateral cephalometric radiographs in orthodontic treatment planning? American Journal of Orthodontics and Dentofacial Orthopedics. 2011;**139**:e175-e181. DOI: 10.1016/j.ajodo.2010.09.021

[63] Gupta A, Kharbanda OP, Sardana V, Balachandran R, Sardana HK. A knowledge-based algorithm for automatic detection of cephalometric landmarks on CBCT images. International Journal of Computer Assisted Radiology and Surgery. 2015;**10**:1737-1752. DOI: 10.1007/s11548-015-1173-6

[64] Duman ŞB, Syed AZ, Celik Ozen D, Bayrakdar İŞ, Salehi HS, Abdelkarim A, et al. Convolutional Neural Network Performance for Sella Turcica Segmentation and Classification Using CBCT Images. Diagnostics (Basel). 2022;**12**(9):2244. DOI: 10.3390/diagnostics12092244

[65] Hiraiwa T, Ariji Y, Fukuda M, Kise Y, Nakata K. A deep-learning artificial intelligence system for assessment of rootmorphology of the mandibular first molar on panoramic radiography. Dento Maxillo Facial Radiology. 2019;**48**(3):20180218

[66] Bruellmann DD, Tjaden H, Schwanecke U, Barth P. An optimized video system for augmented reality in endodontics: A feasibility study. Clinical Oral Investigations. 2013;**17**(2):441-448

[67] Saghiri MA, Asgar K, Boukani KK, Lotfi M, Aghili, H. vd. A new approach for locating the minor apical foramen using an artificial neural network. International Endodontic Journal. 2012;**45**(3):257-265

[68] Aminoshariae A, Kulild J, Nagendrababu V. Artificial intelligence in endodontics: Current applications and future directions. Journal of Endodontic Journal. 2021;**47**:1352-1357

[69] Seidberg BH, Alibrandi BV, Fine H, Logue B. Clinical investigation of measuring working lengths of root canals with an electronic device and with digital-tactile sense. Journal of American Dental Association. 1975;**90**:379-387

[70] Connert T, Hülber JM, Godt A, Löst C, Elayouti A. Accuracy of endodontic working length determination using cone beam computed tomography. International Endodotic Journal. 2014;**47**:698-703

[71] Root ZX, Apex ID, Serna-Peña G, Gomes-Azevedo S, Flores-Treviño J, Madla-Cruz E, et al. In vivo evaluation of 3 electronic apex locators. International Endodotic Journal. 2020;**46**:158-161

[72] Petersson A, Axelsson S, Davidson T, et al. Radiological diagnosis of periapical bone tissue lesions in endodontics: A systematic review. International Endodotic Journal. 2012;**45**:783-801

[73] Tewary S, Luzzo J, Hartwell G. Endodontic radiography: Who is reading the digital radiograph? International Endodotic Journal. 2011;**37**:919-921

[74] Saghiri MA, Garcia-Godoy F, Gutmann JL, Lotfi M, Asgar K. The reliability of artificial neural network in locating minor apical foramen: A cadaver study. International Endodotic Journal. 2012;**38**:1130-1134

[75] Geetha V, Aprameya KS, Hinduja DM. Dental caries diagnosis in digital radiographs using back-propagation neural network. Health Information Science System. 2020;**8**:8. DOI: 10.1007/s13755-019-0096-y

[76] Yu Y, Li Y, Li YJ, Wang JM, Lin DH, Ye WP. Tooth decay diagnosis using back propagation neural network. In: 2006 International Conference on Machine Learning and Cybernetics. IEEE; Aug 2006. pp. 3956-3959

[77] Li W, Kuang W, Li Y. Clinical X-ray image-based tooth decay diagnosis using SVM. Proc Sixth Int Conf Mach Learn Cybern ICMLC. 2007;**3**:1616-1619

[78] Ali RB, Ejbali R, Zaied M. Detection and classification of dental caries in x-ray images using deep neural networks. In: International Conference on Software Engineering Advances (ICSEA). Aug 2016. p. 236

[79] Singh P, Sehgal P. Automated caries detection based on radon transformation and DCT. In: 2017 8th International Conference on Computing, Communication and Networking Technologies (ICCCNT). IEEE; Jul 2017. pp. 1-6

[80] Choi J, Eun H, Kim C. Boosting proximal dental caries detection via combination of variational methods and Convolutional neural network. Journal of Signal Processing System. 2018;**90**:87-97. DOI: 10.1007/s11265-016-1214-6

[81] Moutselos K, Berdouses E, Oulis C, Maglogiannis I. Recognizing occlusal caries in dental intraoral images using deep learning. In: 2019 41st Annual International Conference of the IEEE Engineering in Medicine and Biology Society (EMBC). IEEE; Jul 2019. pp. 1617-1620

[82] Schwendicke F, Elhennawy K, Paris S, Friebertsh.user, P., Krois, J. Deep learning for caries lesion detection in near-infrared light transillumination images: A pilot study. Journal of Dentistry. 2020;**92**:103260

[83] Wang Y, Hays RD, Marcus M, Maida CA, Shen J. Developing children's oral health assessment toolkits using machine learning algorithm. JDR Clinical Transaction Research. 2019;**2019**:23800. DOI: 10.1177/2380084419885612

[84] Kim PJ. Artificial intelligence in anesthesiology: Moving into the future. University of Toronto Medical Journal. 2019;**96**(1):33-37

[85] Kalantarian H, Jedoui K, Washington P, Tariq Q, Dunlap K. Labeling images with facial emotion and the potential for pediatric healthcare. Artificial Intelligence in Medicine. 2019;**98**:77-86

[86] Lygidakis NA, Wong F, Jälevik B, Vierrou AM, Alaluusua S, Espelid I. Best clinical practice guidance for clinicians dealing with children presenting with molar-incisor-hypomineralisation (MIH): An EAPD policy document. European Archives of Paediatric Dentistry. 2010;**11**:75-81. DOI: 10.1007/BF03262716

[87] Lygidakis NA, Garot E, Somani C, Taylor GD, Rouas P, Wong FSL. Best

clinical practice guidance for clinicians dealing with children presenting with molar-incisor-hypomineralisation (MIH): An updated European Academy of Paediatric Dentistry policy document. European Archives of Paediatric Dentistry. 2022;**23**:3-21. DOI: 10.1007/s40368-021-00668-5

[88] Schönewolf J, Meyer O, Engels P, Schlickenrieder A, Hickel R, Gruhn V, et al. Artificial intelligence-based diagnostics of molar-incisor-hypomineralization (MIH) on intraoral photographs. Clinical Oral Investigations. 2022;**26**(9):5923-5930. DOI: 10.1007/s00784-022-04552-4

[89] Shibly O, Rifai S, Zambon JJ. Supragingival dental plaque in the etiology of oral diseases. Periodontology. 2000;**8**:42-59. DOI: 10.1111/j.1600-0757.1995.tb00044.x

[90] Axelsson P, Lindhe J. The effect of a preventive programme on dental plaque, gingivitis and caries in schoolchildren. The results after one and two years. Journal of Periodontology. 1974;**1**(2):126-138

[91] Marsh PD, Moter A, Devine DA. Dental plaque biofilms: Communities, conflict and control. Periodontology. 2000;**55**:16-35. DOI: 10.1111/j.1600-0757.2009.00339.x

[92] You W, Hao A, Li S, Wang Y, Xia B. Deep learning-based dental plaque detection on primary teeth: A comparison with clinical assessments. BMC Oral Health. 2020;**20**(1):141. DOI: 10.1186/s12903-020-01114-6

[93] Rallan M, Rallan NS, Goswami M, Rawat K. Surgical management of multiple supernumerary teeth and an impacted maxillary permanent central incisor. BMJ Case Reports. 2013;**2013**:bcr2013009995

[94] Kaya E, Gunec HG, Aydin KC, Urkmez ES, Duranay R, Ates HF. A deep learning approach to permanent tooth germ detection on pediatric panoramic radiographs. Imaging Science Dental. 2022;**52**(3):275-281. DOI: 10.5624/isd.20220050

[95] Kurniawan A, Chusida A, Atika N, Gianosa TK, Solikhin MD, Margaretha MS, et al. The Applicable Dental Age Estimation Methods for Children and Adolescents in Indonesia. International Journal of Dental. 2022;**2022**:6761476. DOI: 10.1155/2022/6761476

[96] Lossois M, Baccino E. Forensic age estimation in migrants: Where do we stand? WIREs Forensic Science. 2020;**3**:e1408. DOI: 10.1002/wfs2.1408

[97] Swami D, Mishra VK, Bahal L, Rao CM. Age estimation from eruption of temporary teeth in Himachal Pradesh. Journal of Forensic Medicine and Toxicology. 1992;**9**:3-7

[98] Dalitz GD. Age determination of adult human remains by teeth examination. Journal of the Forensic Science Society. 1962;**3**:11-21

[99] Bewes J, Low A, Morphett A, Pate FD, Henneberg M. Artificial intelligence for sex determination of skeletal remains: Application of a deep learning artificial neural network to human skulls. Journal of Forensic and Legal Medicine. 2019;**62**:40-43

[100] Gross GW, Boone JM, Bishop DM. Pediatric skeletal age: Determination with neural networks. Radiology. 1995;**195**:689-695

[101] Avuçlu E, Başçiftçi F. Novel approaches to determine age and gender from dental X-ray images by using multiplayer perceptron neural networks and image processing techniques. Chaos, Solitons and Fractals. 2019;**120**:127-138

[102] Khanagar SB, Vishwanathaiah S, Naik S, et al. Application and performance of artificial intelligence technology in forensic odontology - A systematic review. Legal Medicine (Tokyo, Japan). 2021;**48**:101826

[103] Santosh KC, Pradeep N, Goel V, Ranjan R, Pandey E, Shukla PK, et al. Machine Learning Techniques for Human Age and Gender Identification Based on Teeth X-Ray Images. Journal of Healthcare Engineering. 2022;**2022**:8302674. DOI: 10.1155/2022/8302674

[104] Patil V, Saxena J, Vineetha R, Paul R, Shetty DK, Sharma S, et al. Age assessment through root lengths of mandibular second and third permanent molars using machine learning and artificial neural networks. Journal of Imaging. 2023;**9**(2):33. DOI: 10.3390/jimaging9020033